Futuristic Research Trends and Applications of Internet of Things

Futuristic Research Trends and Applications of Internet of Things

Edited by
Bhawana Rudra
Anshul Verma
Shekhar Verma
Bhanu Shrestha

CRC Press is an imprint of the
Taylor & Francis Group, an **informa** business

First edition published 2022
by CRC Press
6000 Broken Sound Parkway NW, Suite 300, Boca Raton, FL 33487–2742

and by CRC Press
4 Park Square, Milton Park, Abingdon, Oxon, OX14 4RN

CRC Press is an imprint of Taylor & Francis Group, LLC

ISBN: 978-1-032-15561-6 (hbk)
ISBN: 978-1-032-15562-3 (pbk)
ISBN: 978-1-003-24471-4 (ebk)

DOI: 10.1201/9781003244714

Typeset in Times
by Apex CoVantage, LLC

Contents

Acknowledgments

We are extremely thankful to the authors of the 12 chapters of this book, who have worked very hard to bring this unique resource forward in order to help students, researchers, and community practitioners. We feel that it is contextual to mention that as the individual chapters of this book are written by different authors, the responsibility of the contents of each of the chapters lies with the concerned authors.

We would like to thank Randi Cohen, Publisher – Computer Science & IT, and Gabriella Williams, Editor, who worked with us on the project from the beginning, for their professionalism. We also thank Daniel Kershaw, Editorial Assistant at Taylor & Francis Group and his team members, who tirelessly worked with us and helped us in the publication process.

This book is a part of the research funded by 'Seed Grant to Faculty Members under IoE Scheme' under Dev. Scheme No. 6031 at Banaras Hindu University, Varanasi, India, which was awarded to Anshul Verma.

Preface

Overview and Goals

Internet of Things (IoT) is an interconnection of several devices, networks, technologies, and human resources to achieve a common goal. A variety of IoT-based applications are being used in different sectors and have succeeded in providing huge benefits to the users. As a revolution it overtook the entire global landscape with its presence in almost every sector, for example, like Smart cities, Smart grid, intelligent transportation, healthcare, education, and so on. This technological revolution moved to the machines also, converting them into intelligent computers that can take real-time decisions and communicate with each other forming an Internet of Systems/Machines. Finally, we come up with a new paradigm called Internet of Everything wherein anything that becomes a part has sufficient computing and communication resources and can connect. In such an environment where everything can communicate with each other in different clusters based on the application requirement, the data generated and data transmitted is so huge that it cannot be handled by traditional network infrastructure with the same efficiency. The traditional networking based on the TCP/IP model has limitations such as tightly coupled planes, distributed architecture, manual configuration, inconsistent network policies, error-prone, security, and inability to scale.

To overcome these issues a prominent technology was introduced: Software Defined Networking (SDN). This works on the ideas of open flow architecture, which is being widely deployed in different network domains. It is logically centralized software capable of controlling the entire network. It provides solution like network programmability by removing the network intelligence from the forwarding devices. SDN architecture provides better services to manage enterprise, wide area, and data center networks. Data analytics has a significant role to play in the growth and success of IoT applications and investments. The utilization of data analytics shall, therefore, be promoted in the area of IoT to gain improved revenues, competitive gain, and customer engagement. The adoption of Artificial Intelligence (AI) and Machine Learning (ML) to solve the problem of uncertainty, along with the usage of probabilistic models and algorithms to handle predictive analytical problems, is prevalent in a wide range of environments. To make this research area more adaptable for practical and industrial use there is a need to further investigate several research challenges in all aspects of IoT. Therefore, this book will provide theoretical, algorithmic, simulation, and implementation-based research developments related to the fundamentals, applications, and emerging research trends in IoT.

Following are some of the important features of this book, which, we believe, make it a valuable resource for our readers:

- This book is designed, in structure and content, with the intention of making the book useful at all learning levels.

- Most of the chapters are authored by prominent academicians, researchers, and practitioners with solid experience and exposure to research in the IoT. These contributors have been working within this area for many years and have thorough understanding of the concepts and practical applications.
- The authors are distributed in a large number of countries and most of them are affiliated with institutions of worldwide reputation. This gives this book an international flavor.
- The authors of each chapter have attempted to provide a comprehensive bibliography section, which should greatly help the readers interested in further digging into the aforementioned research area.
- Throughout the chapters, most of the core research topics have been covered, making the book particularly useful for industry practitioners working directly with the practical aspects behind enabling the technologies in the field.

We have attempted to make the different chapters of the book look as coherent and synchronized as possible.

Target Audience

This book will be beneficial for academicians, researchers, developers, engineers, and practitioners working in or interested in the fields related to IoT, wireless sensor networks, mobile ad-hoc networks, and ubiquitous computing. This book is expected to serve as a reference book for developers and engineers working in the IoT domain, and for a graduate/postgraduate course in computer science and engineering/information technology/electronics, and communication engineering.

Fundamentals of Internet of Things

1

Sarthak Srivastava[1], Anshul Verma[1], and Pradeepika Verma[2]

[1]*Banaras Hindu University, Varanasi, India*

[2]*Indian Institute of Technology (BHU), Varanasi, India*

Contents

DOI: 10.1201/9781003244714-1

1.1 INTRODUCTION

In this era of growing technologies, modernization is very closely attached to the word 'Automation', which means making things alive, to talk, to think, to see, and take decisions as well. Reducing human intervention and promoting machine-to-machine communication are the key goals that the technical groups are racing for. Making all these things possible, the 'Internet of Things' (IoT) is coming up with a very auspicious class of technologies and computing paradigms that emphasize virtually connecting all the types of physical objects together, forming some sensor networks so that they can exchange data and services among themselves and making the physical world smarter day by day [1]. Sometimes it is weird to hear, and hard to believe, that common things like our house doors, our wrist watches, our tables, chairs, coffee machines, lights, fans, and anything we see around us can work so smartly as well as interact with us without our involvement. But this new paradigm provides us such a class of techniques through which all these things are possible just by embedding some microelectronic devices (sensors, actuators, controllers, etc.) over anything we want, and enjoy visualizing them being alive. Even though it has been just three decades since this revolutionary step of the evolution of internetwork began, but we can clearly see the growth rate of the smart IoT-enabled devices around us. Examples include the smart doors usually seen installed in hotels that open automatically as soon as we come close to them [2–3]; the smart non-contact hand sanitizing machines that automatically spray sanitizing fluid when we put our hands below them [4]; and the health-monitoring machines in hospitals that upload patients' real-time health status to a server so that doctors may remotely observe the data from their homes or cabins [5–6]. Automatic cars, automatic air conditioners, smart lights, smart washbasin taps, and many more things indicate that the world is evolving and allows us to visualize the smart future. In fact, the IoT is developing a worldwide infrastructure and, even today, no field remains untouched by it [7], whether smart cites, transportation, healthcare, home automation, agriculture, defense, or any other sectors. Moreover, according to many projections, such things are expected to reach a few billion in numbers in the coming years.

From years ago, what we know about the Internet is that it is the internetwork of computers (or sometimes network of networks) all over the world through which people can share information and communicate from anywhere [8]. And now if we put some light on the name IoT, that is, Internet of Things, it clearly reflects 'connecting any physical thing to Internet' or 'physical objects embedded with sensors and actuators are linked through wired and wireless networks'. It is just a network of common things around us like our pens, chairs, spoons, etc., made to connect, be controlled, or communicate remotely through the Internet by embedding them with some microelectronic connecting, sensing, and processing devices [9]. The availability of enormous number of sensors [10], like chemical sensors, temperature and humidity sensors, accelerometers, motion sensors, and many more, makes the IoT systems more and more versatile and powerful. They can collect information and send it to servers or to other devices as well, or they can receive information and act accordingly, or they can do both simultaneously. The most prominent thing is these low-cost digital and analog electronic devices which can be used with older technologies to accomplish the intended task.

1.2 HISTORY OF IOT

Embedding physical objects with microprocessors and enabling them to effectively communicate and perform functions, along with minimizing the end users' interaction, has been in practice since the 1970s, when the idea was often called 'embedded internet' or 'pervasive computing' [11–12]. So, today, we can say that the actual history of IoT started in the late 1960s with the invention of INTERNET.

The term 'Internet of Things' was first coined by a British technology pioneer, Kevin Ashton, a co-founder of the Auto-ID Laboratory at MIT in 1999, during his work at Procter & Gamble. Ashton commented on his work in the RFID Journal (June 22, 2009) [13]:

> If we had computers that knew everything there was to know about things – using data they gathered without any help from us – we would be able to track and count everything, and greatly reduce waste, loss and cost. We would know when things needed replacing, repairing or recalling, and whether they were fresh or past their best.

Ashton, in his presentation for Procter & Gamble, described the technology for connecting several heterogeneous devices using RFID tags for supply chain management. Although today's IP-based IoT connectivity differs from what he proposed as RFID-based device connectivity, that breakthrough led to a revolutionary change in the tech world.

In an early 1926 interview in Colliers Magazine (Kennedy) [14], Nikola Tesla described an envisioned world full of advanced connectivity. Tesla said:

When wireless is perfectly applied, the whole Earth will be converted into a huge brain.

At that time, when people were putting various effort into letting the world know the different aspects of smart things and a smart world, ATMs were first considered one of the best and most commonly known smart objects when they went online in early 1974. Moving ahead to the 1980s, the Computer Science Department of Carnegie Mellon University introduced a prototype for a smart automatic Coca-Cola vending machine integrated with some micro-switches and using a very raw form of Internet [15]. The machine facilitated the remote monitoring of coldness of the drinks as well as the availability of the Coke cans within it. Since then, the understanding of possible scope and applications of such technology found a pace and kept increasing its graph.

Vision of Mark Weiser: Mark D. Weiser, a very well-known computer scientist and chief technology officer (CTO) at Xerox PARC, California, is also one the greatest visionaries of the interlinked physical world and considered to be the father of *ubiquitous computing* (coined the term in 1988) [16], which actually is a type of computing that can be done by any device, anywhere, anytime, and in any format and includes all the paradigms of what we call today IoT. Weiser, in 1991, in his essay [17] 'The Computer for the 21st Century', expressed his vision of an interconnected and computerized physical world which was going to be proven so helpful for mankind in its work in a self-effacing way, where the common computationally augmented artifact would interact almost naturally with our senses and spoken words and the most remarkable thing would be their meddlesomeness for users.

Some more introductory efforts:

- In 1991, Cambridge University for the first time proposed using the camera prototype over the Internet for making a smart system to monitor the amount of coffee available in their lab's coffee pot [18]. The camera was programmed to take three pictures per minute and send them to their local computers so that everyone could see if the coffee was available.
- In 2000, the very famous company LG electronics introduced an Internet-connected refrigerator [19]. And in the year 2005, '*Nabaztag*', a small robot rabbit, was developed that was capable of telling the latest news, weather forecasts, and stock market changes [20].
- In 2008, at the first international conference on IoT, held in Switzerland, 23 countries from around the world participated and discussed RFIDs, short-range wireless communications, and sensor networks.

1.3 ARCHITECTURE OF IOT

The growing needs and demands of the wireless and automated systems leads that IoT should be capable of connecting millions of heterogeneous devices across the Internet, and for that a very flexible and coherent layered architecture was required to follow. Architecture is basically a framework that defines the physical components, network configuration, their functional organization, their operational procedures, and data formats as well [21]. Following is a brief discussion about some of the most basic and popular models of IoT architectures.

1.3.1 Three-Layered IoT Architecture

Till now, many architecture models have described IoT connectivity, but none have been accepted as universally agreed to. The most basic and widely accepted one is the three-layer model (Figure 1.1) introduced first with three proposed layers: *perception layer, network layer, and application layer* [7, 21–23].

Perception layer: This is also called the object layer or the physical layer, because of its being responsible for direct and physical interaction with the environment. This is where the physical sensors, actuators, and other connected devices are placed to collect and process the information. The main functionality of this layer is to gather data like locality, temperature, weight, motion, pressure, acceleration, color, humidity, and so on, and transfer the data to the other layers by digitizing it, through secure channels.

Network layer: The data collected at the perception layer needs to be transmitted to other smart devices, servers, or other network devices for being processed. This is the job of the network layer. The main functionality is to provide connectivity and enable communication across the network using technologies like 3G, 4G, UTMS, Wi-Fi, infrared, and so on.

Application layer: This layer is mainly responsible for the user interaction, or we can say providing high-quality smart application specific services to the user. For instance, this layer provides the user with requested information like the temperature and humidity measurements, etc., or any other smart implementation in any domain like activating alarms, sending emails, turning on or off any device, and so on.

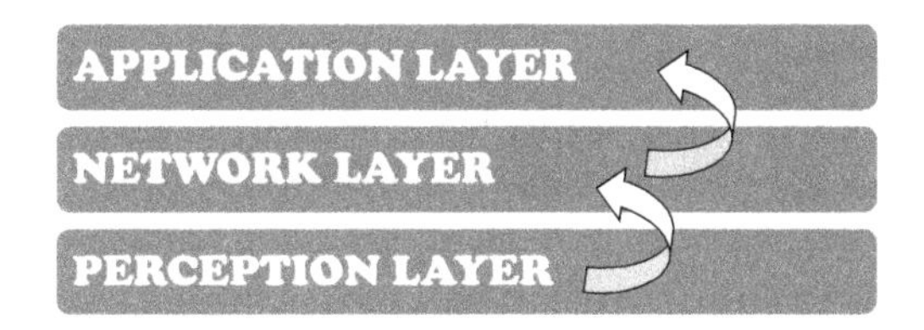

FIGURE 1.1 Three-layered IoT network architecture.

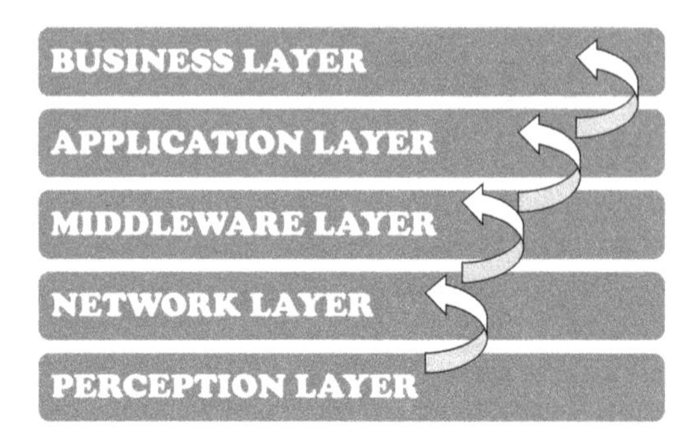

FIGURE 1.2 Five-layered IoT network architecture.

1.3.2 Five-Layered IoT Architecture

This model (also called service-oriented architecture) (Figure 1.2) can be considered as the extension to the basic three-layered architecture because of having just two additional layers. It and has been considered as one of the best proposed architecture models to be followed to study and implement the IoT projects of various devices on broad application areas [7, 21–23].

Middleware layer: This layer, also called the service management layer, came with some advanced features like data storage, computation, processing, and taking actions as well, according to the need. The main functionality of this layer is pairing of the service with its requester, as it stores all the data set obtained from sensors and provides the devices with appropriate data based on their addresses and names.

Business layer: This layer comes with the responsibility of management of overall IoT system services and activities and also with the monitoring of the underlying four layers. The main functionality of this layer is to receive data from the application layer and build the business models, graphs, and flowcharts, etc., based on it. Moreover, it also provides us with the design, implementation, and evaluation of the IoT system-related elements. Big data analysis and decision making on that is the best support of this layer.

1.4 BUILDING BLOCKS OF IOT

The absolute functionality of IoT systems can be understood using the following building blocks [7, 24]. These are the basic elements comprising the IoT network (Figure 1.3).

Identification: There is no need to splash out the point of how much it is needed within a network to provide each object with a unique identity. So as in IoT, it is much needed to identify solitarily the connected devices and match the services with demand. Moreover, the address of the objects within a communication network is also very crucial to assist in the identification of devices. The electronic product codes (EPC) and ubiquitous

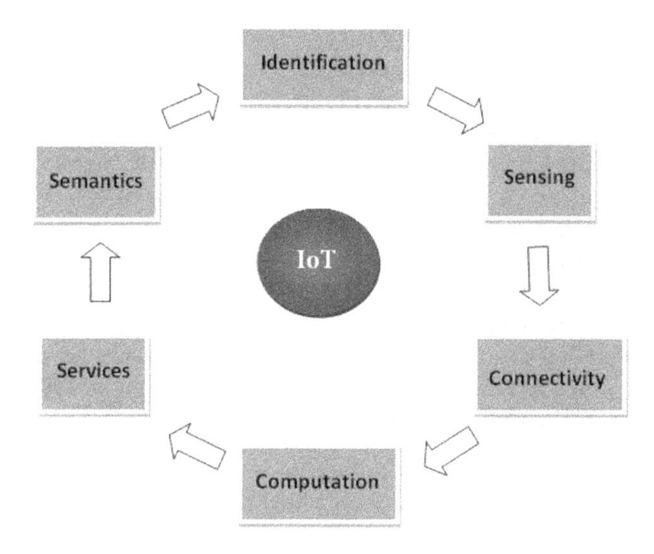

FIGURE 1.3 Basic building blocks of IoT.

codes (uCode) are used very frequently as the IoT identification methods, and IPv4, IPv6, and 6LoWPAN for addressing them.

Sensing: Sensing basically means gathering information or data from the environment or from related objects in a network that can be analyzed further to take actions based on service demand. Data collection and its integrity are the most important parts of the whole process task; that's why sensors are called the backbone of the IoT. Arduino, Raspberry PI, and NodeMCU are some single board computers that come along with the sensors and built in TCP/IP and security functionalities to realize and deploy the IoT products.

Connectivity: Without effective connectivity and communication a network has no meaning. IoT connects heterogeneous objects together to deliver smart devices. The communication protocols and technologies that enable this are Wi-Fi, Bluetooth, LpWANs, Zigbee, RFID, near field communication (NFC), ultra-wide bandwidth (UWB), LTE-A, IEEE 802.15.4, etc.

Computation: IoT computation is mainly connected with three sections, the hardware platforms, the real-time operating systems (RTOS), and cloud platforms. The processing units, like microprocessors, microcontrollers, and Field Programmable Gate Arrays (FPGAs), serve like the brain and heart of the IoT systems. The Arduino, Raspberry Pi, Intel Galileo, Cubieboard, WiSense, T-Mote Sky are some hardware platforms developed for IoT applications. Furthermore, some RTOS are there supporting the IoT systems, for example Contiki RTOS is very popular for IoT scenarios having a simulator COOJA to simulate all wireless sensor network applications. Some other lightweight operating systems designed for the purpose are TinyOS, LiteOS, Open Auto Alliance (OAA), etc. The cloud platform is like the backbone of the IoT computation, providing the facility with big data analysis and processing in real time. All the smart objects send their data to the cloud so that the respective information can be extracted and responses can be made.

Services: Following are descriptions of the main services.

Identification service: This most basic and prior service includes identifying the real-world objects taken into the virtual world to support the other upper-layer processes.

Information-aggregation service: This includes the collection of the data from sensors, summarizing it, and making it ready for processing and reporting to various applications.

Collaborative-aware service: This includes analysis of the obtained data to make decisions and take actions.

Ubiquitous service: This service includes making collaborative services available anywhere for anyone at any time.

Semantics: This block mainly deals with using resources and modeling information to make sense of right decision and provide exact services. Various semantic web technologies are there to support this, such as resource description framework (RDF), Efficient XML Interchange (EXI), Web Ontology Language (OWL), etc.

1.5 HARDWARE OF IOT

The hardware section of IoT includes the controllers, sensors, actuators, routing or bridging devices, and servers. Let's have a brief discussion on each of them and spot some popularly used hardware with their specifications and working.

1.5.1 Microcontrollers and Microprocessors

Microcontroller units (MCU) are basically termed as compact integrated circuits that control the overall working, performance, and functionality of an IoT system. We can say that they are small (sometimes palm sized or even smaller) personal computers designed to control the even larger systems without any operating system front-end interface.

The most prominent point is its contents. First, the MCU has its own fully functional processor (CPU) acting as the brain of the device controlling the peripherals and performing basic arithmetic, logic, I/O, and data transfer operations with them. Second, it comes with inbuilt program and data memory. The program memory (FLASH, EPROM, or EEPROM) of the MCU, being nonvolatile, is responsible for storing the information about the instructions needed for long term and holding it without needing a power source. Whereas the data memory (RAM) being volatile, stores the temporary data needed while the instructions are being executed. Third, it contains some GPIO pins, serial ports, and system bus for receiving the data from peripherals, converting it to binary and sending it to the processor for operation, and transferring the results back to the devices as well, in respective format [25–26].

Moreover, the MCU also includes analog to digital converter (ADC) and digital to analog converter (DAC) as well, to have efficient communication with the external digital and analog devices like sensors and actuators. In the very beginning, this can be programmed only with assembly language, but now the options are extended to C language,

C++, python, and JavaScript too. Next, let's have some focus on some of the basic and most commonly used microcontrollers.

1.5.1.1 Arduino

This is the most basic and commonly used single board microcontroller used for building digital devices [27]. There are a number of variants of Arduino boards (see Figure 1.4), launched with different specifications and sizes, like Arduino Mega, UNO, Nano, Mini, Pro-Mini, Seeduino, etc. Most of them have ATMEL 8-bit AVR microcontroller chip with 5V linear regulator and 16MHz crystal oscillator. These microcontrollers are pre-programmed with a boot loader (Optiboot Bootloader), which facilitates the uploading of programs to the on-chip flash memory from another computer via Universal Serial Bus (USB) or Micro-USB.

Following are some variants of Arduino boards with their basic specifications to be differentiated. Now let's shed some light over different pins and serial ports of the most commonly used Arduino Uno:

(i) Arduino Mega (256 KB flash + 8 KB SRAM, 54 Digital, and 16 Analog I/P pins)

(ii) Arduino UNO (32 KB flash + 2 KB SRAM, 14 Digital, and 6 Analog I/P pins)

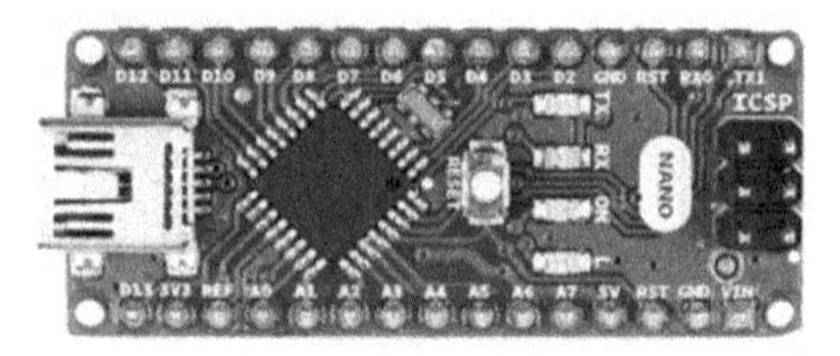

(iii) Arduino Nano (32 KB flash + 2 KB SRAM, 22 Digital, and 8 Analog I/P pins)

(iv) Arduino Pro-Mini (16 KB flash + 1 KB SRAM, 14 Digital, and 6 Analog I/P pins)

FIGURE 1.4 Arduino boards.

- *Digital input pins*: The digital pins (D0, D1, D2, . . ., D13) operate at 5V and are capable of sourcing and sinking 20mA of current. Among these 14 pins, D3, D5, D6, D9, D10, and D11 can produce 8-bit Pulse Width Modulated Signals.
- *Analog input pins*: The analog pins (A0–A5) facilitate to read the analog sensors. These can also provide a 10-bit resolution analog to digital conversion feature.
- *Serial communication pins*: The arduino boards are made to be programmed using serial communication. And to facilitate this, D0 and D1 pins are used as serial pins RX and TX respectively to receive and transmit serial data.
- *SPI (serial peripheral interface) pins*: D10 (SS) is the Slave Select pin that determines to which device the master is communicating with. D11 (SCK) is the Serial Clock pin used for synchronization of data transfer. D12 (MISO) is the Master in Slave Out pin which enables the peripherals to send data to the master device. D13 (MOSI) is the Master Out Slave In pin, which enables the master line to send data to peripherals.
- *I2C or two-wire communication pins*: Analog pins A4 and A5 can be configured to facilitate such communication as SDA, i.e., transmitting data and as SCL, i.e., synchronizing transfers respectively.
- *GND*: Ground pins used for closing of the circuits.
- *RST*: Reset pin.
- *Vin*: Used to provide the Arduino with power using external source.
- *5V and 3.3V*: Respective power supply to external components.

FIGURE 1.5 Raspberry PI board.

1.5.1.2 Raspberry PI

Rather that calling it just a microcontroller, it's better to call it a small-sized computer that can be connected to a monitor or TV. Moreover, it uses a standard keyboard and mouse. It is capable of doing anything that we can expect from a computer to do, e.g., Internet browsing, playing videos, gaming, word processing, and making spreadsheets too [28] (Figure 1.5, Raspberry PI 4 with 4 GB RAM).

Specifications:

- Processor: Broadcom BCM2711, Quad core Cortex-A72 (ARM v8) 64-bit SoC @ 1.5GHz
- SDRAM: 2GB, 4GB or 8GB LPDDR4–3200
- Bluetooth 5.0 + 2.4 GHz and 5.0 GHz IEEE 802.11ac wireless support
- USB ports: 2 USB 3.0 + 2 USB 2.0
- Other ports: 2 micro-HDMI ports + 2-lane MIPI DSI display port + 2-lane MIPI CSI camera port + 4-pole stereo audio and composite video port
- Micro-SD card slot for loading operating system and data storage
- 40 GPIO pins
- Gigabit Ethernet

1.5.1.3 NodeMCU

NodeMCU is another very efficient, micro-USB powered, single board microcontroller that comes with an integrated ESP8266 Wi-Fi module, which makes it one of the best to be used in IoT projects. Its pin specifications and workings are the same as Arduino (may be changed in number), although it has an additional feature of enabling Wi-Fi connectivity through which the system can be made to be used as client as well as server [29] (Figure 1.6, SDRAM: 64KB, Flash memory: 4MB, Wi-Fi built in: 802.11 b/g/n).

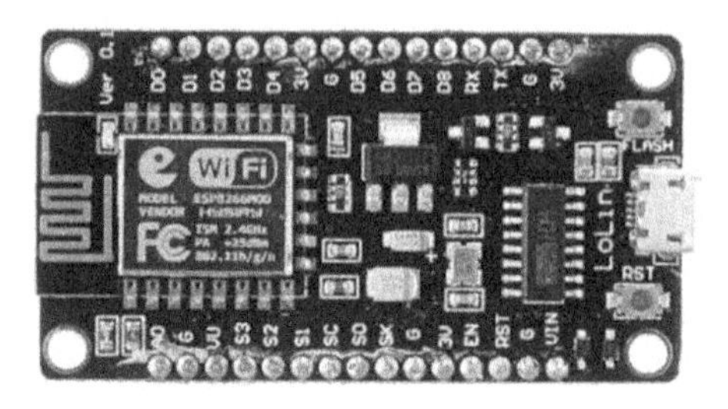

FIGURE 1.6 NodeMCU board.

FIGURE 1.7 Intel Galileo.

1.5.1.4 Intel Galileo

Galileo is the first microcontroller board based on the Intel architecture, having 400 MHz 32-bit Intel® Quark SoC X1000 Application Processor with a full-sized mini-PCI Express slot, 100Mb Ethernet port, Micro-SD slot, RS-232 serial port, USB Host port, and USB Client port. In addition to these, it is also provided with an *Ethernet RJ45 Connector* to get connected with some wired networks. The onboard microSD card reader enables it to extend its storage to 32GB [30] (Figure 1.7, SRAM: 512KB, DRAM: 256KB, EEPROM: 11KB, 14 Digital pins + 6 Analog pins).

1.5.1.5 Cubieboard

This is a very advanced and powerful development board that can be used to assemble small computer system as a whole. In comparison to Raspberry PI, it is better in performance, cheaper in cost, SATA supported, and has a 96 pin extended interface. It has a built-in infrared sensor and supports Ubantu and other Linux distributions with supporting 2.4GHz wireless keyboard and mouse [31] (Figure 1.8, CPU: 1G ARM cortex-A8 processor, NEON, VFPv3, 256KB L2 cache, RAM: 1GB DDR3, FLASH: 4GB NAND, Ports: 1 HDMI 1080P, 2 USB host, 1 micro-SD, 1 SATA, 1 IR).

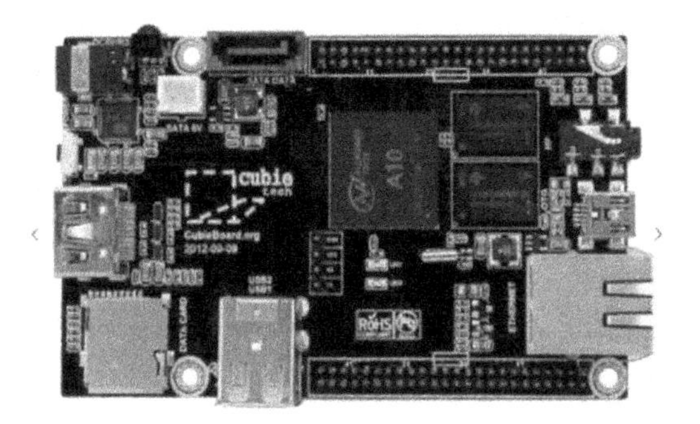

FIGURE 1.8 Cubieboard.

1.5.2 Sensors

Sensors are basically those devices that change their physical state according to the change in some surrounding parameters. More technically, we can say that sensors detect the respective external information or physical phenomenon and replace it with some machine-understandable electrical signals, which are used in IoT systems for smarter decisions. That's why the sensors play a key role in enabling automated systems. While choosing a sensor for any system, it is necessary to ensure that it can sense even the faintest change in the property it is desired to sense and disregards or is unaffected by any other property in its surroundings. It is also very important to mark out that the sensor must not influence the measured property, i.e., it must sense the parameter without affecting it or involving in it [10, 32].

There are various types of sensors available so far enabling the IoT based smart systems. Here are some classes of sensors that are most commonly used:

1.5.2.1 Temperature Sensors

Measure the thermal parameters of the environment or any object (gases, solids, or liquids) and convert it into the electrical signals. These can be used with variety of applications like

1. As voltage devices (*Thermocouple*) giving different output voltage in direct proportion with varying temperature input.
2. As resisting devices (*Thermistor*) giving variable resistance with changing temperatures.
3. As *infrared sensor* measuring the temperature of solids and liquids only using infrared energy of objects.

1.5.2.2 Proximity Sensors

A type of non-contact sensor used to detect the presence of nearby objects by emitting a beam of electromagnetic radiation and converting it to simple electrical signals. These can also be used with a variety of applications like

1. As *inductive sensors* to detect the presence of metallic objects.
2. As *capacitive sensors* for detecting both metallic and non-metallic objects.
3. As *photoelectric sensors* using light beams to detect the presence or absence of devices.
4. As *ultrasonic sensors* to detect the presence as well as measure the distance, very similar to the working of radar and sonar.

1.5.2.3 Pressure Sensors

Measures the pressure and converts it to an electrical signal.

1.5.2.4 Water Quality Sensor

Generally used for monitoring the water quality and presence of ions in a liquid. Some common applications of such modules are

1. *Turbidity sensors* measuring the number of solid particles suspended in water.
2. *Conductivity sensors* providing the ionic concentration in water as output.
3. *pH sensor* measuring the pH level of liquids.
4. *Chlorine residual sensors* detecting the total chlorine molecules in water.

1.5.2.5 Chemical Sensors

Mainly used for monitoring the liquid or air quality for any type of chemical changes. It has a very important role in industries for indicating any accidental or harmful chemical release, radioactive or explosive detection. Many types of chemical sensors are available now; some of them are electrochemical gas sensor, hydrogen sulfide sensor, fluorescent chloride sensor, etc.

1.5.2.6 Gas Sensors

Gas sensors are very little different from chemical sensors, as they can detect the presence of various gases only. Hydrogen sensor, carbon dioxide sensor, oxygen sensor, ozone sensor, hydrometer, nitrogen oxide sensor, smoke sensor, and many more are there to be used accordingly.

1.5.2.7 IR Sensors

These types of sensors use infrared emission for various sensitivity actions. An IR sensor can monitor blood flow and blood pressure and temperature in non-contact means. Also, they have a great application in wearable electronics, optical communication, liquid level monitoring, and house security purposes as well.

1.5.2.8 Level Sensors

As the name clears, these sensors are used to measure or monitor the level or amount of the liquids in any open or closed systems. This level sensing task can be accomplished by various technologies like infrared, ultrasonic, or optical emissions. The level sensors can be used as

1. *Point level sensors* to detect the particular specific level to trigger any type of alarm.
2. *Continuous level sensors* to detect the liquid level within a specified range, like monitoring the fuel level in vehicle.

1.5.2.9 Vision and Imaging Sensors

These sensors are used to convert optical images or their presence into the electrical signals, and also for positioning, orienting, labeling, and inspecting of objects; these are usually found in smart camera modules, night vision equipment, iris devices, thermal imaging systems etc.

1.5.2.10 Motion Detection Sensors

Detects the physical movement of the device and generate electrical signal accordingly. With these modules, it's very easy now to notice anybody's presence or intrusion detection in any intended area and to implement smart door controls, hand dryers, energy management systems, etc. They can be applicable as

1. *PIR sensors (passive infrared sensors).*
2. *Ultrasonic sensors.*
3. *Microwave sensors.*

1.5.2.11 Accelerometer Sensors

As the name defines, these sensors are used to measure the acceleration experienced by any object. Basically, is in intended to convert the mechanical motion into electrical output. They can be used in vibration measurement, free fall detection, making an anti-theft system, etc.

1.5.2.12 Gyroscope Sensors

Gyroscopes are mainly concerned with angular velocity, i.e., measurement of speed of rotation around any axis. They are used for navigation purpose like in drones or orientation of objects like robotic controls, etc.

1.5.2.13 Humidity Sensors

These sensors are used to measure the amount of water vapor present in the atmosphere or in any particular substance. They are more likely found in smart ventilation and air conditioning systems and also in smart agricultural systems for monitoring the humidity level of soil.

1.5.2.14 Optical Sensors

One of the most important sensor modules regarding their applications, optical sensors use light rays for the sensitivity actions. They come in various types, like photo detectors, pyrometers, fiber optics, etc.

Some commonly used sensor modules:

- DHT11, DHT22, HDC1080, LM35, MAX6675 (temperature and humidity sensors)
- ACS712 (current sensor)
- HC-SR04 (ultrasonic distance sensor)
- HC-SR501 (PIR sensor)
- MQ2 (hazardous gas detection sensor)
- TCS3200 / TCS230, ISL29125 (color sensors)
- MAX30102 (heart rate sensor)
- MPU-6050 (gyroscope accelerometer)

1.5.3 Wireless Enablers

Between 1857 and 1894, when electromagnetic waves were discovered, wireless connectivity was born and its first impact was seen in the field of communication when the first ever photophone was invented and patented (in 1880) leading to the first wireless talk that year. So, what wireless technology actually means is, the communication over any distance without using cables and wires but using radio frequency waves and infrared waves depending upon the distance. Semantically, a wireless network is that type of computer network in which the nodes communicate with each other with the data wirelessly [1, 33].

Wireless networks are much integrated with the IoT systems. We can imagine the scenario where anyone lying on the bed wants to know who is knocking at the door outside, and if the person is known then the door will be made to open, without going there, through a handset or a remote. Or if anyone wants to control the whole power system (lights, fans, coolers, AC, or anything) of the house just through a smartphone or any touch screen device, then what can be the main factor affecting this facility? It's the device-to-device communication. Despite the fact that wireless communication requires more power in comparison to wired communication and is sometimes considered un-optimized for event-driven communication, it can be easily deployed, can be made self-managing, and most importantly is simplifying the network connection throughout the system. There is nothing wrong if it has been called to be the blood veins of the automated and remotely controlled IoT devices. There are many wireless connection modules introduced yet for device-to-device and device-to-server communications. Let's have a brief introduction to some of the wireless enablers in IoT systems.

1.5.3.1 Bluetooth

Bluetooth is one of the most commonly known wireless communication protocols, working on data link layer of TCP/IP network model and communication layer of IoT network architecture, used to connect two devices for one-to-one communication for data exchange and to build the personal area networks (PANs) in short-range using short wavelength UFH radio waves (2.4 to 2.485 GHz). Bluetooth 5 is the latest version of this communication protocol, with about 800% more data broadcasting frequency compared to its earlier

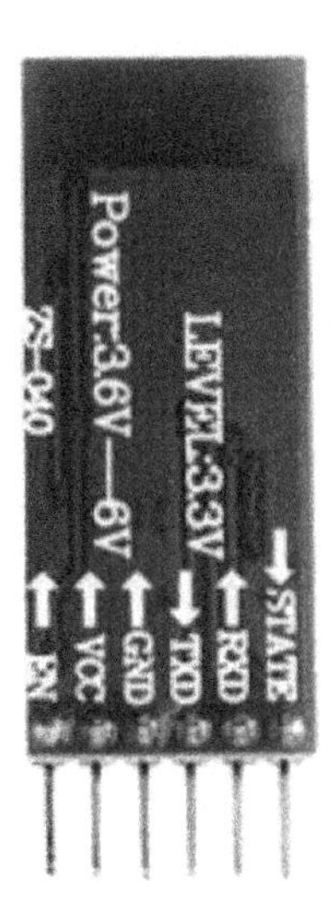

FIGURE 1.9 HC05 Bluetooth module.

version and it's four times longer range and doubled speed make it a very smart choice for the IoT systems. Low power consumption, cheaper cost, fast and secured connection, and low interference characteristics make it a very prominent device to be used for wireless networks [34] (Figure 1.9).

1.5.3.2 Cellular (3G/4G/5G)

In the world of IoT, almost nothing is like that which can be declared impossible. Yes, the name defines itself that, now we have such type of modules, through which we can attach the cellular networks to our smart devices by attaching sim cards to our systems. It facilitates to make some automatic calls (voice/video), text messages, and Internet access as well. In short, this cellular network connects the IoT devices with the existing mobile network, eliminating the need to develop a separate dedicated network infrastructure, which proves this a very cost-effective approach. Through this cellular network, unlike other Internet connection technologies, we can even have the data access across different cities or regions, and it has been found best if the connected device does not use real-time data transfer. As for security, the cellular IoT devices are much secured as a part of GSM standards, and, moreover, they can be configured to use private network technologies like VPN to be more secured and efficient [35] (Figure 1.10).

1.5.3.3 Wi-Fi

Wi-Fi (wireless fidelity) is a technology that uses star topology for network formation, having larger range than Bluetooth, through which the computers and other devices can interface with the Internet, create a network, and exchange information in between (through

FIGURE 1.10 A6-GPRS-GSM module.

FIGURE 1.11 ESP-8266 and ESP-01 Wi-Fi module.

routers). Wi-Fi 6 as the newest of its generation; it provides an enhanced network bandwidth (about 9.6 GBps) and a high throughput data transfer per user even in very congested environments. But, on the other hand, it has always been a less prevalent solution for IoT networks due to its limited coverage and scalability, and also high-power consumption, which makes it less efficiently compatible with battery-operated IoT sensors [36] (Figure 1.11).

1.5.3.4 *Zigbee and Other Mesh Topologies*

Zigbee is a type of wireless local area network (WLAN) designed to share small amounts of data between many battery-powered connected devices using very little power. Zigbee can be considered a better replacement for Wi-Fi as it uses mesh networks in which there are many interconnections between network nodes (between 10–100 meters) (Figure 1.12). Based on IEEE 802.15.4, it operates in radio frequency bands including 868 MHz, 900 MHz, and 2.4 GHz. Some other similar mesh technologies are Z-wave and Thread, etc. [37].

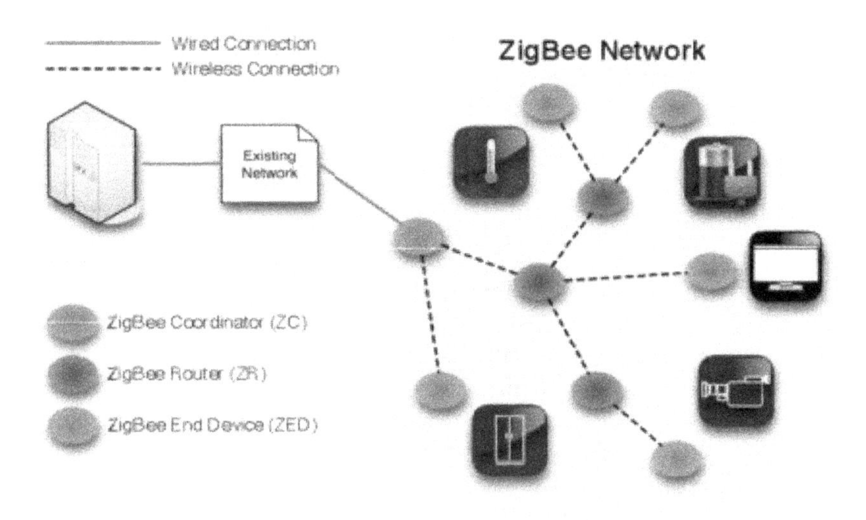

FIGURE 1.12 Zigbee network.

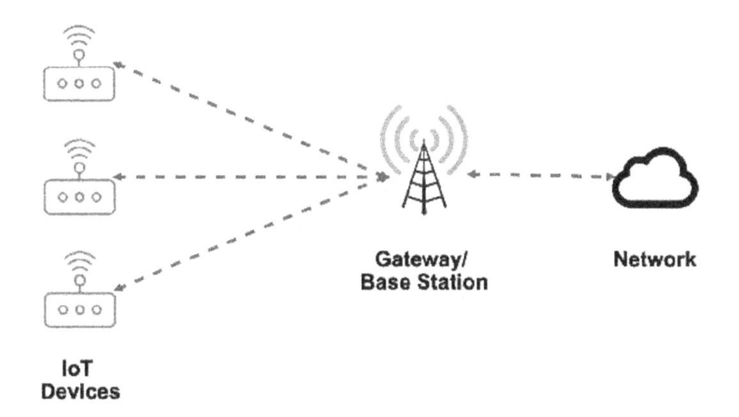

FIGURE 1.13 Low-power WAN technology.

1.5.3.5 LPWAN

These are low-power WAN technology very prominently used for connecting low bandwidth and battery-powered devices that have low bit rates over quite long ranges, from 2 km to 1000 km. If compared to the cellular networks, LPWAN can be operated at lower cost and with greater power efficiency. The number of connected devices supported and the area of connectivity is also bit larger than the cellular connectivity. Most of the LPWANs also use star topology where each end point has a direct connection with the central access point [38] (Figure 1.13). The accommodated packet size lies in the range of 10–10,000 bytes with uplink speed up to 200 kbps. Sigfox, RPMA (random phase multiple access), LoRa, Narrowband-IoT (NB-IoT), DASH7, WAVIoT are some very popular and efficient LPWAN technologies used in different applications according to their specifications.

1.5.3.6 RFID

RFID stands for Radio Frequency Identification system, which is also a very prominent wireless communication technology mainly resembles with the identification of objects in a network and controlling individual targets through radio waves. Moreover, the objects can be identified, tracked, and monitored globally in real time through the unique RFID tags attached with them in the network, which is the main aim of the IoT systems. The RFID systems include the tags having transmitters and responders and the readers having transmitters and receivers (Figure 1.14). The tag is nothing but a microchip attached to every object in the network, which serves as the unique identifier of the objects. These tags can be traced or communicated by the readers by transmitting the radio waves. As we all know that IoT is all about the identification of the objects and their virtual representations across the Internet, the RFID often seems like a prerequisite for that. We can attach our daily life objects with radio tags and operate remotely through computer systems [39].

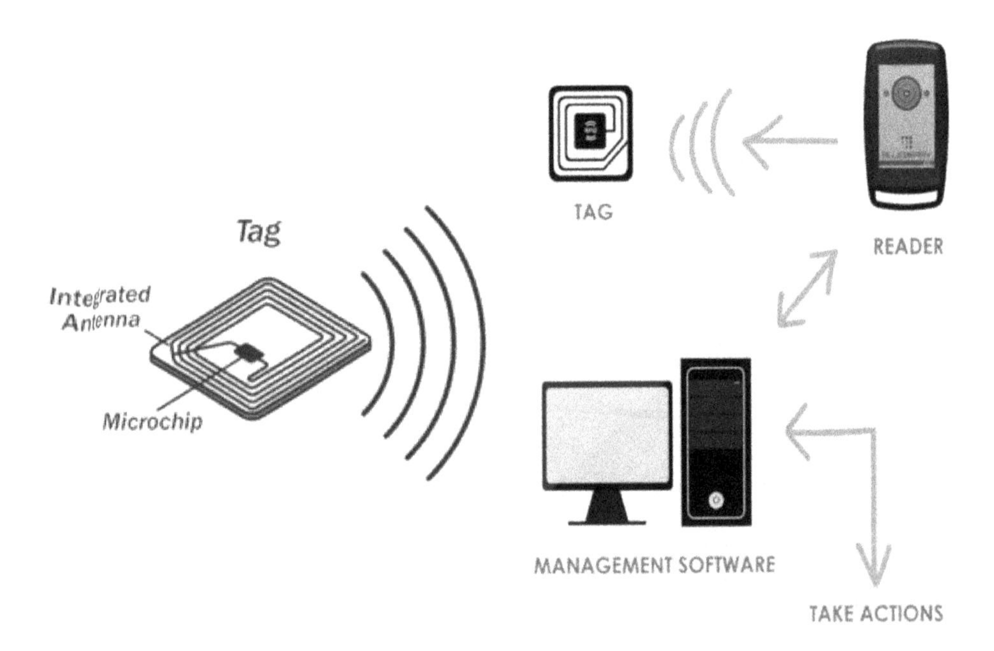

FIGURE 1.14 Radio Frequency Identification.

1.6 IOT APPLICATION PROTOCOLS

CoAP: Constrained Application Protocol (CoAP) is a type of web transfer protocol for machine-to-machine communication applications of smart devices, and it is clear from the name that it can be used with constrained nodes and networks in IoT with low bandwidth and availability. CoAP is featured with URI (uniform resource identifier) and asynchronous message exchange in constrained environment, i.e., in the network of low bandwidth and high congestion or limited connectivity. CoAP can support a network with billions of nodes for a reliable data exchange and can also maintain the satellite communication. It is a basic client server protocol that uses UDP network protocol and methods the same as HTTP [40–41].

MQTT: Message Queuing Telemetry Transport (MQTT) is a TCP-based publish/subscribe protocol, often called a lightweight messaging protocol, much used in industries for communication between devices, servers, and applications. MQTT is one of the protocols with the widest application area because of its low power and bandwidth consumption and also low latency. The publishers are the clients that want to distribute some information on any particular subscriber, and can publish it to the server, which can be referred to as a broker. When any other client (subscriber) wants that information and has subscribed for it, the broker sends that information to the respective subscriber. The complete transmission of data between the clients (publisher and subscriber) is handled by the broker (server). Any client can be a publisher or subscriber, or even both, and moreover they are typically unaware or each other in this protocol [42].

XMPP: Extensible Messaging Presence Protocol (XMPP) is mainly used for instant messaging applications like WhatsApp. For almost real-time exchanging of messages and presence information, it chooses to stream XML elements over the network. XMPP uses the client server architecture where the server serves as an intermediary between clients. But the most important point is that there is no any centralized XMPP server, anyone can run their own XMPP server, hence it is called decentralized [43].

AMQP: The Advanced Message Queuing Protocol (AMQP) is basically a binary application layer protocol for asynchronous message queuing, which involves producers, brokers, and consumers (Figure 1.15). The messages are produced by the publishers and submitted to the broker who ensures that the message will go to the right and intended consumer who comes to pick and process them up. Now, what is advanced in this is, the broker in order to do its job uses two components, exchanges and queues. The publishers actually send the messages to the exchange, which then routes them to appropriate message queues using some pre-defined rules and conditions. From these message queues, the messages get pulled by the consumers or sometimes the queue itself pushes it to the intended consumer depending on the configuration [44].

DDS: Data Distribution Service (DDS) is also based on the publish/subscribe methodology but in addition to that, it is a broker-less architecture and uses multicasting to bring high-quality QoS to the applications (Figure 1.16). In place of the broker, it uses its Global Data Space (GDS) where the applications can read or write data autonomously and asynchronously.

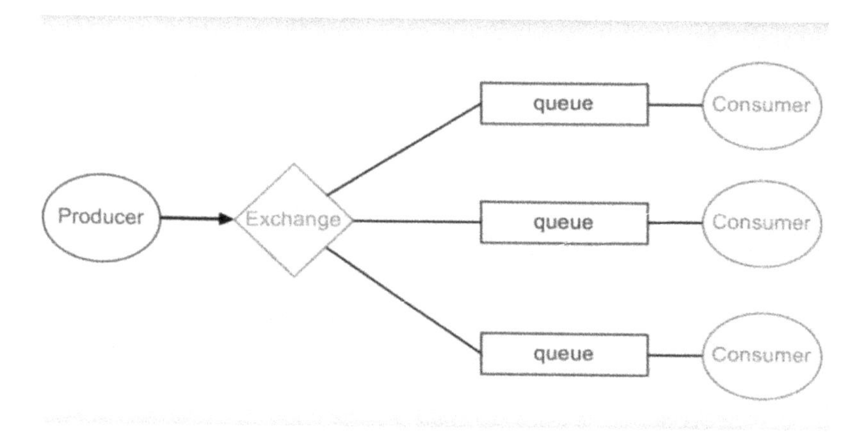

FIGURE 1.15 AMQP queuing.

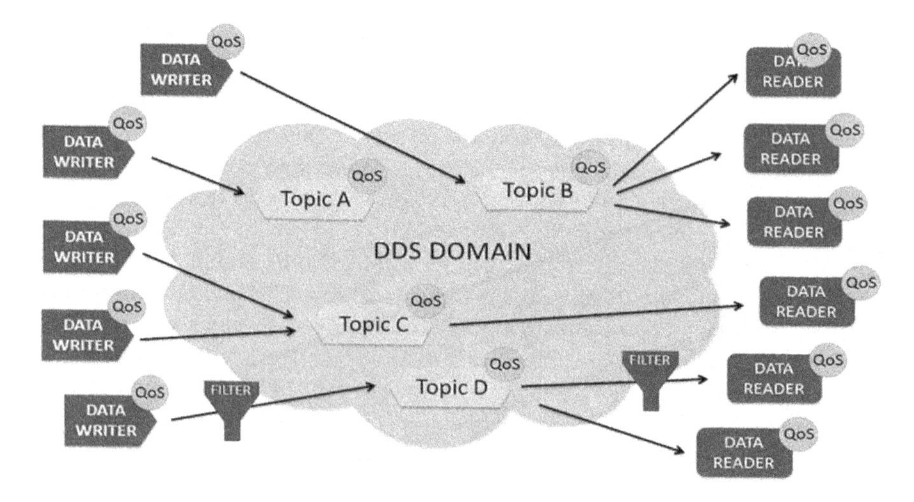

FIGURE 1.16 DDS public/subscribe methodology.

The publishers and subscribers are discovered dynamically and can express their intention of producing or consuming the specific data. The subscriptions are dynamically matched and the data flows from publisher to subscriber [45].

1.7 REAL-TIME OPERATING SYSTEM FOR IOT

In IoT we talk about the embedded systems. But the point is how the firmware can be developed or what can be the basis opted, so that our application can run efficiently. So, there are two methods used most commonly, i.e., Super Loop (Bare Metal) and the RTOS (Real-Time Operating Systems). In *Super Loop*, every task code except the interrupts are written in a single loop. And the firmware is run directly on the microcontroller hardware using memory mapped peripheral registers. There is no need for any operating system or any device drivers (Figure 1.17).

Whereas in the *RTOS* based system, an OS kernel and device drivers provide an interface between actual application code and the hardware (Figure 1.18), and the task is scheduled according to a specific period using a priority-based preemptive scheduling algorithm [46]. Mainly the RTOS are responsible for serving the real-time applications that processes the data in real time without any buffer delays. It forms a time-bounded system with well-defined time constraints, and most remarkably the processing time is the tenths of seconds or even shorter. These are best suited for airline traffic control systems, heart peacemakers, robots, telephone switching equipment, etc.

There are three types of RTOS: hard RTOS, soft RTOS, and firm RTOS.

Hard Real-Time Operating System: In such systems the time deadline is followed very strictly. The task should be started at a scheduled time, not before or after that, and should be completed within the time duration but not much earlier.

Soft Real-Time Operating System: As the name suggest, in such systems, although the deadline has been assigned for the task, they are handled softly and some delays are acceptable, for example, multimedia systems, digital audio systems, and online transaction systems, etc.

Firm Real-Time Operating System: Undoubtedly if it is a real-time operating system, it has to follow the deadlines but, especially in case of firm RTOS, missing the deadline may not have any major impact but could cause some unintended effects in the quality of results, for example, various multimedia applications.

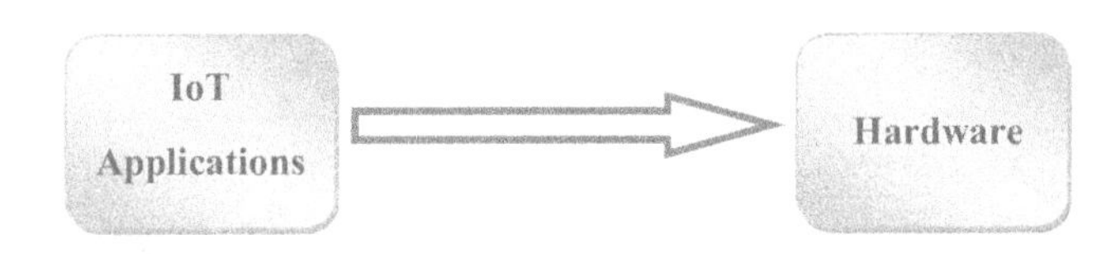

FIGURE 1.17 Super loop.

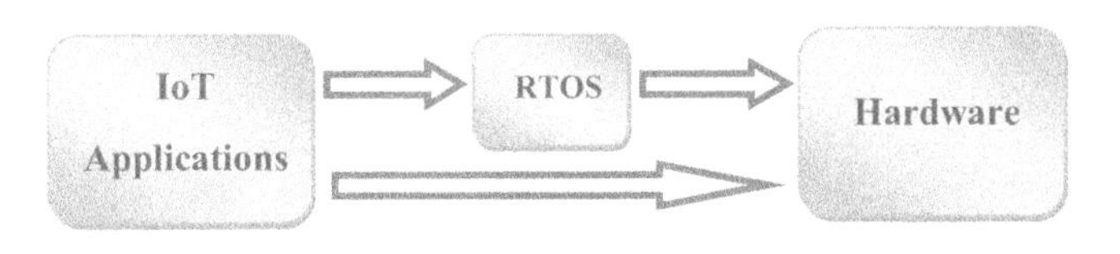

FIGURE 1.18 RTOS-based system.

1.8 IOT AND BIG DATA ANALYSIS

If we talk about the applications of the Internet of Things, we have to reach to a large number of domains, including smart homes, smart healthcare, smart cities, manufacturing, agriculture, and many more. So, IoT analytics is actually the analysis of data from various IoT data sources, including the sensors and actuators and various other objects connected to the Internet (Figure 1.19). The market potential of the IoT mainly depends on this collection and analysis of data from different data streams. But a report from the McKinsey Global Institute [47] [*Unlocking the Potential of Internet of Things*, June (2015)] says that only about 1% of IoT data is currently used, because most of the IoT analytics applications are toward anomaly detection and control. But in addition to that, to provide a greater business value onwards, it should be more oriented toward optimization and prediction.

Proceeding ahead from the simple sensor processing applications and making IoT analytics more widely deployed and used in more sophisticated ways, cloud computing and big data analysis come like a revolution, taking the IoT value to the next evolutionary level. Nowadays, almost all the advanced applications that process IoT data streams are integrated with the cloud computing infrastructures and using big data analytic technology, and come up with better performance, capacity, and scalability.

In this way of adding cherries to the cake, the integration of the edge computing infrastructures very prominently decentralized the processing of data streams at the very edge of the IoT network and the transfer of selected data only to the cloud from the edge devices, making it more robust and accurate. Although there is not much close relationship

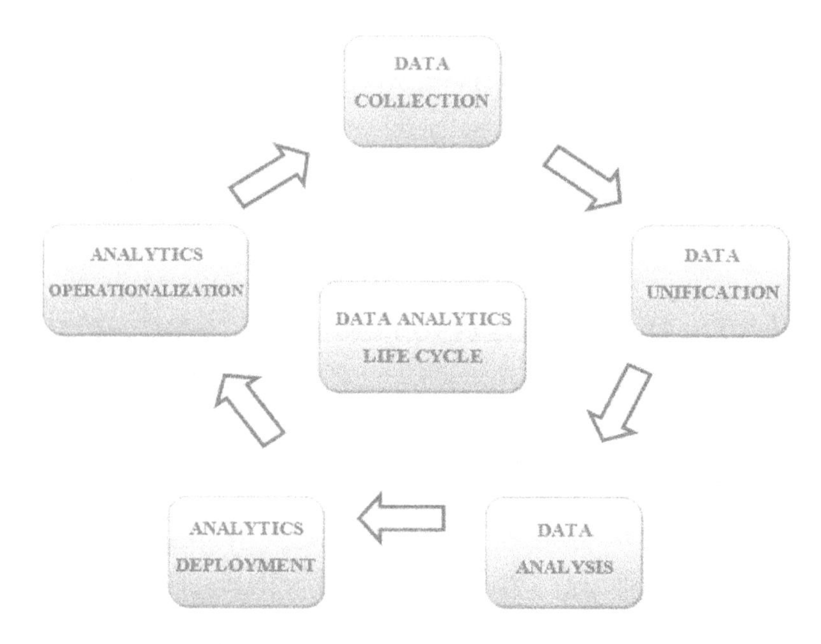

FIGURE 1.19 Data analysis phases.

between IoT analytics and big data analytics, still the IoT data nowadays is considered as big data if splashing out some common characterizing V parameters like Volume, Velocity, Variety, and Veracity.

Volume: The IoT data produced in such a large volume that they may exceed the storage and processing capabilities of conventional database systems.

Velocity: It defines the continuous production of data streams in very large frequencies and high ingestion rates.

Variety: Since we have the IoT devices in such a large number and diversity, obviously we have to deal with heterogeneous data in terms of formats and semantics.

Veracity: There is no way to be amazed if such a huge amount of heterogeneous data contained with noise data and uncertainty.

On the other hand, the big data streams come from large warehouses and numerous data sources rather than like IoT data streams coming from sensors or Internet devices only. For handling the IoT data velocity, streaming engines are needed, but in big data velocity is not the major concern, so map-reduce can be used for that. The variety in IoT data is because of heterogeneity of the sensors, whereas in big data, the variety is for consolidating data sources of different types. The noisy nature of the IoT devices and uncertainty in signals results in the veracity of the data streams, but that in big data is due to the uncertainty in processing the data sources. So, by these things, it can be concluded that with big data analytics tools and techniques like including data warehouses, streaming middleware, and engines and data mining techniques used with IoT data streams, it will be a better way to deal with the analysis [24, 48].

1.9 INDUSTRIAL INTERNET OF THINGS

1.9.1 Industry 4.0

Industry 4.0 was actually initiated by Germany in order to meet the manufacturing and production sectors with the IoT technologies. It was named 4.0 because it was a fourth revolution of the industrial production. Already, it has been widely accepted that before this, the industrial sector had gone through three major revolutions. First was between the 18th and 19th centuries when the production of goods was first facilitated with mechanical ways or machinery. In the second revolution, the productivity was exceptionally increased using the means of electrical energy. And moving forward, toward the third revolution in the industrial sector, in the era of World War II, the information technology together with the mechanical and electrical devices (software plus mechanics) added to the industrial manufacturing and production units led to a drastic change in production rates, efficiency, and decreasing development cost. This Industry 4.0 was mainly launched because many industrial stakeholders thought that it was a perfect time to include the cyber physical systems in the industries to increase the connectivity and communication for making better coordination in complex processes. Also including higher levels of automation by introducing programmable, configurable, and controllable manufacturing units will increase

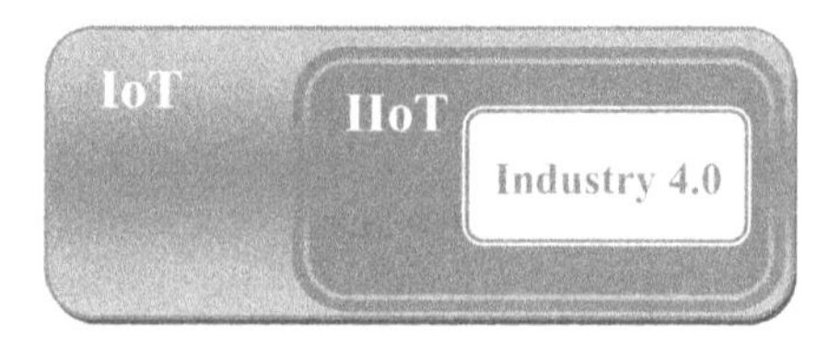

FIGURE 1.20 Domain relation of IoT, IIoT, and Industry 4.0.

the productivity rate of goods, which will give Germany a leading role in manufacturing and productions.

This fourth revolution involves the deployment of very advanced and automated computational and communication resources that have high performance with memories and low power requirements. The goal of making these industries the smart industries was achieved to a great extent, and the smart automated production systems were interconnected in a multilevel hierarchy in order to achieve high flexibility, efficiency, resilience, and safety at low cost. In this strategic evolution, even the customers are able to monitor the development progress of their products, and the manufacturers can optimize the resources used, minimize the environmental impacts, and increase real-time management of productions.

1.9.2 IIoT

What we call Industrial Internet of Things (IIoT) is actually the generalization of Industry 4.0, focusing not only on industrial process efficiency but also on the asset management, maintenance, etc. Although the concept and basic blocks of IoT and IIoT are the same, there are some points that make the IIoT as a subset of IoT (Figure 1.20), i.e., very strong need of continuous operation and the employment of the operational technology to the industrial level.

Another very important term – *Industrial Internet* – was launched by the General Electric Company in 2012 as a leader of IIoT, and it identifies the machine-to-machine communication technologies, industrial data analytics, cybersecurity, programmable logic controllers (PLC), and supervisory control and data acquisition (SCADA) system as the main constituents of IIoT [1, 49].

1.10 CONCLUSION

Going through all these fundamental aspects of IoT and visualizing the future technology, it can be concluded that this class of technology can be called the best till now for improving the quality of life and services. And it is proven that the IoT can automate everything around us. This chapter has presented as the first step knowledge reference for the practitioners working with IoT technologies and embedded systems. The chapter discussed a huge class of wireless communication technology and protocols provided with an enormous range of hardware with amazing versatility. Moreover, the facilitation of big data analysis through cloud and fog computing has also been discussed, giving up the way of deployment at any large stage in the future. Another challenging area is deploying IoT applications in such environments where Internet facility is not available permanently or network is highly intermittent due to difficult geographical terrains (remote areas or hill stations) or existing Internet facility is not working due to a natural disaster. Delay Tolerant Networks and Opportunistic Networks [50–55] are types of mobile ad-hoc networks and are best suitable for such environments. Exploring the possibilities of combining architecture and protocols of such networks with IoT applications is also an emerging research area. If IoT devices are forming a network which is being used for the purpose of distributed computing, then developing failure detection techniques [56–61] for monitoring the health of IoT devices is also an important future research direction.

REFERENCES

[1] Serpanos, D., & Wolf, M. (2017). *Internet-of-Things (IoT) Systems: Architectures, Algorithms, Methodologies*. Springer.

[2] Kodali, R. K., Jain, V., Bose, S., & Boppana, L. (2016, April). IoT based smart security and home automation system. In *2016 International Conference on Computing, Communication and Automation (ICCCA)* (pp. 1286–1289). IEEE.

[3] Ghayvat, H., Mukhopadhyay, S., Gui, X., & Suryadevara, N. (2015). WSN-and IOT-based smart homes and their extension to smart buildings. *Sensors*, 15(5), 10350–10379.

[4] Eddy, Y., Mohammed, M. N., Daoodd, I. I., Bahrain, S. H. K., Al-Zubaidi, S., Al-Sanjary, O. I., & Sairah, A. K. (2020). 2019 novel coronavirus disease (Covid-19): Smart contactless hand sanitizer-dispensing system using IoT based robotics technology. *Revista Argentina de Clínica Psicológica*, 29(5), 215.

[5] Rahaman, A., Islam, M. M., Islam, M. R., Sadi, M. S., & Nooruddin, S. (2019). Developing IoT based smart health monitoring systems: A review. *Revue d'Intelligence Artificielle*, 33(6), 435–440.

[6] Vippalapalli, V., & Ananthula, S. (2016, October). Internet of things (IoT) based smart health care system. In *2016 International Conference on Signal Processing, Communication, Power and Embedded System (SCOPES)* (pp. 1229–1233). IEEE.

[7] Al-Fuqaha, A., Guizani, M., Mohammadi, M., Aledhari, M., & Ayyash, M. (2015). Internet of things: A survey on enabling technologies, protocols, and applications. *IEEE Communications Surveys & Tutorials*, 17(4), 2347–2376.

[8] Luppicini, R. (Ed.). (2013). *Moral, Ethical, and Social Dilemmas in the Age of Technology: Theories and Practice: Theories and Practice*. IGI Global.

[9] Dhanalaxmi, B., & Naidu, G. A. (2017, February). A survey on design and analysis of robust IoT architecture. In *2017 International Conference on Innovative Mechanisms for Industry Applications (ICIMIA)* (pp. 375–378). IEEE.

[10] Deepti, S., & Nasib, S. G. *Smart Sensors: Analysis of Different Types of IoT Sensors*. DOI: 10.1109/ICOEI.2019.8862778.

[11] Satyanarayanan, M. (2001). Pervasive computing: Vision and challenges. *IEEE Personal Communications*, 8(4), 10–17.

[12] Saha, D., & Mukherjee, A. (2003). Pervasive computing: A paradigm for the 21st century. *Computer*, 36(3), 25–31.
[13] Ashton, K. (2009). That 'internet of things' thing. *RFID Journal*, 22(7), 97–114.
[14] Matutinovic, S. F., & Andonovski, J. (2013, May). European newspapers project contribution to the freedom of information: Finding out about Nikola. *Tesla from Historical Newspapers*.
[15] Mutreja, M., Khandelwal, K., Dham, H., & Chawla, P. (2021, July). Perception of IOT: Application and challenges. In *2021 6th International Conference on Communication and Electronics Systems (ICCES)* (pp. 597–603). IEEE.
[16] Lyytinen, K., & Yoo, Y. (2002). Ubiquitous computing. *Communications of the ACM*, 45(12), 63–96.
[17] Weiser, M. (1991). The computer for the 21st century. *Scientific American*, 265(3), 94–105.
[18] López-de-Armentia, J., Casado-Mansilla, D., & López-de-Ipina, D. (2012, July). Fighting against vampire appliances through eco-aware things. In *2012 6th International Conference on Innovative Mobile and Internet Services in Ubiquitous Computing* (pp. 868–873). IEEE.
[19] Hachani, A., Barouni, I., Said, Z. B., & Amamou, L. (2016, November). RFID based smart fridge. In *2016 8th IFIP International Conference on New Technologies, Mobility and Security (NTMS)* (pp. 1–4). IEEE.
[20] Klamer, T., & Allouch, S. B. (2010, March). Acceptance and use of a social robot by elderly users in a domestic environment. In *2010 4th International Conference on Pervasive Computing Technologies for Healthcare* (pp. 1–8). IEEE.
[21] Yang, Z., Yue, Y., Yang, Y., Peng, Y., Wang, X., & Liu, W. (2011, July). Study and application on the architecture and key technologies for IOT. In *2011 International Conference on Multimedia Technology* (pp. 747–751). IEEE.
[22] Wu, M., Lu, T. J., Ling, F. Y., Sun, J., & Du, H. Y. (2010, August). Research on the architecture of internet of things. In *2010 3rd International Conference on Advanced Computer Theory and Engineering (ICACTE)* (Vol. 5, pp. V5–484). IEEE.
[23] Khan, R., Khan, S. U., Zaheer, R., & Khan, S. (2012, December). Future internet: The internet of things architecture, possible applications and key challenges. In *2012 10th International Conference on Frontiers of Information Technology* (pp. 257–260). IEEE.
[24] Soldatos, J. (Ed.). (2016). *Building Blocks for IoT Analytics*. River Publishers.
[25] Gridling, G., & Weiss, B. (2007). *Introduction to Microcontrollers*. Vienna University of Technology Institute of Computer Engineering Embedded Computing Systems Group.
[26] Deshmukh, A. V. (2005). *Microcontrollers: Theory and Applications*. Tata McGraw-Hill Education.
[27] Badamasi, Y. A. (2014, September). The working principle of an Arduino. In *2014 11th International Conference on Electronics, Computer and Computation (ICECCO)* (pp. 1–4). IEEE.
[28] Richardson, M., & Wallace, S. (2012). *Getting Started with Raspberry PI*. O'Reilly Media, Inc.
[29] Kashyap, M., Sharma, V., & Gupta, N. (2018). Taking mqtt and nodemcu to iot: Communication in internet of things. *Procedia Computer Science*, 132, 1611–1618.
[30] Jie, L., Ghayvat, H., & Mukhopadhyay, S. C. (2015, June). Introducing Intel Galileo as a development platform of smart sensor: Evolution, opportunities and challenges. In *2015 IEEE 10th Conference on Industrial Electronics and Applications (ICIEA)* (pp. 1797–1802). IEEE.
[31] Isikdag, U. (2015). Internet of things: Single-board computers. In *Enhanced Building Information Models* (pp. 43–53). Springer.
[32] Dasarathy, B. V. (1997). Sensor fusion potential exploitation-innovative architectures and illustrative applications. *Proceedings of the IEEE*, 85(1), 24–38.
[33] Yacoub, M. D. (2017). *Wireless Technology: Protocols, Standards, and Techniques*. CRC Press.
[34] Bisdikian, C. (2001). An overview of the Bluetooth wireless technology. *IEEE Communications Magazine*, 39(12), 86–94.

[35] Ezhilarasan, E., & Dinakaran, M. (2017, February). A review on mobile technologies: 3G, 4G and 5G. In *2017 Second International Conference on Recent Trends and Challenges in Computational Models (ICRTCCM)* (pp. 369–373). IEEE.
[36] Al-Alawi, A. I. (2006). WIFI technology: Future market challenges and opportunities. *Journal of Computer Science*, 2(1), 13–18.
[37] Ramya, C. M., Shanmugaraj, M., & Prabakaran, R. (2011, April). Study on ZigBee technology. In *2011 3rd International Conference on Electronics Computer Technology* (Vol. 6, pp. 297–301). IEEE.
[38] Mekki, K., Bajic, E., Chaxel, F., & Meyer, F. (2019). A comparative study of LPWAN technologies for large-scale IoT deployment. *ICT Express*, 5(1), 1–7.
[39] Finkenzeller, K. (2010). *RFID Handbook: Fundamentals and Applications in Contactless Smart Cards, Radio Frequency Identification and Near-Field Communication.* John Wiley & Sons.
[40] Karagiannis, V., Chatzimisios, P., Vazquez-Gallego, F., & Alonso-Zarate, J. (2015). A survey on application layer protocols for the internet of things. *Transaction on IoT and Cloud Computing*, 3(1), 11–17.
[41] Shelby, Z., Hartke, K., & Bormann, C. (2014). *The Constrained Application Protocol (CoAP).* RFC Editor. doi: 10.17487/RFC7252. https://www.rfc-editor.org/info/rfc7252
[42] Hunkeler, U., Truong, H. L., & Stanford-Clark, A. (2008, January). MQTT-S – A publish/subscribe protocol for wireless sensor networks. In *2008 3rd International Conference on Communication Systems Software and Middleware and Workshops (COMSWARE'08)* (pp. 791–798). IEEE.
[43] Saint-Andre, P. (2004). *Extensible Messaging and Presence Protocol (XMPP): Core.* RFC Editor. doi: 10.17487/RFC3920. https://www.rfc-editor.org/info/rfc3920
[44] Vinoski, S. (2006). Advanced message queuing protocol. *IEEE Internet Computing*, 10(6), 87–89.
[45] Sanchez-Monedero, J., Povedano-Molina, J., Lopez-Vega, J. M., & Lopez-Soler, J. M. (2011). Bloom filter-based discovery protocol for DDS middleware. *Journal of Parallel and Distributed Computing*, 71(10), 1305–1317.
[46] Hambarde, P., Varma, R., & Jha, S. (2014, January). The survey of real time operating system: RTOS. In *2014 International Conference on Electronic Systems, Signal Processing and Computing Technologies* (pp. 34–39). IEEE.
[47] Manyika, J., Chui, M., Bisson, P., Woetzel, J., Dobbs, R., Bughin, J., & Aharon, D. (2015). Unlocking the potential of the internet of things. *McKinsey Global Institute*, 1.
[48] Malek, Y. N., Kharbouch, A., El Khoukhi, H., Bakhouya, M., De Florio, V., El Ouadghiri, D., . . . Blondia, C. (2017). On the use of IoT and big data technologies for real-time monitoring and data processing. *Procedia Computer Science*, 113, 429–434.
[49] Dalmarco, G., Ramalho, F. R., Barros, A. C., & Soares, A. L. (2019). Providing industry 4.0 technologies: The case of a production technology cluster. *The Journal of High Technology Management Research*, 30(2), 100355.
[50] Verma, A., Verma, P., Dhurandher, S. K., & Woungang, I. (Eds.). (2021). *Opportunistic Networks: Fundamentals, Applications and Emerging Trends.* CRC Press.
[51] Verma, A., & Srivastava, D. (2012). Integrated routing protocol for opportunistic networks. *arXiv preprint arXiv:1204.1658.*
[52] Singh, M., Verma, P., & Verma, A. (2021). Security in opportunistic networks. In *Opportunistic Networks* (pp. 299–312). CRC Press.
[53] Verma, A., Singh, M., Pattanaik, K. K., & Singh, B. K. (2018). Future networks inspired by opportunistic networks. In *Opportunistic Networks* (pp. 230–246). Chapman and Hall/CRC.
[54] Verma, A., & Pattanaik, K. K. (2017). Routing protocols in opportunistic networks. In *Opportunistic Networking* (pp. 123–166). CRC Press.
[55] Verma, A., Pattanaik, K. K., & Ingavale, A. (2013). Context-based routing protocols for oppnets. In *Routing in Opportunistic Networks* (pp. 69–97). Springer.

[56] Ghosh, C., Verma, A., & Verma, P. (2022). Real time fault detection in railway tracks using Fast Fourier Transformation and Discrete Wavelet Transformation. *International Journal of Information Technology*, 14(1), 31–40.
[57] Chaurasia, B., & Verma, A. (2020). A comprehensive study on failure detectors of distributed systems. *Journal of Scientific Research*, 64(2), 250–260.
[58] Verma, A., & Pattanaik, K. K. (2016). Failure detector of perfect P class for synchronous hierarchical distributed systems. *International Journal of Distributed Systems and Technologies (IJDST)*, 7(2), 57–74.
[59] Verma, A., & Pattanaik, K. K. (2015). Multi-agent communication-based train control system for Indian railways: The behavioural analysis. *Journal of Modern Transportation*, 23(4), 272–286.
[60] Verma, A., & Pattanaik, K. K. (2015). Multi-agent communication based train control system for Indian railways: The structural design. *Journal of Software*, 10(3), 250–259.
[61] Verma, A., Singh, M., & Pattanaik, K. K. (2019). Failure detectors of strong S and perfect P classes for time synchronous hierarchical distributed systems. In *Applying Integration Techniques and Methods in Distributed Systems and Technologies* (pp. 246–280). IGI Global.

Issues and Challenges Associated with Fog Computing in Healthcare 4.0

2

Jeetendra Kumar[1] and Rashmi Gupta[1]
[1]*Assistant Professor, ABV University, Bilaspur (C.G.), India*

Contents

DOI: 10.1201/9781003244714-2

2.1 INTRODUCTION

It has taken decades for the industrial sector to evolve from Industry 1.0 to Industry 4.0; it hasn't happened overnight. As Industry 4.0 and Logistics 4.0 have grown, so has the healthcare sector. It is enabling technologies that are responsible for the evolution of industry and healthcare. This chapter will discuss Industry 4.0, Healthcare 4.0, fog computing, fog computing in healthcare, various issues of fog computing in healthcare, and research challenges associated with fog computing in healthcare.

2.1.1 Industry 4.0

In recent years, Industry 4.0 has become a hot topic. As a result, it has attracted a lot of attention because it makes human existence much easier than it was previously. In 2011, the German government first used the term 'Industry 4.0'. In Industry 4.0, emphasis is on smart sensors, real-time connected technologies, and artificial intelligence-based industry automation.

> Industry 4.0, also sometimes referred to as IIoT or smart manufacturing, marries physical production and operations with smart digital technology, machine learning, and big data to create a more holistic and better-connected ecosystem for companies that focus on manufacturing and supply chain management [1].

As information and communication technology exploded during this decade, we saw a shift in the way we shared and traded information. For the same reason, it shifted perspectives in the assembly industry and in traditional creativity activities, bringing physical and virtual worlds closer together. There are a number of ways that cyber-physical systems (CPSs) have disguised this barrier in the commercial world. CPSs allow the machines to communicate more intelligently with one another, essentially without any physical or geological hindrances, thanks to their ability to communicate. Industry 4.0 is focused on creating 'more smart machines/devices'. With CPSs, Industry 4.0 shares, analyzes, and guides intelligent actions for a variety of industrial processes. Industry 4.0 makes devices smarter and smarter, which helps you to do your work easily.

It's a radical shift that involves digitalization and automation of every aspect of the business, including production, inventory management, human resource management, and raw material management, among others. From little to big, and public to global, organizations that utilize the ideas of persistent improvement and have elevated requirements for innovative work will acknowledge the idea of Industry 4.0 and make themselves significantly more serious on the lookout. Self-advancement, self-awareness, and self-customization can be introduced into the business to achieve this. However, instead of operating them, the makers will want to communicate with PCs or other computers instead of operating them.

During the COVID-19 pandemic, we have seen the importance of digital technology in our lives. The era has seen the increasing trend of online shopping, online teaching, video conferencing, virtual exhibition, online payment, etc. All these technologies take a lot of changes in the industrial process to the next level, that is, Industry 4.0. The term 'Industry 4.0' has been visualized due to adoption of digital technologies in manufacturing industries. The idea of Industry 4.0 expects obscuring the contrasts between crafted by individuals and crafted by machines. Figure 2.1 shows the empowering technologies of Industry 4.0. Industry 1.0 brought manually working machines in industry, Industry 2.0 brought power into the business, and Industry 3.0 robotized the uniform undertakings of line laborers. Now Industry 4.0 applies artificial intelligence with fewer human interventions. [2–3] Industry 4.0 varies from the past eras in the way that it identifies with all fields of life. Industries are still in the early stages of adopting Industry 4.0, which is still in its infancy. For industries to remain competitive and profitable, they must implement the new methods as quickly as feasible. [4]

Industry 4.0 is revolutionizing the manufacturing industry by automating processes that were previously done by hand or with the help of animals. Mechanization through water and steam power; mass production in assembly lines; and automation utilizing digital technology were the previous industrial revolutions that led to paradigm shifts in manufacturing [5]. In the late 18th century, mechanized production facilities were introduced to the world along with Industry 1.0. For large manufacture of commodities, water and steam-powered machines were created. In 1784, the first weaving loom was invented. Larger organizations with more owners, managers, and employees were created for manufacturing efficiency, and scale increased. Quality, efficiency, and scalability were all equally important in Industry 1.0 [6].

After World War I, the second industrial revolution, which is known as Industry 2.0, arrived. The introduction of electrically powered machines played a major role in this revolution. In the past, electrical energy had been used as a key power source. At this time, industry culture was also transformed into management programs to improve the efficiency of manufacturing facilities [7]. Worker, workplace practices, and effective resource

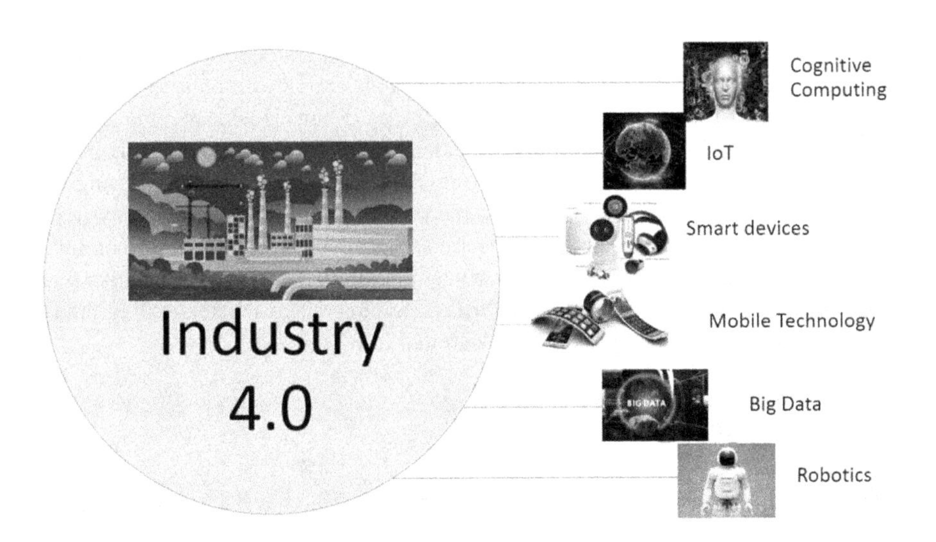

FIGURE 2.1 Empowering technologies of Industry 4.0.

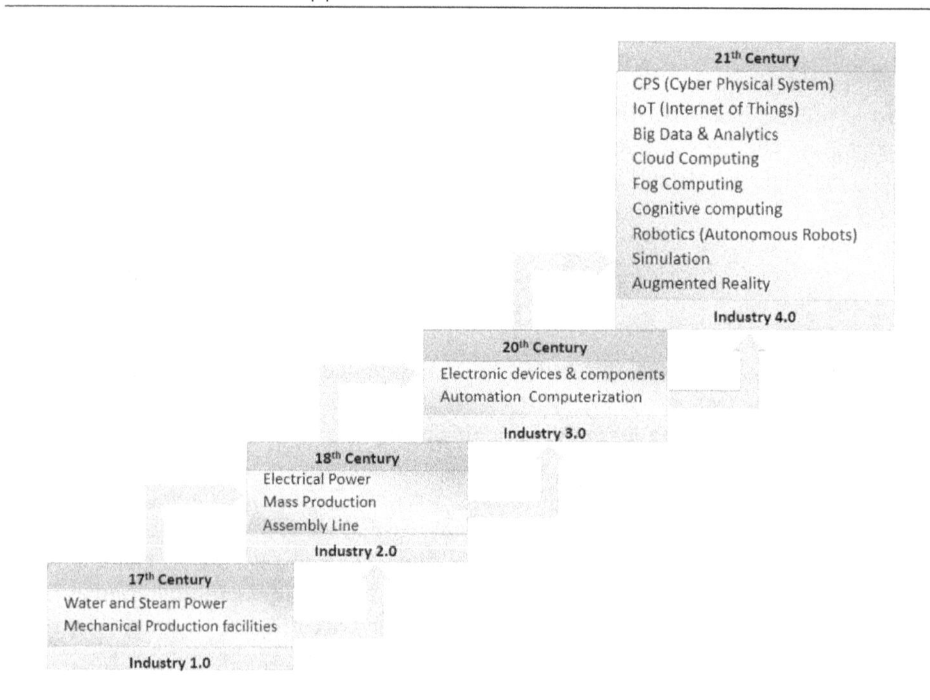

FIGURE 2.2 Evolution from Industry 1.0 to Industry 4.0.

allocation were pioneered by American mechanical engineer Fredrick Taylor. General mass production became the primary method of production in Industry 2.0. Because of vast steel production, railroads were introduced into the industrial system, which in turn led to mass manufacturing.

It was in the late 20th century when the digital revolution began and Industry 3.0 was born. Many electronic devices, such as transistors and integrated circuits (IC), have been invented and manufactured in recent years. This has lowered the amount of work required by machines and enhanced their speed as well as precision. Nowadays, the majority of individuals are familiar with the production sector's reliance on digital technologies [8–9]. Industry 4.0 is the fourth industrial revolution. Thanks to the Internet of Things (IoT), real-time data, and cyber-physical systems, the emphasis on digital technology in recent decades has been elevated to a new level. There are many benefits to adopting an Industry 4.0 strategy to production. There is a better way to collaborate across departments, partners, providers, products, and people with this technology. The schematic diagram of the overview for the industrial revolutions is illustrated in Figure 2.2.

2.1.2 Healthcare 4.0

Nowadays, health issues are a major concern. A busy life or a hard workload prevents people from having enough time to contact a doctor about their health difficulties. Due to a lack of attention to their health, routine medical checkups are not carried out. Indirectly, one's health has a direct impact on one's productivity. Having a healthy population is crucial to a country's development since it has a direct impact on the country's progress. In general, it is duty of the country to emphasize public health policies [10]. COVID-19 is a highly infectious disease with pneumonia-like symptoms caused by the novel, severe, acute respiratory syndrome corona viruses (SARS-CoV). [7] COVID-19 not only affects people physically but mentally as well. This pandemic gave a new impetus to the evolution in the health sector. Prior to COVID-19, evolution was happening in the health sector, but its pace was slow. After COVID-19 the healthcare sector has seen the biggest advantages of digital healthcare [11–12].

A global transformation to Industry 4.0 (digital, fully automated settings, and cyber-physical systems) is underway. Technology and innovation are being implemented in many different industries as part of Industry 4.0. With the evolution in industrial technologies (from Industry 1.0 to 4.0), the whole healthcare system is also transformed from Healthcare 1.0 to Healthcare 4.0. Due to the impact of technical and innovative approaches, the healthcare sector is moved toward a Healthcare 4.0 paradigm. Industry 4.0 concentrates on making smart devices through the application of digital technology. Smart (intelligent) devices involving data analytics capabilities are empowered with technological advancements such as Machine Learning (ML), Deep Learning (DL), Artificial Intelligence (AI), Big Data Analytics (BDA), IoT, etc. The application areas of smart devices in healthcare are assisted living [13–14], physiological signal monitoring [15–16], disease prediction [17], etc. Figure 2.3 shows a schematic diagram of Healthcare 4.0.

With advances in medical technology, the healthcare sector has evolved from Healthcare 1.0 to Healthcare 4.0, as shown in Figure 2.4.

Healthcare 1.0: In Healthcare 1.0, basic mechanical and manual instruments came into the existence. These instruments included hand-operated blood pressure checking machines, syringes, thermometers, stethoscopes, hand-operated surgery instrument, medical oxygen cylinders, etc. At that time, some of the devices were invasive like invasive EEG. In Healthcare 1.0 doctors were manually checking the patients and suggesting prescriptions.

Healthcare 2.0: During the transition from Healthcare 1.0 to 2.0 some advanced medical instruments were used, for example X-ray machines, ECG machines,

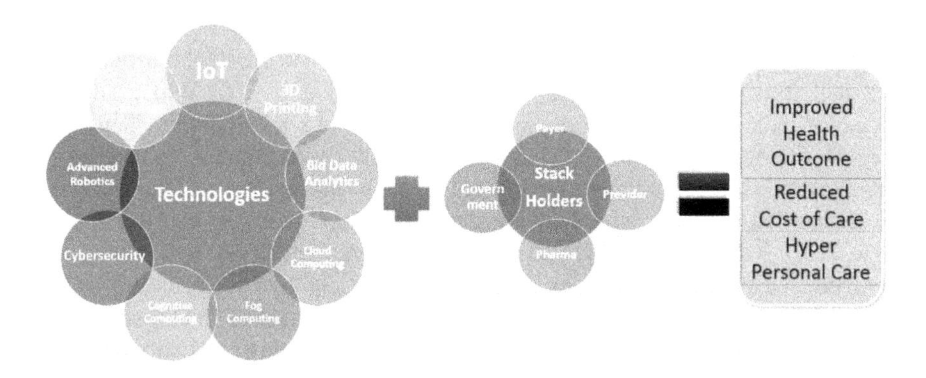

FIGURE 2.3 Schematic diagram of Healthcare 4.0.

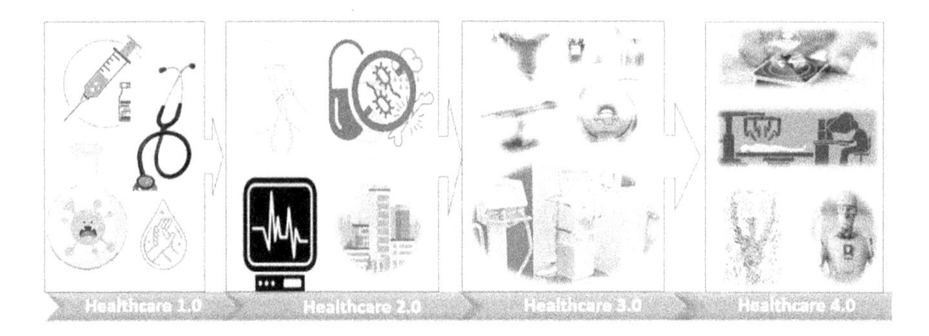

FIGURE 2.4 Evolution from Healthcare 1.0 to Healthcare 4.0.

sphygmomanometers, and systems for continuous monitoring of heart rate, blood pressure, respiration rate. Data of the patient was manually entered into the computer system by the operator. Doctors were checking their patients manually and based on their previous records.

Healthcare 3.0: In Healthcare 3.0, some more complex devices like MRI machines, CT scan machines, heart echo machines, implanted pacemakers, and ultrasonography were used. These medical devices were sophisticatedly designed with high accuracy. These devices were controlled by computer software.

Healthcare 4.0: Penetration of artificial intelligence techniques in the healthcare sector has revolutionized the healthcare sector. Nowadays the use of IoT and fog computing-based technology have allowed the automatic recording and analyzing of patient data in real-time and the automatic ordering of prescriptions. Many computerized expert systems are also available, which analyzed the patient data in realtime and predict the suggested diagnosis. Doctors can monitor their patients from remote locations. Advanced smart robots can perform surgeries without doctor invention [18–19].

2.1.3 Empowering Tools for Healthcare 4.0

Healthcare 4.0 is the reinvention of the care delivery ecosystem to directly benefit patients and their loved ones, by equipping the care parties with technologies that drastically transformed their ability to create and maintain a healthy community [20]. Without the use of advanced technology, the concept of Healthcare 4.0 would not be possible. Sensors play an important role in Healthcare 4.0. The sensor senses the information and transfers the sensed data to fog/cloud for analytics purposes. In the healthcare sector, these sensors can be biosensors that measure physiological signals like blood pressure, heart rate, respiratory rate, EEG, ECG, EMG, activity monitoring, speech recognition, etc., or they can be environment sensor that measure environmental conditions like temperature, humidity, air quality, water quality, etc. With these sensors, data can be sensed and transferred automatically without human intervention. Many smart healthcare devices are available in the market, for example Fitbit smart bands, Mi smart band, GOQii fitness tracker, envirochip for radiation protection, etc.

Another pillar of Healthcare 4.0 is artificial intelligence. The raw data is not useful until it is processed. Many machine learning algorithms are being used continuously to make predictions using healthcare data. Neural networks are one more step ahead of machine learning. Neural networks are high-level algorithms that can take a decision as a human being takes. Many critical healthcare problems have been solved using these highly trained artificial networks. Advanced robots are also doing surgeries with human intervention. They are using high-level artificial intelligence-based methods and smart sensors. Neoguide colposcopy system [21], minimally invasive neurosurgical intracranial robot (MINIR) [22], and STIFF-FLOP [23] soft robot are some examples of robotics systems being used for minimally invasive surgery. Management of a huge amount of healthcare data generated by healthcare sensors is a very cumbersome task but thanks to new technologies like big data and blockchain, techniques that take care of data in all three dimensions – volume, velocity, and variety. Real-time response in the healthcare sector and management for networks was only possible due to the cloud/IoT/fog computing facilities. Cloud computing ensure 24/7 availability of high-speed computing servers and storage facilities. Figure 2.5 shows digital tools for empowerment of Healthcare 4.0.

FIGURE 2.5 Digital tools for empowerment of Healthcare 4.0.

2.2 FOG COMPUTING

Fog computing is the extension of the Internet of Things (IoT) and cloud computing where computing facilities have been extended toward the edge of the network. According to the U.S. National Institute of Standards and Technology, 'Fog computing decentralizes management, data analytics, and applications within the network itself. This, in turn, minimizes the intervention time between requests and responses, providing local computing resource to devices and network connectivity to centralized services' [24]. In recent years, fog computing has gained the attention of networking professionals because it decreases the load on the cloud. Inclusion of a fog node in-between sensor nodes reduces the load on the cloud as well as on the network. Any device having computing facilities and Internet connectivity can work as a fog node. Cisco has introduced the term 'fog computing' in 2012 to manage the high load on the cloud. Small boards like Arduino having computing facilities has increased the demand for fog computing. The fundamental purpose of fog computing is to reduce latency and bandwidth consumption by incorporating intelligence into the access point to optimize common IoT situations. Fog servers communicate with mobile users via off-shelf wireless interfaces, including Wi-Fi, mobile, and Bluetooth, via single-hop wireless connections. The fog server can provide cloud-like services to mobile users without the help of other fog servers or the remote cloud for predefined applications. On the other hand, the fog server can be connected to the cloud via the Internet to use rich cloud computing and content resources. Many protocols, like MQTT, CoAP, 6LowPAN, etc., have provided the basis for fog computing.

2.2.1 Fog Computing Architecture

Fog computing is represented as a hierarchical architecture, which extends the cloud computing facilities to the edge of the network. Figure 2.6 shows traditional cloud computing

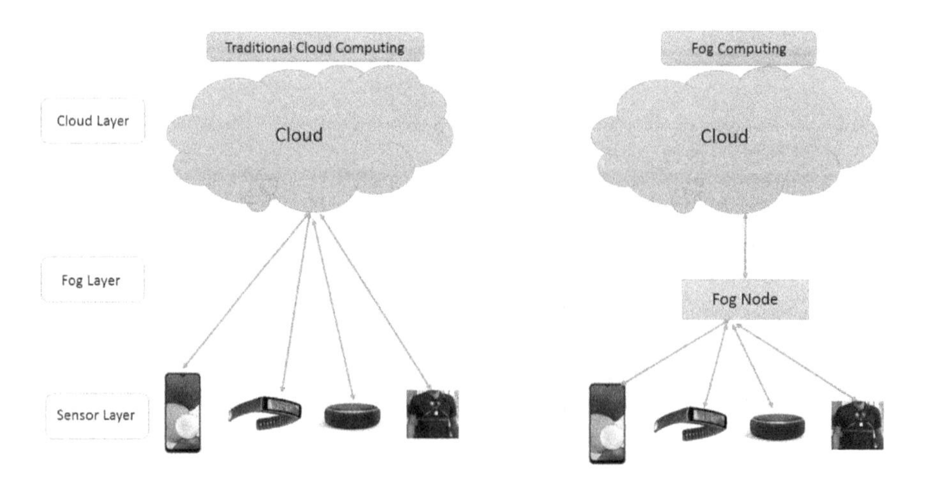

FIGURE 2.6 Traditional cloud computing versus fog computing.

versus fog computing. In traditional cloud computing, data is collected by sensors, and raw data is directly transferred to the cloud node where data visualization and data analysis are performed. In contrast to traditional cloud computing, fog computing incorporated one middle layer in between device node and cloud node, i.e., fog node. Fog computing architecture has three layers: (1) devices layer or sensor layer or terminal layer, (2) fog layer, and (3) cloud layer. The bottom-most layer is the device layer where many sensors exist. These sensors continuously sense data from their surrounding environment and transfer the data to nearby fog node using a wired/wireless data transmission method. Fog node cleans and preprocesses the raw data to make it ready for performing data analytics. After preprocessing, reduced data is transferred to the cloud for high-level data analysis. The system designer defines the distribution of data processing between cloud node and fog node.

2.2.2 Fog Computing in Healthcare 4.0

In Healthcare 4.0, many smart devices have been incorporated for bringing the revolution in the healthcare industry. The healthcare industry has seen a change in the strategy of treatment from manually checking of patients by a doctor to remote monitoring of patients by AI-based expert systems. The role of IoT is increasing day by day to improve patient monitoring, quality of care, early diagnosing of diseases, etc. 'The reliance of healthcare on IoT is increasing to improve access to care, increase the quality of care, and most importantly, reduce the cost of care' [25]. In real-life healthcare services, a speedy response is needed. Thus in healthcare services, the fog computing concept is being used widely, and data is analyzed at fog nodes rather than cloud nodes.

The use of fog computing has been incorporated in the healthcare sector to reduce the load to cloud and network bandwidth. The healthcare sector always requires a real-time response from the computer. While using traditional cloud computing, the computers were always heavily loaded and they took some time for giving responses, but in fog computing, computation work can also be performed on the fog node and response can also be generated from the fog computing in less time, which is suitable for the healthcare sector. Justas in the industry sector, in the healthcare sector, fog computing follows a three-layer architecture. Figure 2.7 shows fog computing architecture for Healthcare 4.0.

1. **Device node:** The device layer consist of various sensors, and sensor sense the data. These sensors can be fixed on the room to sense surrounding environmental conditions like temperature, humidity, air quality; or can be worn on the human body to form body area networks like EEG sensor, ECG sensor, BP sensor, glucose sensor, etc. These sensors sense the data and transfer all raw data to the fog node for further processing.
2. **Fog node:** The fog node is responsible for performing the initial computation of the raw data. Data cleaning, segmentation, removal of missing values, converting unstructured data to structured data, data compression, etc., steps are performed on the fog node. Fog nodes are generally placed in the proximity of the sensors, like building, hospital campuses, etc., and have high-speed Internet connectivity with the cloud.

3. **Cloud node:** Cloud nodes comprise a high level of computing facilities, large storage, and high-speed servers. At the cloud node, an AI-based expert system analyzes the incoming data and makes predictions, like early disease prediction, patient condition prediction, medicine prescription, bed availability prediction in hospitals, etc. The secret of prediction lies behind the highly efficient machine learning models. These models are trained on very huge data to learn the pattern of healthcare data, and after training these models can predict the trend of new incoming data.

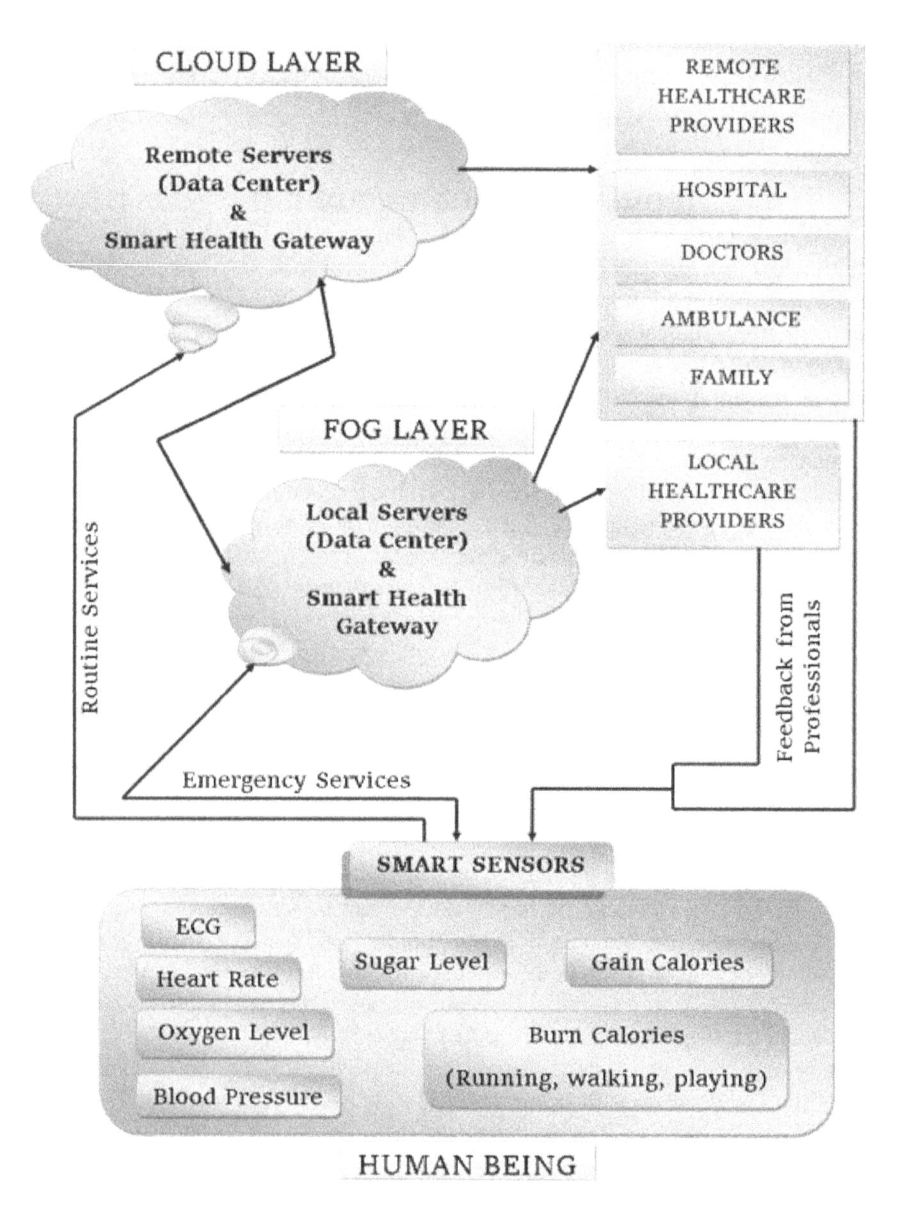

FIGURE 2.7 Fog computing architecture for Healthcare 4.0.

2.2.3 Fog Computing Framework for Healthcare 4.0 in Recent Research Works

The market for medical sensors has been growing constantly. According to one report of MORDOR intelligence, 'The medical sensors market was valued at USD 5.69 billion in 2020 and expected to reach USD 10.11 billion by 2026 and grow at a CAGR of 10.07% over the forecast period (2021–2026)' [26]. Parallelly with an increase in the market of the sensor, research on medical sensors has also increased. Many researchers are continuously working on smart sensors along with fog computing. We have selected some recent research work for discussion. Table 2.1 shows recent work done in the area of Healthcare 4.0 in fog computing.

TABLE 2.1 Recent Work Done in the Area of Healthcare 4.0 in Fog Computing

AUTHOR	*YEAR*	*OBJECTIVE*	*SENSOR USED*	*METHODOLOGY*	*RESULTS*
Pravin et al. [27]	2019	Prediction of Dengue disease using secure healthcare framework.	Temperature, blood pressure, heart rate, respiratory rate, body pain, skin symptoms for collecting the data of body condition of patient; water quality, humidity, rainfall, temperature and degree of mosquito for locating mosquito breeding sites.	Data is collected from IoT sensors and sent to fog node. Based on the results, alert signal is sent to the people, which contains the information of severity of disease and suggested treatment. Then final data is sent to the cloud to be considered as a base for future prediction.	They found that their processing time, delay time, and energy were less than the existing system.
Sanjay et al. [28]	2021	Prediction and prevention of Zika virus in fog computing.	Mobile phone and mosquito sensor	They used three-layer fog architecture and Fuzzy KNN for diagnosis purpose.	They achieved 94.2% accuracy for the initial diagnosis of patients.

(Continued)

TABLE 2.1 (Continued)

AUTHOR	*YEAR*	*OBJECTIVE*	*SENSOR USED*	*METHODOLOGY*	*RESULTS*
Tuli et al. [29]	2020	Smart healthcare system for diagnosis of heart disease.	Spo2, heart rate, respiration rate, EEG, ECG, EMG, blood pressure, glucose level, and activity level sensor.	They integrated ensemble deep learning at the edge device.	The model performed better in terms of arbitration time, jitter, latency, execution time, network usage, and power consumption.
Priydarshani et al. [30]	2018	Prediction of stress type, diabetes, and hypertension.	EDA, HR, Spo2, temperature, SBP, DBP, TC, HDL, LDL, PGC, HR, DGC, serum insulin, BMI, age, pedigree function.	They used three-layer fog computing architecture and deep neural networks	91.63% accuracy.
Rahmani et al. [31]	2018	Development of smart E-health gateways	Blood pressure, temperature, SPo2, sleep, ECG, heart rate, fall detector, steps, posture, activity, surrounding temperature, light, pressure.	They used three-layer fog computing architecture in which analysis is performed on cloud and preprocessing is performed on fog node.	Better functionality.
Hassan et al. [32]	2019	Development of E-Health system for monitoring of elderly.	Mysignals HW V2 platform havingSpo2, pulse, airflow, glucometer, ECG, temperature, BP, Galvanic skin response sensors.	In the proposed system Mysignals HW V2 was used for data collection, Android app was used as fog node which preprocesses the data and sends it to the cloud node.	83 points on the system usability scale.

AUTHOR	*YEAR*	*OBJECTIVE*	*SENSOR USED*	*METHODOLOGY*	*RESULTS*
Dhillon et al. [33]	2020	To develop health information system for predicting alcohol addiction in Punjab.	Blood pressure, body temperature, sound.	They used three-layered fog computing architecture, and for prediction purposes, they used Bayesian network, neural network and KNN algorithm.	Their framework performed better in terms of latency, network bandwidth, energy, and response time.
Muhammad et al. [34]	2020	To analyze medical data from the body area network.	Number of steps taken in a day, calories burned, cholesterol level, sleeping hours, sleep quality, calorie intake.	Data fusion enabled a multi-sensor approach to analyze the data.	98% accuracy.
Ajay Dev et al. [35]	2021	To monitor and predict stroke disease in fog computing scenario.	**Health Parameter:** numbness, dizziness, headache, cholesterol, blood pressure, slurring speech, paralysis, difficulty in seeing, sudden vision loss. **Environment Parameter:** air pollution, physical activity, smoke, dietary habit, working environment. **Location Parameter** **Personal Parameter**	They used three-layered fog architecture and RFB, SVM, ANN, DT, and NB for classification.	Highest sensitivity and specificity with random forest algorithm with boosting (RFB) algorithm.

2.2.4 Advantages of using Fog Computing in Healthcare

The trend of implementing smart healthcare in fog computing-based systems has increased since the last decade. Even in the era of the COVID-19 pandemic, it has been seen that many patients have been monitored by doctors from remote locations because many corona virus-infected patients were at home in quarantine. This COVID-19 pandemic has shown the importance of fog computing in remote monitoring. Fog computing has penetrated the Healthcare 4.0 market due to ease and real-time response. Following are some advantages of using fog computing in Healthcare 4.0.

1. **Real-time response:** The main reason behind the penetration of fog computing in Healthcare sector 4.0 is the real-time response. In the healthcare sector, based on the critical situation of patients, real-time response is also required because the late response can cause severe health problems, even death. In fog computing, rather than sending data to the cloud, data is analyzed at a nearby fog node so that immediate response can be collected. Extending the computing facility from cloud to edge is the main reason for real-time response. Even the fog node has less storage and computing facility than the cloud but initial processing can be easily performed on the data efficiently.
2. **Reduction of load on the cloud:** As it has been already discussed that some intermediary calculation on data can be performed at fog node and preprocessed and cleaned data is transferred to the cloud, hence the load on the cloud is reduced. If the load on the cloud will be reduced then the cloud can handle more requests from many nodes. The work distribution between the cloud and fog node is dependent on developers' designs. Sometimes even the analytics part and prediction are also performed at the fog node and data is stored in the cloud.
3. **Efficient network bandwidth utilization:** Employing the fog node between the cloud node and sensor node helps to utilize the network bandwidth very efficiently. Consider an example: Suppose a sensor generates 1MB of data per second. In traditional cloud computing whole 1MB of data is sent to the cloud but in fog computing suppose after preprocessing its size is reduced to 0.2MB. So only 0.2MB of data is sent to the cloud for further processing, which reduces the load on the network. In the healthcare sector, a large amount of data is continuously generated by many sensors like BP, glucose, Spo2, HR, etc. So preprocessing of this data at the fog node will reduce the load on the network.
4. **Energy efficiency:** Fog computing also provides energy efficiency by reducing the load on the cloud and network. Nowadays many energy-efficient sensors and fog devices are available that can work even with less energy availability.
5. **Privacy and security:** As in the Healthcare Internet of Things (HIoT), the health data generated by various body sensor networks are mobilized on the Internet. Data privacy and security is one of the major threats when data is exploded over the Internet. By employing the fog node rather than sending the raw data over the Internet, most of the data is analyzed at the fog node. From the fog node, compressed and encrypted data is transferred to the cloud so it cannot be patched.

2.3 ISSUES OF FOG COMPUTING IN HEALTHCARE 4.0

The use of fog computing in healthcare is very helpful for the healthcare sector, but many issues are associated with it. Resolving these issues will be very important but if these issues persist they will negatively affect the concept of fog computing in healthcare rather than support it. The healthcare sector is very critical and even a single mistake cannot be tolerated, so the healthcare devices should be very efficient and less prone to failure. Sometimes the cloud and fog service providers are small companies and they don't have policies to handle many issues. It creates a negative impact of technology on the mind of customers. It looks like the technology is more dangerous than useful. So organizations and companies should emphasize resolving the various critical issues to make the hassle-free working of healthcare devices. Following are some issues of implementing Healthcare 4.0 with the help of fog computing.

2.3.1 Technical Issues

1. **Heterogeneity &interoperability:** As the market for biosensors has increased, many companies have started to make their products. Each product has its configuration and operating standards. In Healthcare 4.0 it is expected that many patients, hospitals, and doctors are connected in a network so that the data and information can be easily transferred. Due to the variety of sensor devices, networking devices, and computing devices, their consolidated working in a single system is a very challenging task. Continuous development in these devices also creates a big interoperability challenge because in some devices old versions do not support the functionality of new devices. In 2018 Redovan et al. proposed an interoperable fog-based IoT healthcare system using the iFogSim simulator. Performance was improved in terms of instance cost, distribution energy, and network delay.
2. **Power consumption:** In fog computing, it has been seen that many sensor devices are attached to the network. The size of the sensor devices is small and generally operated in wireless mode using dry batteries. So it is very true that now day demand for energy-efficient devices has been increased. It is required that devices can work for a long time with low battery usage. Even for smartphone devices many companies are putting their effort into increase battery life. To date, the Samsung Galaxy S31 smartphone has the largest battery life. The sensors are continuously sensing the surrounding information. Unnecessary sensing the data drains the battery faster, so smart sensing period policies should be proposed to save the battery life.
3. **Data management:** In the healthcare sector, many biosensors are continuously generating a large amount of data per second. This data is generally in unstructured form. The International Data Corporation (IDC) estimated that 'till 2020 digital universe has been expanded to 40,000 Exabytes' [36]. For

processing the huge amount of data various techniques like big data and blockchain are also available. The 3V's – volume, velocity, and variety – define the big data. Volume means a large volume of data, velocity means data generated at very high speed, and variety means data generated in various formats. Big data technologies are handling the three V's of data but more techniques are also needed to reduce the load on big data clouds. Preprocessing raw sensory data is one of the most efficient ways to reduce the load of big data in cloud computing. Data should also be stored in some reliable storage devices so that in any of the situations like cyber-attacks, fire, or even natural disasters data will not be lost. Many researchers are also working on data management in fog computing. In 2016, Farhoud et al. [37] proposed reshaping of raw and passive forms of data to intelligent data cell for big data management. In 2019, Wang et al. [38] proposed a fog computing with data integration model that was able to reduce the data. In 2021,Saeed et al. [39] proposed fault-tolerant data management schemes for healthcare IoT.

4. **Scalability:** Scalability refers to the capability of a system, network, or process to handle the growing number of systems and have the potential to meet growing demand in the future. As the new computing devices are being connected to the Internet, connecting these devices to the existing system will be very difficult [40]. According to one report of Statista 'The total installed base of Internet of Things (IoT) connected devices worldwide is projected to amount to 30.9 billion units by 2025, a sharp jump from the 13.8 billion units that are expected in 2021' [41]. As the number of devices is growing, the network should be able to provide addressing, processing capability, network bandwidth, storage resources, and computing devices to all newly connected devices [42]. To handle the growing demand, hardware as well as software systems should be able to support scaling. For example, IPv4 address was updated to IPv6 address for providing a unique address to more devices.

2.3.2 Security and Privacy Issues

As much as data has exploded over the Internet, the issue of security and privacy will increase. Since the data is generated by sensor nodes and transferred to the cloud using various gateways, so many security breaches are in the way of data transfer. The security and privacy issues can be defined in terms of trust, authentication, and end-user privacy. Trust means there should be a level of trust between two communication devices. Trust should be a two-way process between two nodes in fog computing. A well-defined trust model should be applied to fog computing. Authentication is very important in the healthcare sector because a large amount of personal information is flooded over the Internet. There should be a well-defined authentication structure so that only authenticated persons can access the data. Various security protocols are also needed to be implemented at each node so that malicious programs cannot affect the system. Leakage of any personal information should be prohibited over fog computing. Many researchers are working to resolve security and privacy issues [43–45] in fog computing. Pengfie et al. [43] proposed face authentication method; Chien et al. [44] and Wazid et al. [45] proposed secure key exchange-based scheme. Research work is also going on for multitier authentication schemes [46], mutual authentication schemes [47], filtering schemes [48], etc.

2.3.4 Social Issues

The concept of fog computing has been invaded in every sector nowadays. Still many devices are lacking of stable standards. Many companies have started to make many devices. In small businesses, people are borrowing cloud facilities from some small cloud providers. Sometimes these providers don't have clear policies regarding any problems. For example, some companies don't have clear policies regarding data loss, 24/7 computing facility availability, etc. In these cases, people will lose their trust over cloud and fog providers and if a trend of mistrust about any technology flows in society then this negatively affects its usage among people.

2.3.5 Regulatory Issues

Nowadays many companies and providers are providing cloud computing and fog computing facilities. So some uniform standards should be developed so that each product will be aligned with these standards. Sensors, fog nodes, and clouds should follow the standard protocols during data acquisition, processing, and transmission. If uniform standards are used all over the world it will solve all other issues to some extent. In 2014, IEEE [49] defined Open Fog Consortium (Open Fog Reference Architecture) for the adoption of standard fog architecture. In 2018, National Institute of Standards and Technology (NIST) [50] also defined the standards for fog computing architecture.

2.3.6 Ethical Issues

While using fog computing, companies should ensure that from their end any personal data will not be given to another firm for commercial purposes without user consent. This unethical process can even become the biggest complexity between two countries. So every company should provide the services ethically. General Data Protection Regulation (GDPR) and the EU-US Privacy Shield are some standards that maintain ethical standards [51].

2.4 RESEARCH CHALLENGES OF FOG COMPUTING IN HEALTHCARE 4.0

All the previously discussed issues posed various research challenges before the researchers, and there are some possibilities that can be considered while doing research work in the considered section. Table 2.2 shows the challenges and possibilities of fog computing.

TABLE 2.2 Challenges and Possibilities of Fog Computing

CHALLENGES	*POSSIBILITIES*
Interoperability of devices	• There should be some standards at each level (sensor layer, fog layer, and cloud layer), which should be followed by every manufacturer. • All devices should be manufactured in a way so that they can operate with previous and new versions also. • Companies should declare their configurations so that other manufacturers can align their devices accordingly.
Data Management	• New techniques of data management like big data or blockchain should be used to handle a huge amount of data. • High-speed hard disks should be used. • Cloud/fog service providers should ensure 24/7 availability in realtime. • Cloud/fog service providers should ensure data reliability without any data loss.
Security	• High-level antivirus software/firewall should be at each node so that data cannot be breached. • High-level authentication schemes should be implemented at each level. • Security should be ensured from the lower layer to the upper layer.
Privacy	• Without end-user consent, no one can access the data. • Personal healthcare data should not be used for commercial purposes without user consent.
High computing power	• High computing power servers should be deployed at cloud nodes or fog nodes. • Cloud providers who are serving a huge number of nodes can use multiple servers. • Computing servers can be located geographically all over the world.
End-user portability	• Battery-operated sensors should be used so that patients can move around wearing these devices.

2.5 CONCLUSION

With the increase of smart devices, the healthcare sector has also progressed toward Healthcare 4.0. In Healthcare 4.0 online patient monitoring, expert systems, and online medicine prescription will take the place of manual monitoring of patients. To make the real-time prediction in healthcare, the fog computing concept has been entering towards inside of healthcare. With the help of fog computing, the concept of Healthcare 4.0 is being visualized in real life. Industry 4.0, Healthcare 4.0, fog computing for Healthcare 4.0, enabling technologies of Healthcare 4.0 have been discussed rigorously in this chapter. We have also discussed issues and challenges of implementing the fog computing concept in Healthcare 4.0.

REFERENCES

[1] "What is Industry 4.0? | The Industrial Internet of Things (IIoT) | Epicor," [Online]. Available: www.epicor.com/en/resource-center/articles/what-is-industry-4-0/. [Accessed: 07–Sep–2021].

[2] M. Ghobakhloo, "Industry 4.0, digitization, and opportunities for sustainability," *J. Clean. Prod.*, vol. 252, p. 119869, Apr. 2020.

[3] S. Vaidya, P. Ambad, and S. Bhosle, "Industry 4.0 – A glimpse," *Procedia Manuf.*, vol. 20, pp. 233–238, Jan. 2018.

[4] A. G. Frank, L. S. Dalenogare, and N. F. Ayala, "Industry 4.0 technologies: Implementation patterns in manufacturing companies," *Int. J. Prod. Econ.*, vol. 210, pp. 15–26, Apr. 2019.

[5] K. Vinitha, R. Ambrose Prabhu, R. Bhaskar, and R. Hariharan, "Review on industrial mathematics and materials at industry 1.0 to industry 4.0," *Mater. Today Proc.*, vol. 33, pp. 3956–3960, Jan. 2020.

[6] D. P. Penumuru, S. Muthuswamy, and P. Karumbu, "Identification and classification of materials using machine vision and machine learning in the context of industry 4.0," *J. Intell. Manuf. 2019 315*, vol. 31, no. 5, pp. 1229–1241, Nov. 2019.

[7] Y. Yin, K. E. Stecke, and D. Li, "The evolution of production systems from industry 2.0 through industry 4.0," vol. 56, no. 1–2, pp. 848–861, Jan. 2017.https://doi.org/10.1080/00207543.2017.1403664.

[8] K. H. Tantawi, A. Sokolov, and O. Tantawi, "Advances in industrial robotics: From industry 3.0 automation to industry 4.0 collaboration," *TIMES-iCON 2019–2019 4th Technol. Innov. Manag. Eng. Sci. Int. Conf.*, pp. 1–4, Dec. 2019.

[9] D. A. Zakoldaev, A. V Shukalov, I. O. Zharinov, and O. O. Zharinov, "Modernization stages of the industry 3.0 company and projection route for the industry 4.0 virtual factory," *IOP Conf. Ser. Mater. Sci. Eng.*, vol. 537, no. 3, p. 032005, May 2019.

[10] P. P. Jayaraman, A. R. M. Forkan, A. Morshed, P. D. Haghighi, and Y.-B. Kang, "Healthcare 4.0: A review of frontiers in digital health," *Wiley Interdiscip. Rev. Data Min. Knowl. Discov.*, vol. 10, no. 2, p. e1350, Mar. 2020.

[11] A. Haleem and M. Javaid, "Medical 4.0 and its role in healthcare during COVID-19 pandemic: A review," vol. 5, no. 4, pp. 531–545, Nov. 2020.https://doi.org/10.1142/S2424862220300045.

[12] M. Javaid, A. Haleem, R. Vaishya, S. Bahl, R. Suman, and A. Vaish, "Industry 4.0 technologies and their applications in fighting COVID-19 pandemic," *Diabetes Metab. Syndr. Clin. Res. Rev.*, vol. 14, no. 4, pp. 419–422, Jul. 2020.

[13] R. Shrivastava and M. Pandey, "Real time fall detection in fog computing scenario,"*Cluster Comput.*, pp. 1–10, Jan. 2020.

[14] R. Shrivastava and M. Pandey, "Adaptive window based fall detection using anomaly identification in fog computing scenario,"*Multiagent Grid Syst.*, vol. 17, no. 1, pp. 15–37, Jan. 2021.

[15] S. Xia, S. Song, F. Jia, and G. Gao, "A flexible, adhesive and self-healable hydrogel-based wearable strain sensor for human motion and physiological signal monitoring," *J. Mater. Chem. B*, vol. 7, no. 30, pp. 4638–4648, Jul. 2019.

[16] D. Wang et al., "Self-adhesive protein/polypyrrole hybrid film for flexible electronic sensors in physiological signal monitoring," *Int. J. Biol. Macromol.*, vol. 181, pp. 160–168, Jun. 2021.

[17] P. G. Shynu, V. G. Menon, R. L. Kumar, S. Kadry, and Y. Nam, "Blockchain-based secure healthcare application for diabetic-cardio disease prediction in fog computing," *IEEE Access*, vol. 9, pp. 45706–45720, 2021.
[18] A. K. Sarangi, A. G. Mohapatra, T. C. Mishra, and B. Keswani, "Healthcare 4.0: A voyage of fog computing with IOT, cloud computing, big data, and machine learning," *Signals Commun. Technol.*, pp. 177–210, 2021.
[19] D. Sharma, G. S. Aujla, and R. Bajaj, "Evolution from ancient medication to human-centered healthcare 4.0: A review on health care recommender systems," *Int. J. Commun. Syst.*, p. e4058, 2019.
[20] A. A. Mutlag, M. K. Abd Ghani, N. Arunkumar, M. A. Mohammed, and O. Mohd, "Enabling technologies for fog computing in healthcare IoT systems," *Futur. Gener. Comput. Syst.*, vol. 90, pp. 62–78, Jan. 2019.
[21] Jannick P. Rolland, and Peter J. Delfyett Jr. *Three dimensional optical imaging colposcopy* (U.S. Patent No. 5,921,926, issued 13 July, 1999.
[22] "MINIR surgical robot may revolutionize brain tumor removal | Maryland robotics center," [Online]. Available: https://robotics.umd.edu/news/story/minir-surgical-robot-may-revolutionize-brain-tumor-removal. [Accessed: 08–Sep–2021].
[23] "STIFF-FLOP | Centre for robotics research, department of informatics, king's college London," [Online]. Available: https://nms.kcl.ac.uk/core/?page_id=105. [Accessed: 08–Sep–2021].
[24] "Fog computing market to witness a robust CAGR of 65% over 2017–2024," [Online]. Available: www.comparethecloud.net/articles/fog-computing-market/. [Accessed: 04–Sep–2021].
[25] "How the internet of things is revolutionizing healthcare," [Online]. Available: www.nxp.com/docs/en/white-paper/IOTREVHEALCARWP.pdf. [Accessed: 20–Jul–2020].
[26] "Medical sensors market | 2021–26 | industry share, size, growth – mordor intelligence," [Online]. Available: www.mordorintelligence.com/industry-reports/medical-sensors-market. [Accessed: 07–Sep–2021].
[27] A. Pravin, T. P. Jacob, and G. Nagarajan, "An intelligent and secure healthcare framework for the prediction and prevention of Dengue virus outbreak using fog computing," *Heal. Technol. 2019 101*, vol. 10, no. 1, pp. 303–311, Mar. 2019.
[28] S. Sareen, S. K. Gupta, and S. K. Sood, "An intelligent and secure system for predicting and preventing Zika virus outbreak using Fog computing," vol. 11, no. 9, pp. 1436–1456, Oct. 2017.https://doi.org/10.1080/17517575.2016.1277558.
[29] S. Tuli *et al.*, "HealthFog: An ensemble deep learning based smart healthcare system for automatic diagnosis of heart diseases in integrated IoT and fog computing environments,"*Futur. Gener. Comput. Syst.*, vol. 104, pp. 187–200, Mar. 2020.
[30] R. Priyadarshini, R. K. Barik, and H. Dubey, "DeepFog: Fog computing-based deep neural architecture for prediction of stress types, diabetes and hypertension attacks," *Comput.*, vol. 6, no. 4, p. 62, Dec. 2018.
[31] A. M. Rahmani *et al.*, "Exploiting smart e-Health gateways at the edge of healthcare internet-of-things: A fog computing approach," *Futur. Gener. Comput. Syst.*, vol. 78, pp. 641–658, Jan. 2018.
[32] H. Ben Hassen, W. Dghais, and B. Hamdi, "An E-health system for monitoring elderly health based on internet of things and fog computing," *Heal. Inf. Sci. Syst. 2019 71*, vol. 7, no. 1, pp. 1–9, Oct. 2019.
[33] A. Dhillon, A. Singh, H. Vohra, C. Ellis, B. Varghese, and S. S. Gill, "IoTPulse: Machine learning-based enterprise health information system to predict alcohol addiction in Punjab (India) using IoT and fog computing," 2020.https://doi.org/10.1080/17517575.2020.1820583.
[34] M. Muzammal, R. Talat, A. H. Sodhro, and S. Pirbhulal, "A multi-sensor data fusion enabled ensemble approach for medical data from body sensor networks," *Inf. Fusion*, vol. 53, pp. 155–164, Jan. 2020.
[35] Ajay Dev et al. IoT and fog computing based prediction and monitoring system for stroke disease. *Turkish J. Comput. Math. Educ. [Internet]*, vol. 12, no. 12, pp. 3211–3223, 23 May 2021 [cited 2021 Sep 4]. Available from: https://www.turcomat.org/index.php/turkbilmat/article/view/7998

[36] S. Dash, S. K. Shakyawar, M. Sharma, and S. Kaushik, "Big data in healthcare: Management, analysis and future prospects," *J. Big Data 2019 61*, vol. 6, no. 1, pp. 1–25, Jun. 2019.
[37] F. Hosseinpour, J. Plosila, and H. Tenhunen, "An approach for smart management of big data in the fog computing context,"*Proc. Int. Conf. Cloud Comput. Technol. Sci. CloudCom*, vol. 0, pp. 468–471, Jul. 2016.
[38] J. Wang, P. Zheng, Y. Lv, J. Bao, and J. Zhang, "Fog-IBDIS: Industrial big data integration and sharing with fog computing for manufacturing systems," *Engineering*, vol. 5, no. 4, pp. 662–670, Aug. 2019.
[39] W. Z. A. I. N. A. I. J. Saeed, "A fault tolerant data management scheme for healthcare internet of things in fog computing,"*KSII Trans. Internet Inf. Syst.*, vol. 15, no. 1, pp. 35–57, Jan. 2021.
[40] R. Barik, H. Dubey, S. Sasane, C. Misra, N. Constant, and K. Mankodiya, "Fog2Fog: Augmenting scalability in fog computing for health GIS systems," *Proc. – 2017 IEEE 2nd Int. Conf. Connect. Heal. Appl. Syst. Eng. Technol. CHASE 2017*, pp. 241–242, Aug. 2017.
[41] "Global IoT and non-IoT connections 2010–2025 | statista," [Online]. Available: www.statista.com/statistics/1101442/iot-number-of-connected-devices-worldwide/. [Accessed: 06–Sep–2021].
[42] M. Ghobaei-Arani, A. Souri, and A. A. Rahmanian, "Resource management approaches in fog computing: A comprehensive review," *J. Grid Comput. 2019 181*, vol. 18, no. 1, pp. 1–42, Sep. 2019.
[43] P. Hu, H. Ning, T. Qiu, H. Song, Y. Wang, and X. Yao, "Security and privacy preservation scheme of face identification and resolution framework using fog computing in internet of things,"*IEEE Internet Things J.*, vol. 4, no. 5, pp. 1143–1155, Oct. 2017.
[44] C.-M. Chen, Y. Huang, K.-H. Wang, S. Kumari, and M.-E. Wu, "A secure authenticated and key exchange scheme for fog computing," 2020.https://doi.org/10.1080/17517575.2020.1712746.
[45] M. Wazid, A. K. Das, N. Kumar, and A. V. Vasilakos, "Design of secure key management and user authentication scheme for fog computing services," *Futur. Gener. Comput. Syst.*, vol. 91, pp. 475–492, Feb. 2019.
[46] A. Manzoor, M. A. Shah, H. A. Khattak, I. U. Din, and M. K. Khan, "Multi-tier authentication schemes for fog computing: Architecture, security perspective, and challenges," *Int. J. Commun. Syst.*, p. e4033, 2019.
[47] G. Rahman and C. C. Wen, "Mutual authentication security scheme in fog computing," *Int. J. Adv. Comput. Sci. Appl.*, vol. 10, no. 11, pp. 443–451, 2019.
[48] J. Wu, M. Dong, K. Ota, J. Li, and Z. Guan, "FCSS: Fog-computing-based content-aware filtering for security services in information-centric social networks," *IEEE Trans. Emerg. Top. Comput.*, vol. 7, no. 4, pp. 553–564, Oct. 2019.
[49] "IEEE 1934–2018–IEEE standard for adoption of openfog reference architecture for fog computing," [Online]. Available: https://standards.ieee.org/standard/1934-2018.html. [Accessed: 08–Sep–2021].
[50] M. Iorga, L. Feldman, R. Barton, M. J. Martin, N. Goren, and C. Mahmoudi, "Draft SP 800–191, The NIST definition of fog computing," 2018.
[51] "GDPR–General data protection regulation," [Online]. Available: www.privacytrust.com/guidance/gdpr.html. [Accessed: 08–Sep–2021].

Modeling of an IoT System in Healthcare Subject to Gateway Failure

3

Shensheng Tang
Department of Electrical and Computer Engineering
St. Cloud State University

Contents

DOI: 10.1201/9781003244714-3

3.1 INTRODUCTION

The Internet of Things (IoT) is affecting our daily lives by realizing the interactions of human-to-human, machine-to-machine, and human-to-machine. As one of the most promising IoT applications, the healthcare IoT technology can capture data streaming in real time from the biosensors, wearables, and other connected medical devices that monitor physiological data and other health information. The captured data streaming, along with the power of cloud computing and data analytics, can enable precise diagnoses and remote treatments and improve patient safety and outcomes.

However, applications of new technologies often bring certain risks, including failures of components, devices, and infrastructure, which may cause disastrous results for patients [1]. To minimize such risks and ensure the required system performance and availability, the system modeling and performance evaluation for a health IoT system is an inevitable development step before the practical IoT system is deployed.

There are quite a few research works in the literature on the modeling and performance evaluation techniques for the IoT systems [2–11], which are summarized as follows. In [2], a model was constructed by utilizing feedback control theory to introduce a method for representing the 'things' as the controlled objects. Then the overall system model of the real-time IoT with the controlled objects was represented and validated through the numerical analysis. In [3], the impact of factors on the performance in IoT networks was studied using simulation based models. Further, an analytical framework was developed to model the impact of individual node behavior on overall performance using Markov chains. In [4], a framework was introduced that autonomously chooses the best solution for the application given the current deployed environment. The framework uses a performance model to predict the expected performance of a particular solution and then chooses an apt solution for the application from a set of available solutions.

In [5], a binding update scheme on Proxy Mobile IPv6, which reduces signal traffic during location updates by Virtual LMA (VLMA) on the top original Local Mobility Anchor (LMA) Domain, was proposed to reduce the total cost, where the Mobile Access Gateways (MAGs) eliminate global binding updates for the mobile nodes between LMA domains and significantly reduce the packet loss and latency by eliminating the handoff between LMAs. In [6], an IoT system was modeled by a queueing network and detailed quantitative model analyses under different request arrival distributions were presented along with some performance metrics. In [7], a simulation model was presented for the performance analysis of the traffic generated by devices connected by the IoT, aiming at dimensioning the capacity of an IoT mediator. The simulation model is based on both discrete event and Random Waypoint simulation for an ad-hoc network that accommodates clusters and the effects of mote mobility. The mean-queuing time and the CPU utilization were analyzed through a case study for the cluster-heads in each cluster, given the probability of connectivity of the motes resulting from their mobility and transmission power.

In [8], the authors introduced the software defined networking (SDN) into narrow band (NB)-IoT and investigated the traffic modeling and performance evaluation of the SDN-based NB-IoT access network. Different types of queuing models were introduced to evaluate the network performance in different environments. The influence of different network parameters on the network performance was investigated through simulations. In [9], the uplink scheduler for the NB-IoT access network was modeled using the state-machine modeling methodology on the Simulink environment. Then the NB-IoT uplink scheduler was optimized to exploit the periodicity nature of smart-meter applications. An integrated scheduling protocol was proposed to rearrange the transmission times of different smart-meter utilities and draw a map for the transmission schedule of them to better utilize the sparse time resources.

In [10], the reliability of wireless link under multiple factors, such as fading, shadowing, interference, and noise, was analyzed by capturing the effect of different parameters through compound probability distributions. Expressions for link and node outage were obtained and measured through network simulations. In [11], simulation-based queueing models were presented for the representation of various QoS settings of IoT interactions, including message-drop probabilities, intermittent mobile connectivity, message availability or validity, the prioritization of important information, and the processing or transmission of messages.

More introduction of healthcare IoT systems and technologies can be found in some review works [12–14]. In [12], the authors surveyed advances in IoT-based healthcare technologies and reviewed the state-of-the-art network architectures/platforms, applications, and industrial trends in IoT-based healthcare solutions. In [13], a systematic review was performed for the studies of the current advancements in wearable sensors and IoT-based monitoring applications to support independent living for older adults. Most studies were found to focus on the system aspects of wearable sensors and IoT monitoring solutions, including advanced sensors, wireless data collection, communication platform, and usability. In [14], IoT-based distributed healthcare systems were studied, where all available medical resources are interconnected to provide effective and efficient healthcare services to those in need of medical assistance.

In this chapter, we propose a queueing model of a healthcare IoT system subject to the gateway failure and analyze the performance of the system based on different system parameters such as data packet arrival rate, packet processing time, system failure rate, and system recovery time. By using the generating function technique, we derive the analytical solution for the global balance equations of the queuing model. From the explicit expressions of the steady-state probabilities, we develop some performance metrics of interest and perform the numerical evaluation for the metrics for the verification of the analytical results. The proposed system modeling and performance analysis method can be extended to different application fields of IoT systems.

The major contribution of the chapter is summarized as follows:

- Describe an IoT system in healthcare subject to gateway failure and build a queueing model for the system.
- Solve the system model by using the generating function technique and obtain analytical solution.
- Develop some important performance metrics of interest.
- Perform the numerical evaluation for verifying the analytical results.

The remainder of the chapter is organized as follows. Section 3.2 presents a healthcare IoT system description. Section 3.3 develops the mathematical model of the IoT system by using the queueing theory and provides explicit solution of the system. Section 3.4 derives some system performance metrics based on the analytical solution. Section 3.5 presents the numerical evaluation. Finally, the chapter is concluded in Section 3.6.

3.2 A HEALTHCARE IOT SYSTEM DESCRIPTION

The main components of the healthcare IoT system include biosensor nodes installed on wearables, a portable gateway device, and a cloud server, as shown in Figure 3.1. A brief description of each component follows.

A group of biosensors installed on different parts of the wearables for a patient sense physiological data and send them to the portable gateway device wirelessly, which again wirelessly communicates with the cloud server for data transfer. Different types of medical sensors and sensor devices can be used in a healthcare IoT system. For example, a blood pressure (BP) sensor can collect BP data regularly. A body temperature sensor can monitor body temperature. An electroencephalogram (EEG) sensor module can detect minute electrical activity of brain cells. An accelerometer sensor can capture the intensity of physical activity for human movement by attaching the sensor on a person wrist or ankle or to a person waist with a belt clip. Data collected by biosensors are transmitted to the portable gateway device using Bluetooth or Zigbee [15] protocol. The Zigbee protocol is a low-power wireless specification that uses physical (PHY) and media access control (MAC) layers based on IEEE 802.15.4 standard [16]. The Zigbee technology operates at the 2.4 GHz ISM (industrial, scientific, and medical) bands. The protocol allows sensor

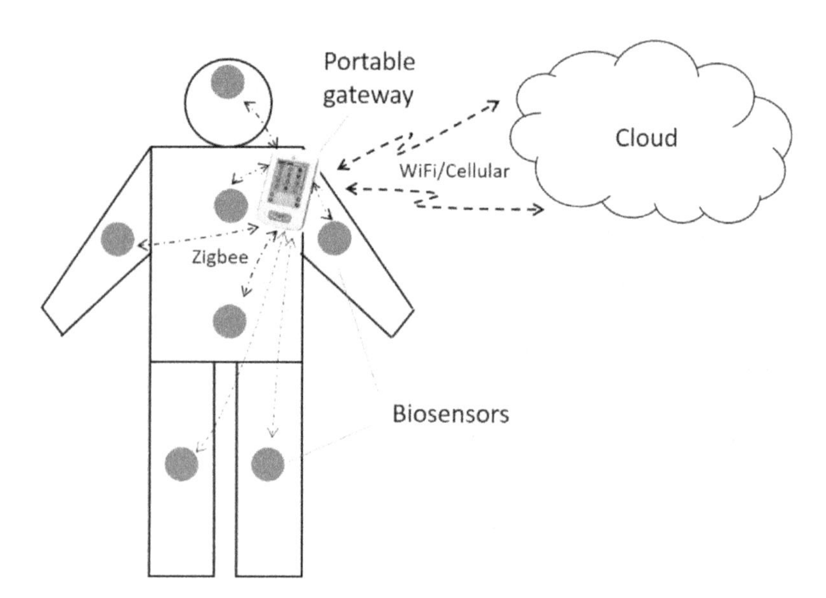

FIGURE 3.1 A typical healthcare IoT system infrastructure.

nodes to communicate in a variety of network topologies and the battery life to last for a long time.

The gateway device is a base station that collects data from the sensors and sends the data to the cloud server; it also monitors the sensor status, changes parameter configuration, and updates software. The gateway connects to the cloud server through Wi-Fi or cellular wireless technologies and to the sensor nodes through Zigbee or Bluetooth technology. Therefore, the gateway device is critical to the IoT system. Assume that the gateway is a single channel server. When it is processing the data from one sensor, it cannot process the data from another sensor. However, it can put the subsequent data packets in its queue. The queue size is assume to be infinite. When its current service is done, the gateway will serve the data packets in the queue according to first-come-first-served (FCFS) mode.

The cloud server is a virtual server running in a cloud computing environment over Internet. It provides a web-based information infrastructure, including computing power, storage space, and network technology. Health providers can easily process, exchange, secure, and manage data in either web-based or location-independent styles. They perform comprehensive data analysis and send commands from the cloud to the gateway device or further, from the gateway to do necessary software update or other commands to the specific sensor nodes.

In order to ensure the successful data transfer in the IoT system, the two types of communication links must both be operatable: the links between the sensors and the gateway device, and the link between gateway and the cloud. Link failures are prone to occur in wireless environments. Many factors can cause wireless connection failures such as insufficient battery power, physical hardware defect due to aging, software fault due to software bug, etc. However, among the related three components – biosensors, gateway, and cloud server – the gateway is critical to the realibility of the IoT system due to its unity. When a gateway fails, the IoT system will be completely down. It cannot receive data from the biosensors; it cannot send the collected data to the cloud for processing. On the contrary, the IoT system has multiple biosensors installed on the wearable equipment for the patient. If some sensors fail, the rest of the sensors can still transfer data to the gateway to maintain an operational wireless link. Similarly, the cloud server generally has very high reliability compared with the sensors and the gateway. Moreover, the cloud server usually has many backup servers in the cloud; the gateway switch to communicate a backup server in case the current server accidently fails. Therefore, in this work we only focus on the gateway failure condition for the IoT sytem. In general, the gateway failure immdiately triggers a repair process and a random recovery time is needed.

3.3 MODELING OF THE IOT SYSTEM

In this section, we build the mathematical model for the healthcare IoT system considering the gateway failure condition. The gateway serves as a single server with a queue to hold the subsequent data packets from sensors when it is in service. Assume that the queue size is infinite.

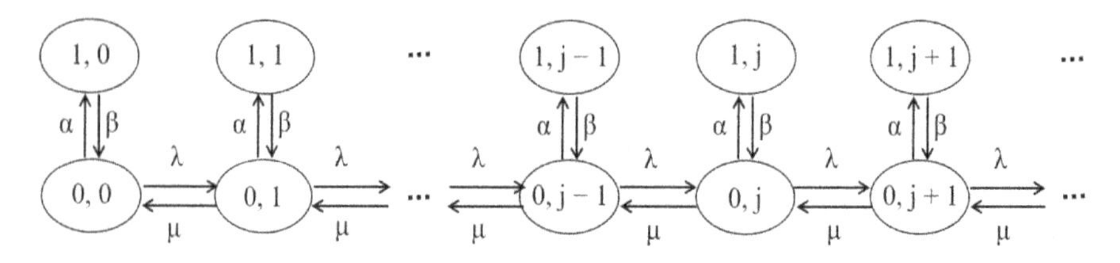

FIGURE 3.2 State transition diagram of the IoT system.

Let the arrival process of the data packets be a Poisson with rate λ and data packet service time be exponentially distributed with mean 1/μ. Since the gateway may be subject to failure, we assume that the gateway remains operational for a period, which is exponentially distributed with rate α. When the gateway fails, a recovery process starts, which is exponentially distributed with rate β. Note that the the gateway does not receive or transmit data in failure state.

Let {X(t), Y(t); t ≥ 0} represent the state of the IoT system at time t, where X(t) is set to 0 if the system is in operational state serving the data packets and 1 if the gateway is in failure state; and Y(t) is the number of data packets in the gateway including the one in service. Then (X(t), Y(t)) is a two-dimensional Markov process with the transition rate diagram shown in Figure 3.2 and state space {(i, j) | i = 0, 1; j = 0, 1, 2, . . .}.

Therefore, state (0, 0) represents that the gateway system is empty without any failure, state (1, 0) represents that the gateway is empty with failure, state (0, *j*) represents that the gateway is operational and has *j* data packets in the system, and state (1, *j*) represents that the gateway is in failure condition and has *j* data packets in the system.

Let $\pi(i,j)$ denote the steady-state probability that the gateway system is in state (*i*, *j*) with state space {(*i*, *j*) | *i* = 0, 1; 0 ≤ *j* ≤ ∞}. The global balance equations of the system [17] are derived as follows.

$$(\lambda+\alpha)\pi(0,0)=\mu\pi(0,1)+\beta\pi(1,0) \tag{3.1}$$

$$(\lambda+\mu+\alpha)\pi(0,j)=\lambda\pi(0,j-1)+\mu\pi(0,j+1)+\beta\pi(1,j),\ \text{for}\ j\geq 1 \tag{3.2}$$

$$\beta\pi(1,j)=\alpha\pi(0,j),\ \text{for}\ j\geq 0 \tag{3.3}$$

We shall use the generating function technique [18] to solve the preceding system equations. Define the partial generating functions as follows.

$$G_i(z)=\sum_{j=0}^{\infty}\pi(i,j)z^j,\ |z|\leq 1,\ i=0,1. \tag{3.4}$$

Applying $G_i(z)$ to Equation (3.2) and utilizing Equation (3.1), we can easily find the following equation:

$$\left[\lambda z^2-(\lambda+\mu+\alpha)z+\mu\right]G_0(z)+\beta zG_1(z)=(1-z)\mu\pi(0,0) \tag{3.5}$$

Similarly, applying $G_i(z)$ to Equation (3.3), we find:

$$\alpha G_0(z) = \beta G_1(z) \tag{3.6}$$

Solving Equations (3.5) and (3.6) with a series of mathematical arrangements, we have

$$G_0(z) = \frac{1}{1-\rho z}\pi(0,0), \tag{3.7}$$

$$G_1(z) = \frac{\gamma}{1-\rho z}\pi(0,0), \tag{3.8}$$

where $\rho = \frac{\lambda}{\mu}$ which is the traffic intensity of the system, a measure of how busy a system is; $\gamma = \frac{\alpha}{\beta}$, which is the ratio of the mean time during which a server is repaired $(1/\beta)$ to the mean time between server failures $(1/\alpha)$. Then we apply the normalization condition

$$\sum_{i=0}^{1} G_i(z)|_{z=1} = 1 \tag{3.9}$$

and obtain the probability that gateway system is empty without any failure as

$$\pi(0,0) = \frac{1-\rho}{1+\gamma} \tag{3.10}$$

It turns out that the system has a steady state solution only if $\rho < 1$. From Equation (3.2), we have the probability that gateway system is empty and under failure condition as

$$\pi(1,0) = \gamma\pi(0,0). \tag{3.11}$$

Similarly, from Equation (3.1), we derive

$$\pi(0,1) = \rho\pi(0,0). \tag{3.12}$$

Repeatedly apply the equations of (3.1), (3.2), and (3.3), we derive the following general results:

$$\pi(0,j) = \rho^j\pi(0,0) = \frac{(1-\rho)\rho^j}{1+\gamma}, j \geq 0 \tag{3.13}$$

$$\pi(1,j) = \gamma\rho^j\pi(0,0) = \frac{\gamma(1-\rho)\rho^j}{1+\gamma}, j \geq 0 \tag{3.14}$$

Note here $\pi(i,j)$ represents the steady-state probability that the system is in state (i, j), i = 0 means the system is operational, i = 1 means wireless connection failure in the system, and j denotes the number of data packets in the gateway system.

Note also that the preceding general results have verified the indepedence property of the data packet arrival/departure processes and the gateway failure/recovery processes, by which the same results can be easily derived.

3.4 PERFORMANCE METRICS

3.4.1 Mean Number of Data Packets in the System

The mean number of data packets in the system when the system is operable, $E[L_0]$, can be expressed as

$$E[L_0]=\sum_{j=1}^{\infty} j\pi(0,j)=\frac{dG_0(z)}{dz}|_{z=1}$$

Substituting Equation (7) into the above formula, we can calculate the deriative of $G_0(z)$ as follows.

$$\frac{dG_0(z)}{dz}|_{z=1}=\frac{\rho}{(1-\rho z)^2}\pi(0,0)|_{z=1}=\frac{\rho}{(1-\rho)^2}\pi(0,0)=\frac{\rho}{(1-\rho)(1+\gamma)}$$

Therefore we have

$$E[L_0]=\frac{\rho}{(1-\rho)(1+\gamma)} \tag{3.15}$$

Similarly, the mean number of data packets in the system when the system is under failure condition, $E[L_1]$, can be obtained as

$$E[L_1]=\sum_{j=1}^{\infty} j\pi(1,j)=\frac{dG_1(z)}{dz}|_{z=1}=\frac{\rho\gamma}{(1-\rho)(1+\gamma)} \tag{3.16}$$

Finally, the mean number of data packets in the system (regardless of the system status), $E[L]$, can be obtained as

$$E[L]=E[L_0]+E[L_1]==\frac{\rho}{1-\rho} \tag{3.17}$$

3.4.2 Mean Sojourn Time of Data Packets in the System

The mean sojourn time of data packets in the system, $E[S]$, is the amount of time a data packet is expected to spend in the system before leaving the system and can be calculated by Little's Law,

$$E[S]=\frac{E[L]}{\lambda}=\frac{1}{\mu(1-\rho)} \tag{3.18}$$

3.4.3 Steady-State Availability of the System

The steady-state availability of the system, denoted by A, is defined as a percentage measure of the degree to which the system is in an operable and committable state at the point in time when it is needed. It can be calculated as the sum of all the probabilities in steady state that the system is operable.

$$A=\sum_{j=0}^{\infty}\pi(0,j)=G_0(1)=\frac{1}{1+\gamma} \tag{3.19}$$

Similarly, the steady-state unavailability of the system, denoted by U, can be calculated as the sum of all the probabilities in steady state that the system is under the wireless cloud connection failure.

$$U=\sum_{j=0}^{\infty}\pi(1,j)=G_1(1)=\frac{\gamma}{1+\gamma}. \tag{3.20}$$

Actually, U can be directly obtained from the relationship $U = 1 - A$. The above derivation of Equation (3.17) verifies the correctness of the result.

3.4.4 Steady-State Throughput of the System

The steady-state throughput of the system, denoted by T, is defined as the number of data packets processed per unit time in the system. It can be calculated as the percentage of the system processing rate to which the system is processing data packets.

$$T=\mu\sum_{j=1}^{\infty}\pi(0,j)=\mu[G_0(1)-\pi(0,0)]=\frac{\lambda}{1+\gamma} \tag{3.21}$$

3.4.5 Mean Busy Period

The mean busy period (BP) is defined as the time duration between the arrival of a data packet to the empty system and the first epoch when the system becomes empty again. The time duration that the system is empty is referred to as an idle period (IP). Due to the memoryless property of the exponential distribution, an IP is exponentially distributed with mean $1/\lambda$. It is clear that the ratio of the mean busy period (E[BP]) of data packets in the system to the probability that the system is not empty is proportional to the ratio of the mean idle period (E[IP]) to the probability that the system is empty. Therefore, we have

$$E[BP] = \frac{1-\pi(0,0)-\pi(1,0)}{\pi(0,0)+\pi(1,0)} E[IP] = \frac{1}{\mu(1-\rho)} \tag{3.22}$$

3.5 NUMERICAL EVALUATION

In this section, we present numerical evaluation for the IoT system under the gateway failure condition by studying selected metrics with respect to (w.r.t.) various parameters. The typical parameter settings are given in Table 3.1. Other values of related parameters are set separately in the figures to study the performance of the relevant metrics.

Figure 3.3 shows the probability distribution of $\pi(0, j)$ with respect to the number of data packets j and other parameters. We observe that under the specific settings, $\pi(0, j)$ decreases with j at the small values of j and tends to be constant at the large values of j. We also observe that when the packet arrival rate (λ) increases, the probability $\pi(0, j)$ will decrease at the small values of j and increase at the large values of j. This is because the component (1-ρ) in Equation (3.12) dominates at the small values of j while the component ρ^j dominates at the large values of j. The increase of the packet processing rate (μ) has the same impact as that of the decrease of λ. As expected, when the gateway failure rate (α) or the recovery time ($1/\beta$) increases, the probability $\pi(0, j)$ will decrease. The increase of α will increase the opportunity from the states of $\pi(0, j)$ to enter the states of $\pi(1, j)$ and thus tend to reduce the value of $\pi(0, j)$.

Figure 3.4 shows the probability distribution of $\pi(1, j)$ with respect to the number of data packets j and other parameters. As expected, $\pi(1, j)$ will increase when the gateway failure rate (α) or the recovery time ($1/\beta$) increases. A larger failure rate will cause more opportunity of entering the states of $\pi(1, j)$. Similarly, a large recovery time ($1/\beta$) will put off the opportunity of leaving the states of $\pi(1, j)$. We also observe that when the packet arrival rate (λ) or the packet processing time ($1/\mu$) increases, the $\pi(1, j)$ will decrease at the small values of j and increase at the large values of j. The reason is the same as that mentioned in Figure 3.3.

Figure 3.5 shows the performance of the mean number of data packets in the operable system $E[L_0]$. We observe that $E[L_0]$ will increase with the increase of the packet arrival rate (λ) or the decrease of the packet processing rate (μ). In a single server queueing system, it is obvious that the increase of the packet arrival rate will lead to the increase of

TABLE 3.1 Typical Parameter Configuration for Numerical Evaluation

PARAMETER	*VALUE*	*UNIT*	*DESCRIPTION*
j	varies	packets	# of packets
λ	varies	packets/unit time	arrival rate
μ	11, 13, 15	packets/unit time	service rate
α	5, 8, 10	failures/unit time	failure rate
β	10, 16, 25	recoveries/unit time	recovery rate

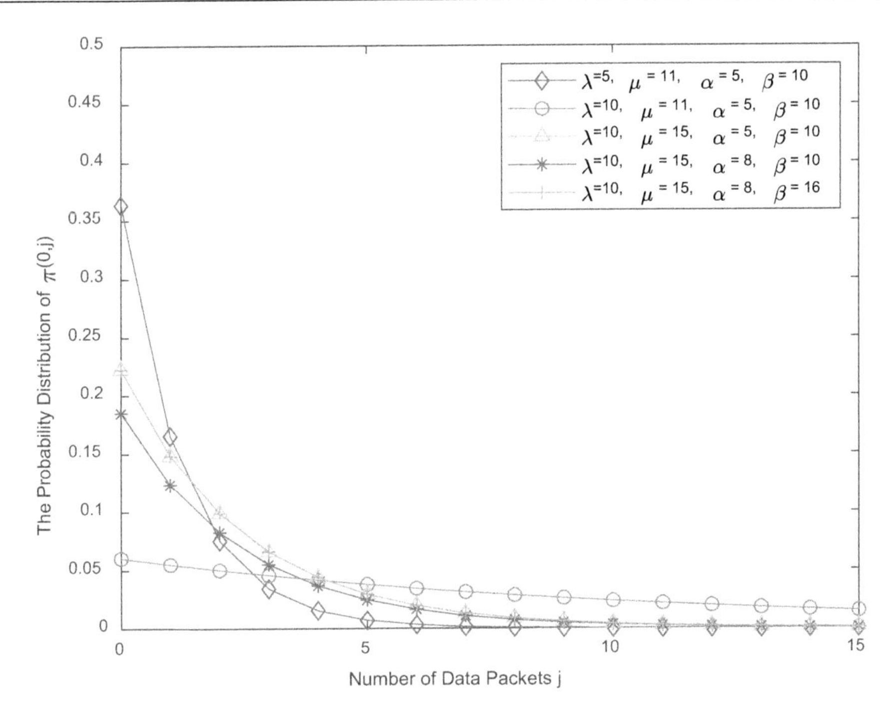

FIGURE 3.3 The probability distribution of π(0, j) w.r.t. parameters.

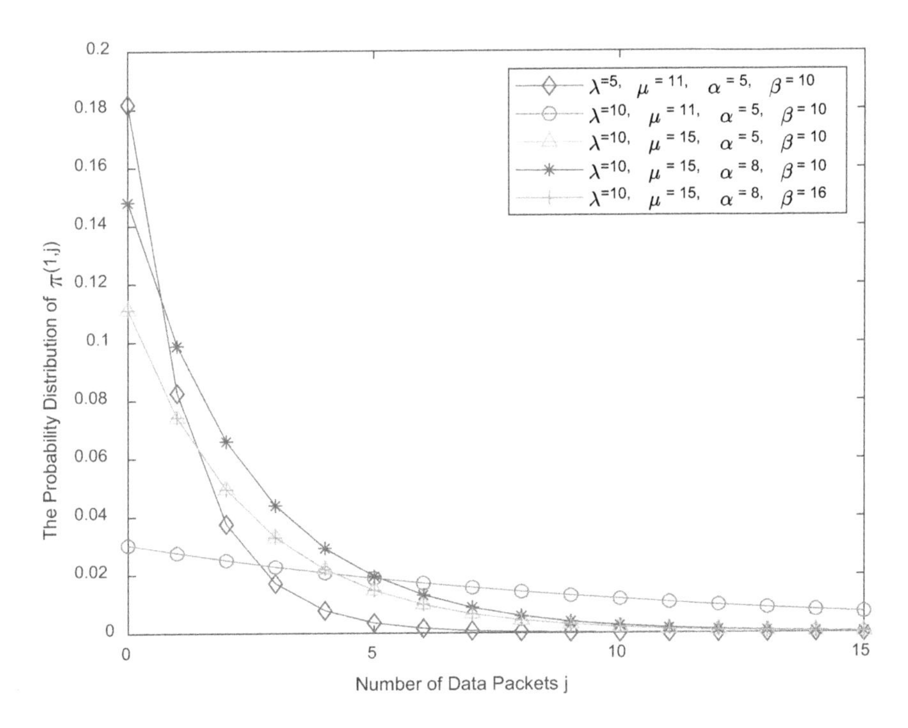

FIGURE 3.4 The probability distribution of π(1, j) w.r.t. parameters.

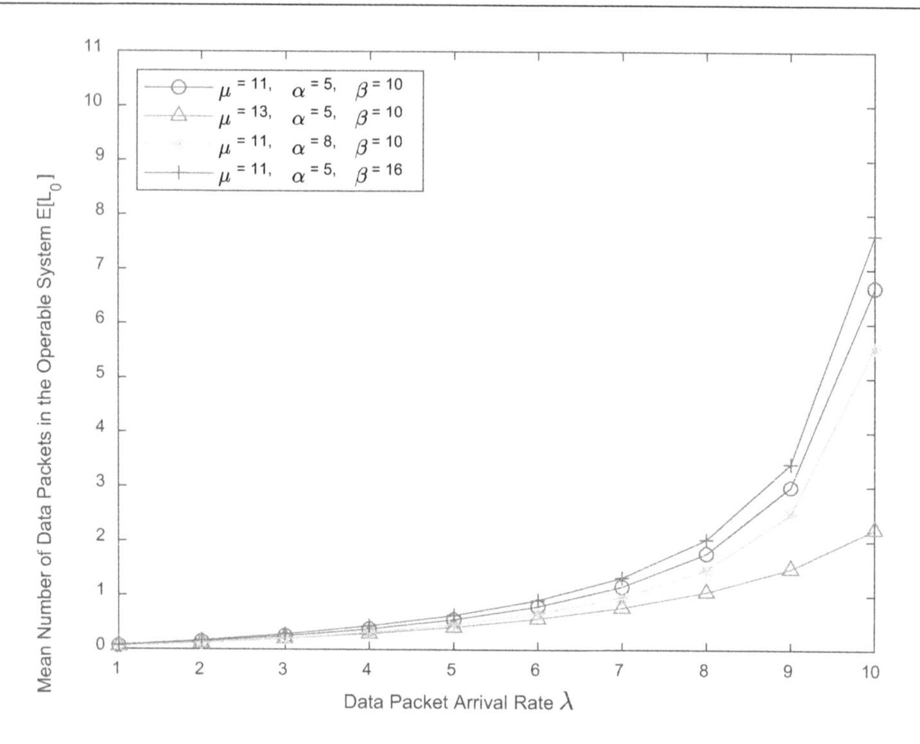

FIGURE 3.5 The mean number of data packets in the operable system w.r.t. parameters.

the queueing length. Equivalently, the decrease of the packet processing rate will cause more packets to wait in the queue and thus lead to the increase of queueing length. We also observe that $E[L_0]$ will decrease with the increase of the gateway failure rate (α) or the gateway recovery time ($1/\beta$). The reason is when the gateway failure rate (α) increases, more operable states will be changed to the failure states, leading to the decrease of $E[L_0]$. The equivalent situation will occur when the gateway recovery time ($1/\beta$) increases.

Figure 3.6 shows the performance of the mean number of data packets in the failed system $E[L_1]$. We observe that $E[L_1]$ will increase with the increase of the packet arrival rate (λ) or the decrease of the packet processing rate (μ), given the fixed gateway failure rate (α) and the gateway recovery time ($1/\beta$). This is obvious since the increase of α or decrease of μ directly lead to the increase of the queueing length. We also observe that $E[L_1]$ will increase with the increase of the gateway failure rate (α) or the gateway recovery time ($1/\beta$). When the gateway failure rate (α), or equivalently, the gateway recovery time ($1/\beta$) increases, more operable states will be changed to the failure states, leading to the increase of $E[L_1]$.

Figure 3.7 shows how the mean total number of data packets in the system $E[L]$ changes with respect to different parameters. We observe that $E[L]$ will increase with the increase of the packet arrival rate (λ) or the decrease of the packet processing rate (μ). The reason has been explained in the part of Figure 3.5. We also observe that $E[L]$ does not change with the gateway failure rate (α) or the gateway recovery time ($1/\beta$). Clearly the

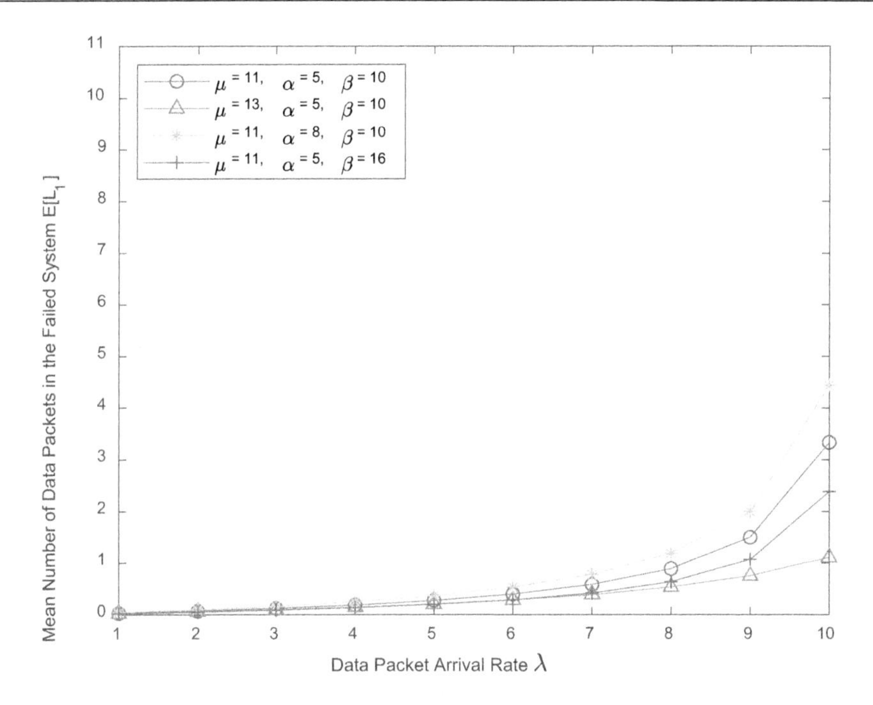

FIGURE 3.6 The mean number of data packets in the failed system w.r.t. parameters.

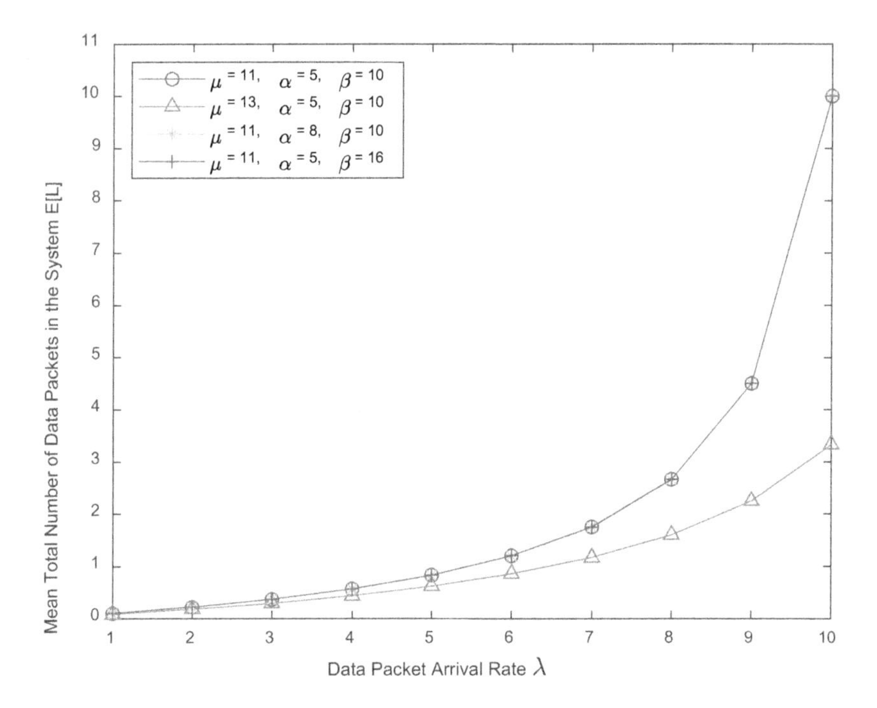

FIGURE 3.7 The mean total number of data packets in the operable system w.r.t. parameters.

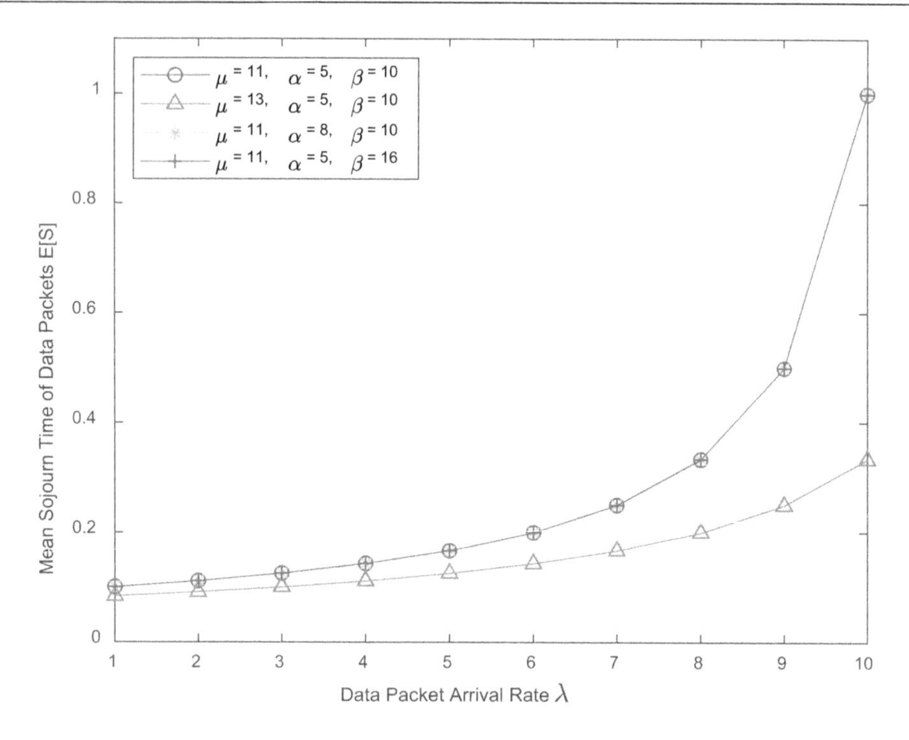

FIGURE 3.8 The mean sojourn time of packets w.r.t. parameters.

total number of data packets in the system is not affected by the gateway failure rate or the gateway recovery time.

Figure 3.8 shows the mean sojourn time of data packets E[S] with respect to the change of different parameters. We observe that E[S] has the same trend to the E[L]. The larger the packet arrival rate (λ), the more the packets waiting in the queue. This leads to a larger mean sojourn time of the packets in the system. Similar to Figure 3.7, E[S] does not change with the change of the gateway failure rate or the gateway recovery time.

Figure 3.9 shows the steady-state throughput of the system T with respect to the change of different parameters. We observe that, in steady state, T will increase with the increase of the packet arrival rate (λ), but does not change with the change of processing time ($1/\mu$). This is intuitive since it is assumed that $\lambda < \mu$ in our system configuration. The larger the data packet arrival rate, the higher the system throughput. On the contrary, the increase of the processing rate μ does not affect the system throughput, just waste some system processing ability. We also observe that T will decrease with the increase of the gateway failure rate (α) or equivalently, the recovery time ($1/\beta$). The gateway failure makes the data processing completely stop in the system.

Figure 3.10 shows the steady-state availability with respect to the change of different parameters. We observe that the availability will decrease with the increase of the gateway failure rate (α) or the decrease of the gateway recovery time (β). This is because the increase of either the gateway failure (α) rate or recovery time ($1/\beta$) will cause a larger opportunity of an unavailable gateway. We also observe that the availability does not

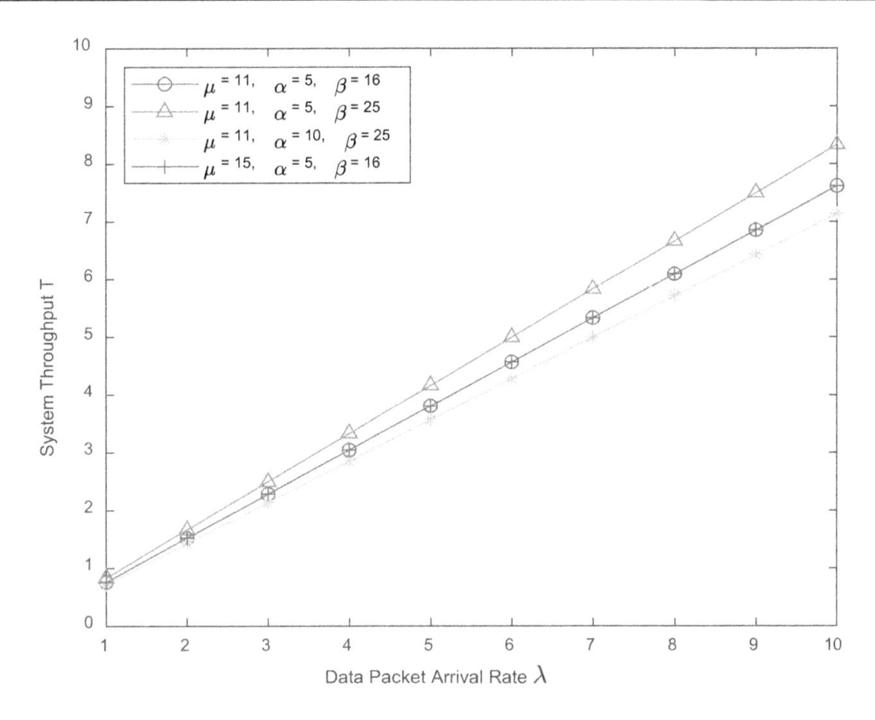

FIGURE 3.9 The steady-state throughput of the system w.r.t. parameters.

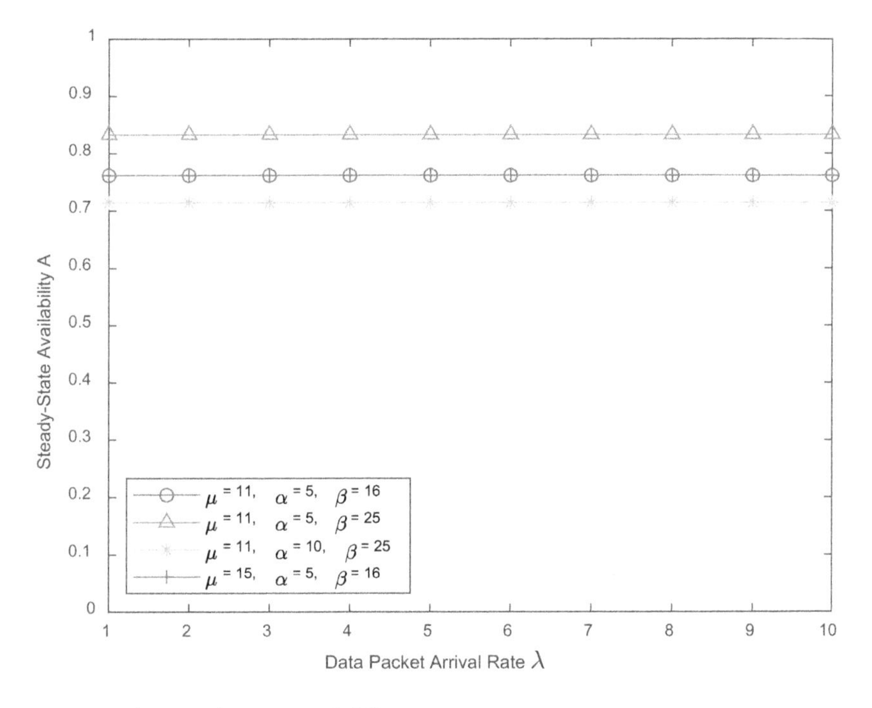

FIGURE 3.10 The steady-state availability w.r.t. parameters.

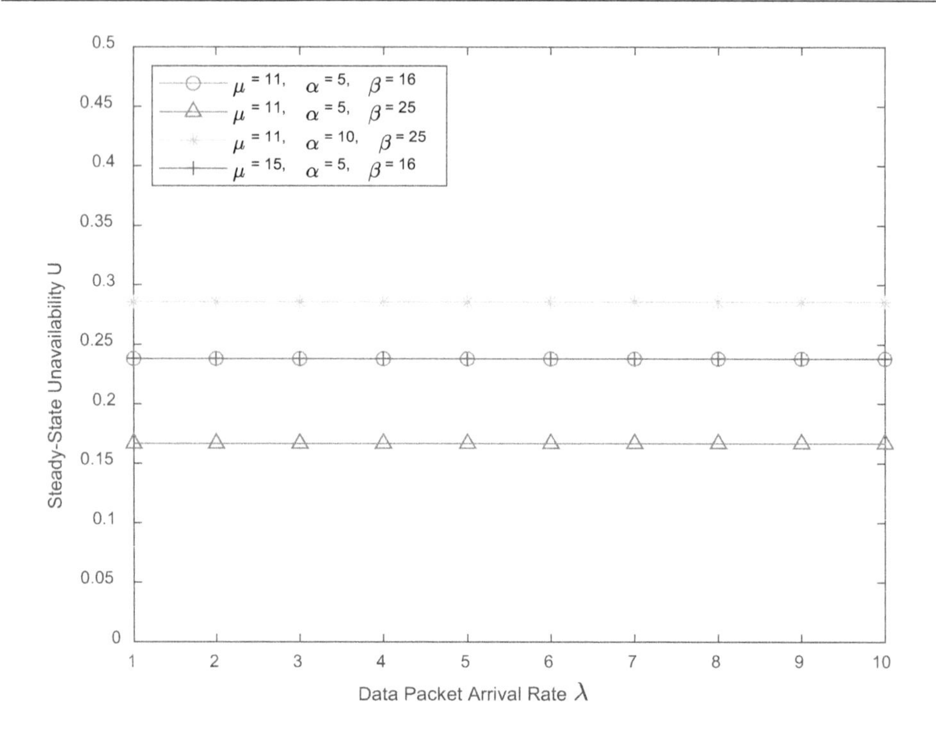

FIGURE 3.11 The steady-state unavailability w.r.t. parameters.

change with the change of packet arrival rate (λ) and processing time ($1/\mu$). This is because the gateway system is assumed to have an infinite queue, which is always available to the data packets regardless of the packet arrival rate and processing time. The corresponding steady-state unavailability is shown in Figure 3.11.

3.6 CONCLUSION

We proposed a queueing model of a healthcare IoT system subject to the gateway failure and did the performance analysis of the system based on different system parameters. By using the generating function technique, we derived the analytical solution for the system balance equations. The explicit solution of the system operational probability $\pi(0, j)$ and the system failure probability $\pi(1, j)$ should be useful for the development of algorithm and scheduling method for different applications, which will be considered for future research. Some performance metrics of interest were then developed, and detailed numerical evaluation for the metrics were performed for the verification of the analytic results. The proposed system modeling and performance analysis method can serve as a useful reference for the general IoT system design and evaluation.

3.7 ACKNOWLEDGMENT

The author would like to acknowledge the support of the project development from St. Cloud State University, MN, USA, through the Early Career Grant (No. 211129).

REFERENCES

[1] S. Tang and Y. Xie, "Availability modeling and performance improving of a healthcare internet of things (IoT) system", *Journal IoT*, Vol. 2, No. 2, pp. 310–325, 2021.
[2] K. Sato, Y. Kawamoto, H. Nishiyama, N. Kato and Y. Shimizu, "A modeling technique utilizing feedback control theory for performance evaluation of IoT system in real-time", 2015 International Conference on Wireless Communications & Signal Processing (WCSP), pp. 1–5, 2015. doi: 10.1109/WCSP.2015.7341303.
[3] S. Sankaran, "Modeling the performance of IoT networks", 2016 IEEE International Conference on Advanced Networks and Telecommunications Systems (ANTS), pp. 1–6, 2016. doi: 10.1109/ANTS.2016.7947807.
[4] D. T. Delaney and G. M. O'Hare, "A framework to implement IoT network performance modelling techniques for network solution selection", *Sensors (Basel, Switzerland)*, Vol. 16, No. 12, p. 2038, Dec. 2016. https://doi.org/10.3390/s16122038.
[5] C. Cho, J.Y. Choi, J. Jeong, and T.M. Chung, "Performance analysis of inter-domain handoff scheme based on virtual layer in PMIPv6 networks for IP-based internet of things", *PLoS One,* Vol. 12, No. 1, p. e0170566. https://doi.org/10.1371/journal.pone.0170566.
[6] J. Huang, S. Li, Y. Chen, and J. Chen, "Performance modelling and analysis for IoT services", *International Journal of Web and Grid Services*, Vol. 14, No. 2, pp. 146–169, 2018.
[7] J. R. E. Leite, E. L. Ursini and P. S. Martins, "Performance analysis of IoT networks with mobility via modeling and simulation", 2018 International Symposium on Performance Evaluation of Computer and Telecommunication Systems (SPECTS), pp. 1–13, 2018. doi: 10.1109/SPECTS.2018.8574144.
[8] X. Chen, Z. Li, Y. Chen, and Y. Zhang, "Traffic modeling and performance evaluation of SDN-based NB-IoT access network", *Concurrency and Computation Practice and Experience*, 2019. https://doi.org/10.1002/cpe.5145.
[9] A. M. Abbas, K. Y. Youssef, I. I. Mahmoud, and A. Zekry, "NB-IoT optimization for smart meters networks of smart cities: Case study", *Alexandria Engineering Journal*, Vol. 59, No. 6, pp. 4267–4281, 2020. https://doi.org/10.1016/j.aej.2020.07.030.
[10] A. Sehgal, R. Agrawal, R. Bhardwaj, and K. K. Singh, "Reliability analysis of wireless link for IOT applications under shadow-fading conditions", *Procedia Computer Science*, Vol. 167, pp. 1515–1523, 2020. https://doi.org/10.1016/j.procs.2020.03.362.
[11] G. Bouloukakis, I. Moscholios, N. Georgantas, and V. Issarny, "Performance analysis of internet of things interactions via simulation-based queueing models", *Future Internet*, Vol. 13, No. 4, p. 87, 2021. https://doi.org/10.3390/fi13040087.
[12] S. M. R. Islam, D. Kwak, M. H. Kabir, M. Hossain, and K. Kwak, "The internet of things for health care: A comprehensive survey", *IEEE Access*, Vol. 3, pp. 678–708, 2015. doi: 10.1109/ACCESS.2015.2437951.
[13] M. M. Baig, S. Afifi, H. GholamHosseini, and F. Mirza, "A systematic review of wearable sensors and IoT-based monitoring applications for older adults – a focus on ageing population and independent living", *Journal of Medical Systems*, Vol. 43, Article No. 233, 2019. https://doi.org/10.1007/s10916-019-1365-7.

[14] M. N. Birje and S. S. Hanji, "Internet of things based distributed healthcare systems: A review", *Journal of Data, Information and Management*, vol. 2, pp. 149–165, 2020. https://doi.org/10.1007/s42488-020-00027-x.
[15] The ZigBee Alliance, "ZigBee Document 05–3474–21", ZigBee Specification. Available online: https://zigbeealliance.org/wp-content/uploads/2019/11/docs-05-3474-21-0csg-zigbee-specification.pdf.
[16] N. Salman, I. Rasool, and A. H. Kemp, "Overview of the IEEE 802.15.4 standards family for low rate wireless personal area networks," 7th International Symposium on Wireless Communication Systems, New York, UK, 2010, pp. 701–705, doi: 10.1109/ISWCS.2010.5624516.
[17] P. G. Harrison and N. M. Patel, *Performance Modelling of Communication Networks and Computer Architectures*. Addison-Wesley, Boston, 1993.
[18] U. Yechiali and P. Naor, "Queuing problems with heterogeneous arrivals and service", *Operations Research*, vol. 19, pp. 722–734, 1971.

LoRa for Long-Range and Low-Cost IoT Applications

4

Mayur Rajaram Parate[1] and Ankit A. Bhurane[2]
[1]*Indian Institute of Information Technology, Nagpur*
[2]*Visvesvaraya National Institute of Technology, Nagpur*

Contents

DOI: 10.1201/9781003244714-4

4.1 INTRODUCTION TO LORA AND SCOPE OF THE CHAPTER

Long range (LoRa) is a patented frequency modulation technology developed by Semtech Corporation that provides communication over a wide area networks finding applications majorly in Internet of Things (IoT).

Some salient features of LoRa are,

1. Long range: LoRa can facilitate communication range up to as long as 15 km and provides penetrated coverage indoors without GPS.
2. Low power: LoRa is designed for low-power applications and ensures extended battery life for IoT applications. The node power is adapted as per the distance to the gateway (GW) and spread.
3. High capacity: LoRa is capable to handle multiple thousands of message transactions in a day thus making it suitable for public wide networks.
4. Open-source software: The software being open source keeps the software overhead to minimal.
5. Security: LoRa is protected by AES-128 encryption technique.
6. Robust communication: LoRa signals are immune to in and out of the band interference, time and frequency shifts.

LoRa is based on chirp spread spectrum (CSS) technique where the data bits are bandwidth spread by multiplying the data bits by a spreading code (also called as chip sequence) [1]. The spectrum spreading of the signal is controlled by various spreading factors (SF), which also determine the range of communication. The signals with different SF, if transmitted in same time slot, do not overlap with each other.

4.2 LORA STANDARD, TOPOLOGIES (LORAWAN), FREQUENCY BANDS (ASIA)

A typical single and multi-node topologies of a LoRaWAN network is shown in Figures 4.1 and 4.2. The general operation of a LoRaWAN network is as follows.

1. **End device:** An end device/node can be any wireless device capable of sensing or actuating a signal. These are usually battery-operated devices. The data from the end device (say sensor) is up-linked to the hearing GW. There can be multiple GW deployed to ensure high reception rate.
2. **GW:** GW are responsible for reception and forwarding of the messages sent by the end nodes. The messages are further sent to central network server. Entire transaction of messages through GW happen at the lowest (physical) layer.

FIGURE 4.1 A typical LoRaWAN network.

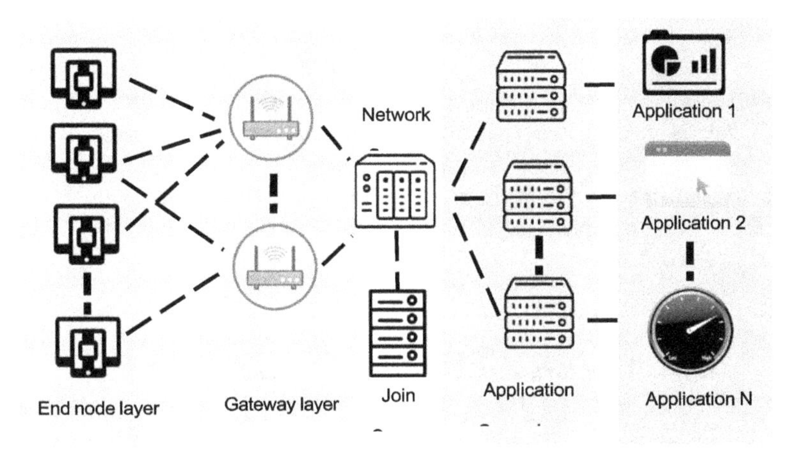

FIGURE 4.2 A typical LoRaWAN network with multiple end nodes, gateways, and applications

3. **Network servers (NS):** NS handles the end-to-end networking, relaying to application server, securing, authenticity, acknowledgment, and transport of the data.
4. **Application servers (AS):** The user end management, application packet downloads, delivery, and connectivity are all managed by the respective AS.

In addition to network and application servers, the overall join-request-accept of uploading and downloading, storage, and verification of encryption keys is managed by an additional *join server* (JS).

4.3 BASIC STEPS OF END-TO-END OPERATION OF A LORA NETWORK

Step 1: An end node transmits a *send request* signal to JS.

Step 2: Post authentication, the JS accepts and acknowledges request to the respective end node.

TABLE 4.1 Region-Wise LoRa Bands

REGION	*INDIA*	*ASIA*	*EUROPE*	*UNITED STATES*	*AUSTRALIA*	*SOUTH KOREA*
Frequency band (MHz)	865–867	923	863–870	902–928	915–928	920–923

Stem 3: Both end node and join server generates session keys.
Step 4: The connection between the end node, network server, and application server is established using respective session keys. The sessions being secured with 128-bit AES ensures data confidentiality with NS and GW.
Step 5: Data is exchanged between end node and intended application.

LoRa operates at different frequencies in different regions. For some regions the bands are shown in Table 4.1. A complete list of LoRa bands can be obtained in [2].

4.4 OVERVIEW OF ESP32-BASED LORA PLATFORM

Internet-of-things (IoT) is best defined by the network connecting the physical objects ('things') that employ sensors, communication capabilities, and software to communicate with other devices over the Internet and exchange information. IoT has evolved as an amalgamation of multiple technologies that need powerful computing capabilities, good communication range, and low power requirement. Here, we introduce ESP32 as a computing element and LoRa as a communication technology in an IoT device.

4.4.1 Introduction to ESP32

ESP32 is a series of low-power, low-cost system-on-chip (SOC) microcontrollers Espressif Systems, a Shanghai-based company. ESP32 is manufactured using 40 nm technology by TSMC. Being SOC, it is integrated with on chip dual-mode Bluetooth and full stack Wi-Fi. The ESP32 is enabled with either a Tensilica Xtensa LX6 microprocessor in dual-core or single-core variations or Xtensa LX7 dual-core microprocessor or single-core RISC-V microprocessor. It also includes power amplifier, low-noise receive amplifier, built-in antenna switches, RF balun, filters, and power-management modules [3].

ESP32 can perform as a complete stand-alone system or can act as a slave device for a host microcontroller to reduce communication stack overhead on the main application processor. Further, ESP32 can also be used to provide Wi-Fi and Bluetooth capabilities through its SPI/SDIO or I2C/UART interfaces. The ESP32 are designed to sustain industrial environment and enabled with advanced calibration circuitries. It is capable of dynamically reducing imperfections in the external circuit and adapting to changes in external conditions.

Moreover, it can sustain the temperature in industrial scenario, i.e., from –40°C to +125°C.

ESP32 achieves low power consumption employing its various power modes and dynamic power scaling. It also has a ultra-low power core which is ideal for mobile devices, wearable electronics, and IoT applications.

4.4.2 Introduction to Heltec LoRa Module

LoRa is low-power, long-range, wireless platform that became ideal for Internet of Things (IoT). LoRa devices and networks such as the LoRaWAN® stands in a Low Power, Wide Area Network (LPWAN) category that enable smart IoT applications to communicate over a long distance using very less power. An ESP32 based LoRa module by Heltec Automations is a complete development board with GPIOs and have surface mount 0.96-inch Blue OLED display [4] as shown in Figure 4.3. Due to its onboard display and ease of programming using Arduino, it is a very simple and time efficient option for prototyping, program validation, and product development.

The heart of ESP32 is Tensilica LX6 dual-core processor operating at a clock speed of 240 MHz. The chip has built-in 520 KB SRAM, full stack (802.11 b/g/N) HT40 Wi-Fi and Lightweight IP (LWIP) Bluetooth enabled with Bluetooth low power technology (BLE). Further, the chip is powerful enough to execute 600 million instructions per second.

To add extra functionality in the ESP32, Heltec added SX1278 chip, which is a LoRa remote modem and support -148 dBm high sensitivity at +20 dBm output power. The main objective behind this integration is to achieve highly reliable long-distance transmission with all the features supported by ESP32. The communication range achieved using Heltec ESP32 LoRa development board measured in open area is 2.6 km.

The ease of programming is achieved by integrating CP2102 USB to serial chip on the board. It support Arduino integrated development environment and can be used with all other available libraries for sensors and interfaces. The pin diagram for the ESP32 based LoRa module by Heltech Automations is shown in Figure 4.4.

FIGURE 4.3 Heltec ESP32 LoRa module.

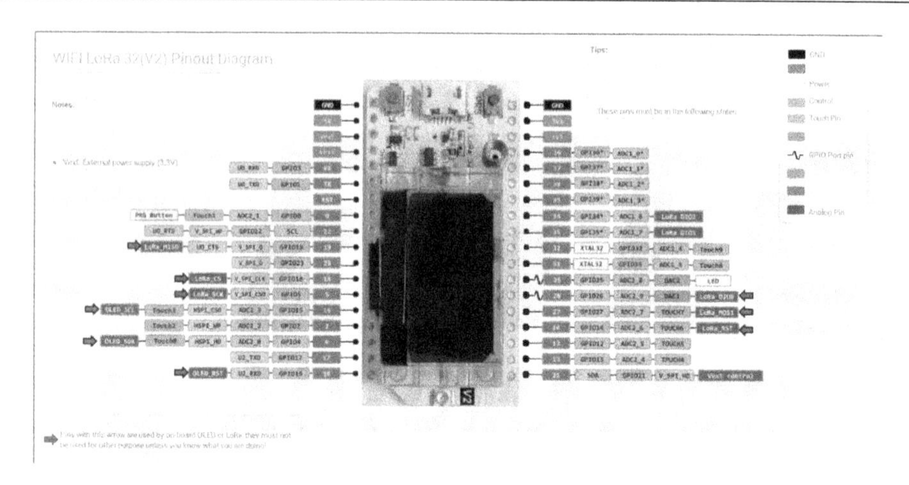

FIGURE 4.4 Pin diagram for Heltec ESP32 LoRa module [4].

4.5 GET-STARTED 'HELLO/ WELCOME!' PROGRAM

In order to make programming and development user friendly, we illustrate development of IoT nodes using 'Arduino' Integrated Development Environment (IDE). We are using 'Heltec esp32-LoRa Development Board' to demonstrate integration of ESP32 and LoRa with Arduino IDE. To do this, we need to install Arduino IDE and required libraries and dependencies.

4.5.1 Arduino IDE Installation

Arduino is an open source IDE for programming various microcontrollers of Atmel, STM32, ESP32, etc. [5]. This section illustrates the process of installing Arduino IDE and required dependencies for ESP32 with LoRa.

- Download Arduino IDE as shown in Figure 4.5 from www.arduino.cc/.
- Install the IDE suitable for your operating system.
- Open the IDE, go to 'File' and then to 'Preferences' as shown in Figure 4.6.
- In Preferences window, go to 'Additional Boards Manager URLs' and paste the following URL and click 'OK'.
  https://github.com/Heltec-Aaron-Lee/WiFi_Kit_series/releases/download/0.0.5/package_heltec_esp32_index.json
  Adding this URL allows Arduino IDE to capture libraries and boards from the repository.
- Now, go to 'Tools' and select the 'Boards Manager' option as shown in Figure 4.7.

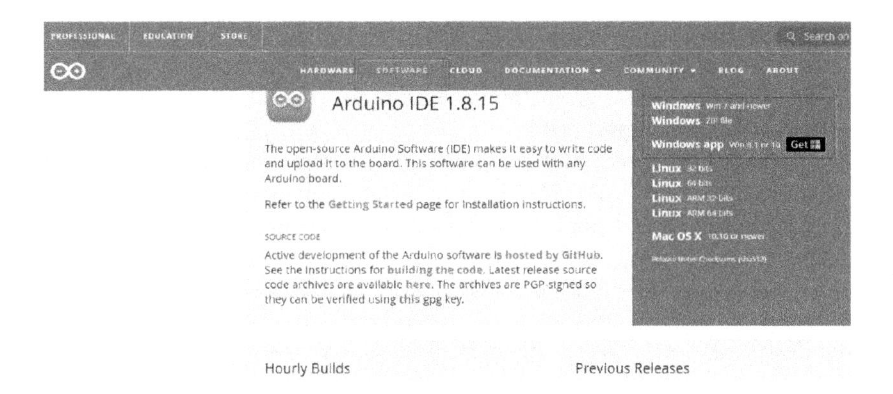

FIGURE 4.5 Arduino website.

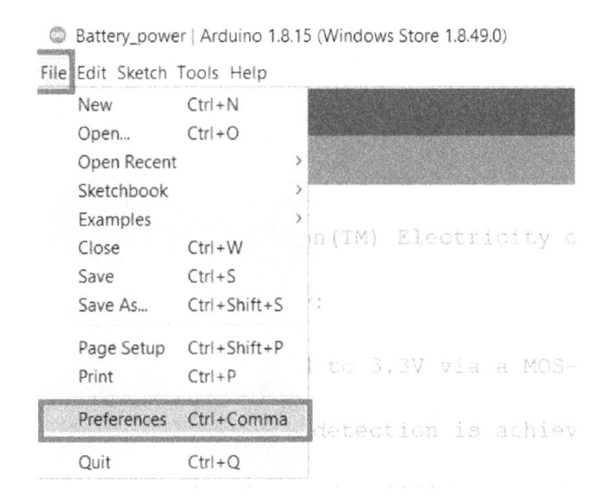

FIGURE 4.6 Go to 'File' and then to 'Preferences'.

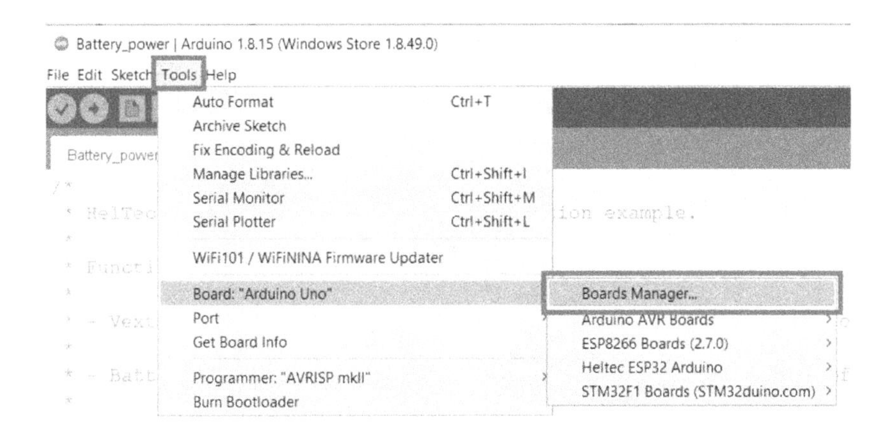

FIGURE 4.7 Go to 'Tools' and then 'Boards Manager'.

- In Boards Manager window, search for 'heltec ESP32' and install the latest version of boards by 'Heltec ESP32 Series Dev-Boards' as shown in Figure 4.8.
- Now, again in 'Tools', go to 'Manage Libraries' option to add corresponding libraries from the repository.
- In 'Manage Libraries' search for 'heltec ESP32' and install the latest version of libraries by 'Heltec ESP32 Dev-Boards' as shown in Figure 4.9. At this point all the dependencies and libraries are installed in Arduino IDE for ESP32-LoRa Development Board.

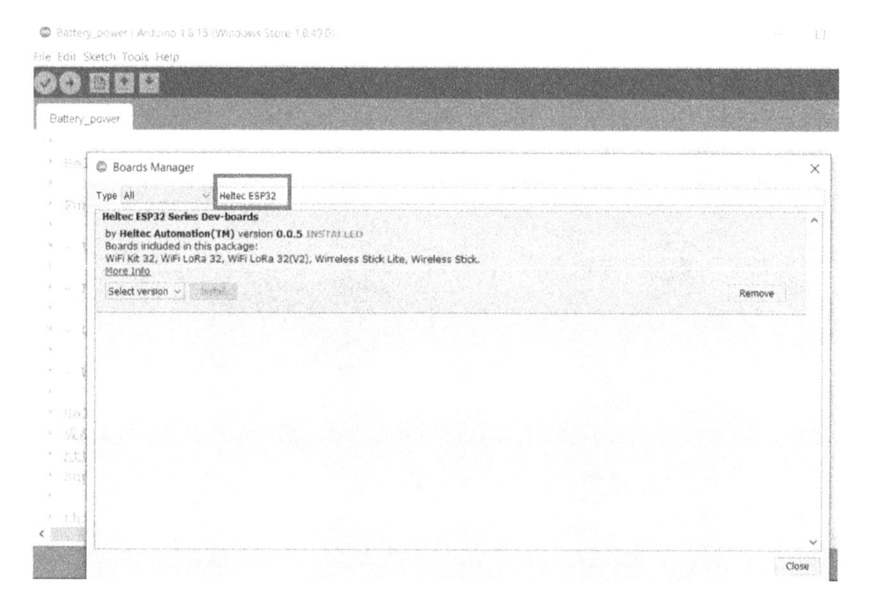

FIGURE 4.8 In Board Manager window search for 'Heltec ESP32' and 'Install'.

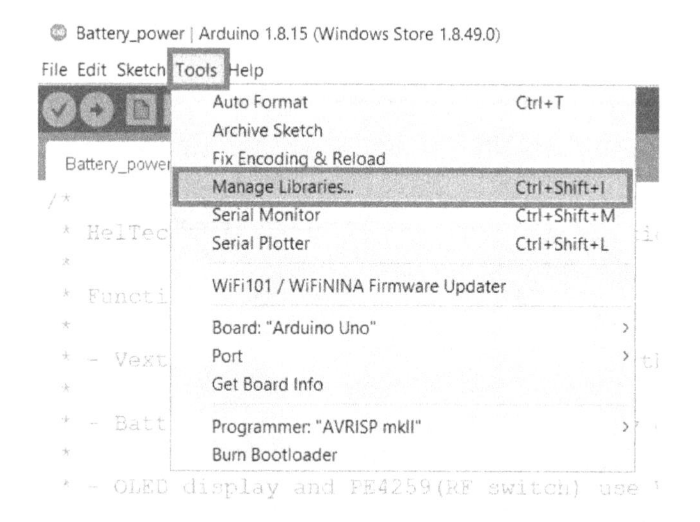

FIGURE 4.9 In 'Tools' go to 'Manage Libraries'.

- After completing all installations, we select the appropriate board for programming.

Go to Tools -> Board -> Heltec ESP32 Arduino -> WiFi LoRa 32(v2). Here, we are using 'WiFi LoRa 32(v2)', if you have any other version then choose accordingly.

- Arduino IDE have facility to program a supported device by serial communication through USB using CP210X USB to UART bridge. It creates a virtual serial COM port in a computer. The COM port corresponding to the connected device can be verified at 'Device Manager' in a Windows machine as shown in a Figure 4.12. To program a connected device, select the same 'COM Port' in arduino IDE at Tools -> Port.

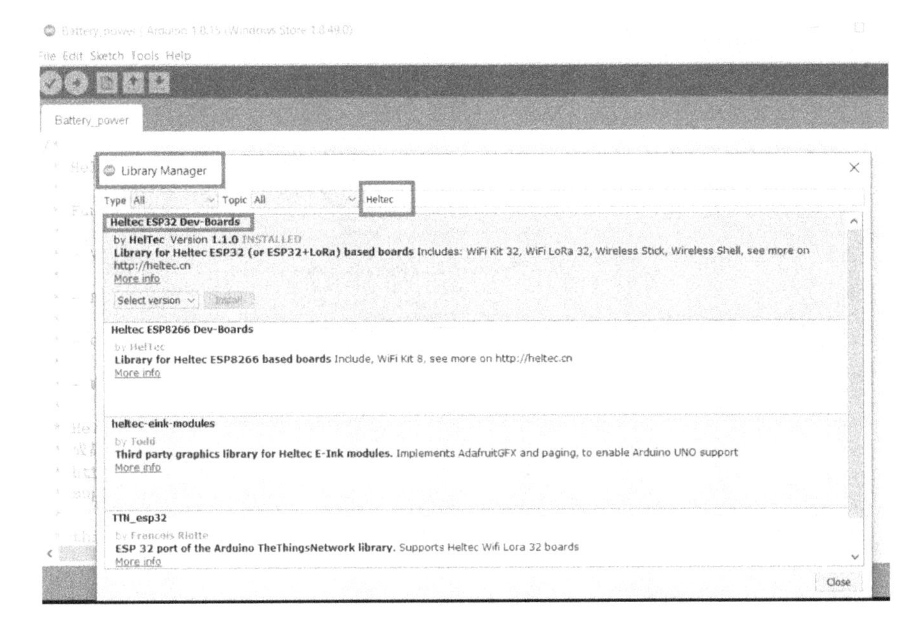

FIGURE 4.10 In 'Library Manager' window search for 'Heltec ESP32' and click 'Install'.

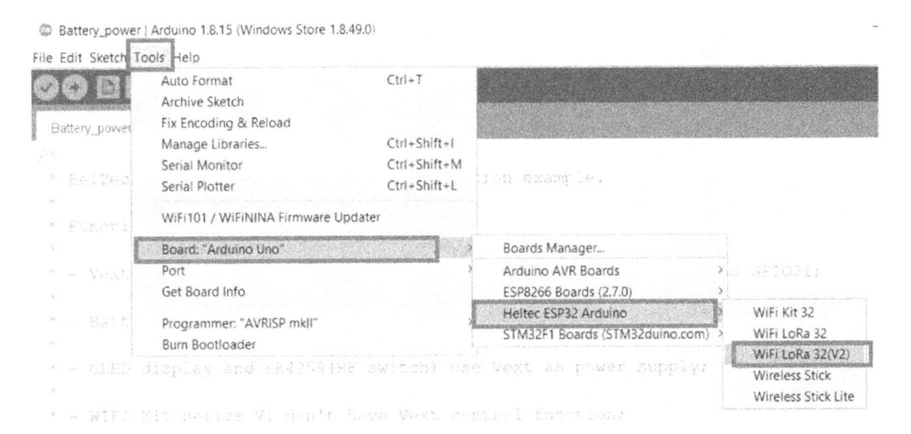

FIGURE 4.11 Select the appropriate board.

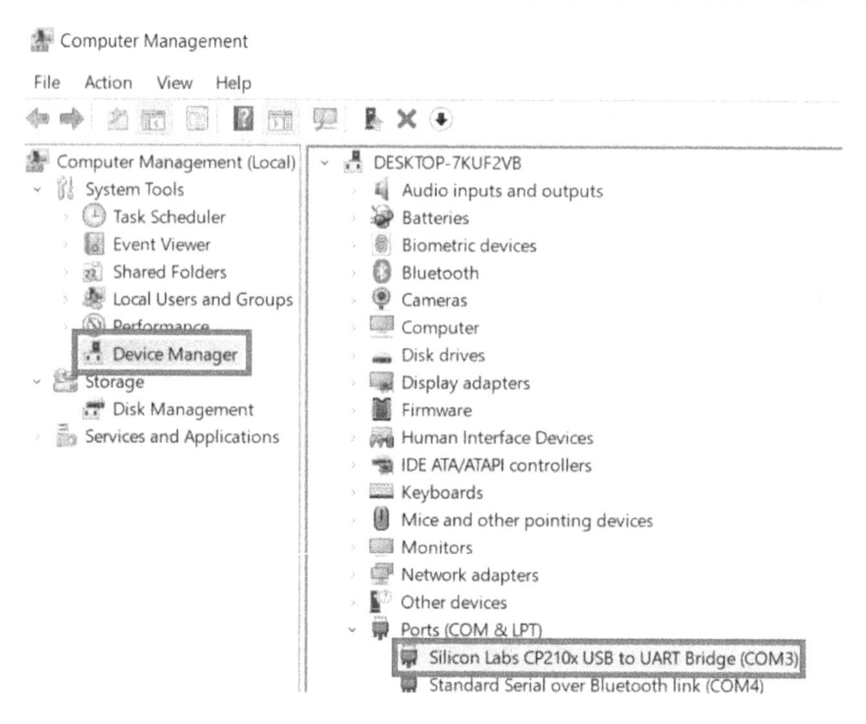

FIGURE 4.12 Verify COM port at 'Device Manager' in Windows machine.

4.5.2 Testing the Development Board

With this setting, the development board can be tested for its various functionalities [6].

4.5.2.1 Testing the Basic Functionalities

The board supports multiple functionalities and needed to be tested before using in a complex development. The factory test program test the development board for the following

- Basic Wi-Fi function testing
  We test the Wi-Fi functionalities of the development board by setting Wi-Fi in a station mode. If the board is already used and we have SSID and password saved in it, then it connects automatically, otherwise use provided SSID and password for a new connection.
  Listing.1 presents the code for Wi-Fi-initialization and testing its functionalities, #include 'heltec.h' and #include 'WiFi.h' include the header files 'heltec.h' and 'WiFi.h'. These files are developed by Heltec Automation and provide definitions for the functions we will be using in the setup(). WiFi.mode (WIFI_STA), on line 8, sets the Wi-Fi in station mode so that the development board can connects to provided SSID and password in WIFI. begin ('SSID', 'Password').

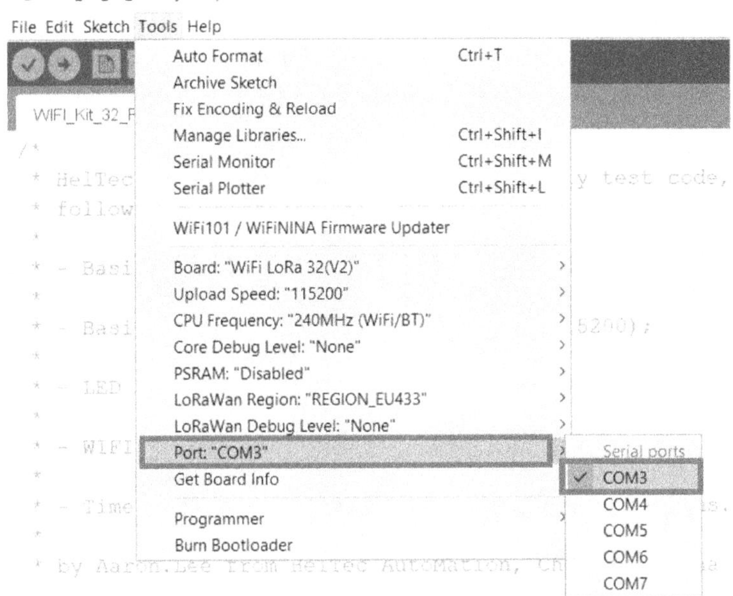

FIGURE 4.13 Select respective COM Port for connected device.

```
#include "heltec.h"
#include "WiFi.h"
void setup()
{
  // Setup a fresh WiFi station
  WiFi.disconnect(true);
  delay(2000);
  WiFi.mode(WIFI_STA); //select station mode
  WiFi.setAutoConnect(true);
  WiFi.begin("SSID","Password");
  delay(200);
}

void loop()
{
  // your main code will come here.
}

```

- Basic OLED function test
  We initialize OLED display and print 'Hello!' on the screen. It is mandatory to initialize the Heltec Object as follows.
  Heltec.begin(Enable-Display, Enable-LoRa, Enable-Serial-Communication, Enable-Antenna-Power-Boost, BAND);
  The argument 'true' and 'false' provided to a function Heltec.begin() either enable or disable the corresponding functionality in the heltec development board. The first four arguments are Boolean whereas the last argument is the LoRa band.

The lines 8 to 10 indicate, the initialization of OLED display, flipping the screen vertically for horizontal view, setting font to 'ArialMT_Plain' and text size to 10. Line 11, display the string 'It's an OLED Display' starting from line position 0 and character position 0. The required simple setup is shown below in Listing.2.

```
1           #include "heltec.h"
2           #include "images.h"
3           #define BAND     865E6  // Set LoRa band (here Indian)
4           void setup()
5           {
6             Heltec.begin(true, true, true, true, BAND);
7             // Here we setup the settings for OLED Display
8             Heltec.display->init();
9             Heltec.display->flipScreenVertically();
10            Heltec.display->setFont(ArialMT_Plain_10);
11            Heltec.display->drawString(0, 0, "It's an OLED
      Display");
12            Heltec.display->display();
13            delay(1000);
14          }
15
16          void loop()
17          {
18              // your main code will come here.
19          }
20
```

- Basic LoRa function test
  Basic functionalities of the LoRa can be tested using the following code. Initially, the frequency band allowed as per the region needs to be selected for communication over the LoRa. 'Heltc.begin()' function initializes LoRa and other features in the development board as explained in 'Basic OLED function test'. The data packet of the LoRa needs initialization and termination as provided by 'LoRa.beginPacket()' and 'LoRa.endPacket()' and data to be sent is placed in-between these function using 'LoRa.print(data)'.
  The example code is shown in Listing.3. In this, the data we are sending over the LoRa-Network is 'hello'.

```
1           # include "heltec.h"
2           #define BAND     865E6  // Set LoRa band (here Indian)
3           void setup()
4           {
5             Heltec.begin(true, true, true, true, BAND);
6             delay(2000);
7           }
8           void loop()
9           {
10            // send packet
11            LoRa.beginPacket(); // starts the data packet
      formation
12            LoRa.setTxPower(14,RF_PACONFIG_PASELECT_PABOOST); //
      sets the transmission power
13            LoRa.print("hello ");// incapsulate the data into
      the LoRa packet
14            LoRa.endPacket(); // ends and transmit the data
      packet
15          }
16
```

- Basic LoRa transmitter and receiver
  We studied the simple LoRa configuration in previous topic, whereas, how to set LoRa module into transmitter and receiver is explored here. The transmitter and receiver can be obtained by configuring one LoRa kit in transmitter mode using LoRa.setTxPower(txPower,RFOUT_pin) as in line 13 in Listing.4. and other LoRa kit in the receiving mode using LoRa.read() commands as in line 13 in Listing.5. During transmission, the RFOUT_pin could be RFO to attain a maximum output power of 14 dBm or PABOOST to achieve maximum of 20 dBm. A simple working example for transmitter and receiver is given. In transmitter, we are sending data as 'Sent Packet', followed by the packet number.
  At receiver, we parse the received packet for data at line 9 and print the same data on serial port at line 13. The listings and the output for the same is shown in Figure 4.14.

```
1    #include "heltec.h"
2    #define BAND    868E6 // Set LoRa band (here Indian)
3    int Packet_Number = 0;
4    void setup() {
5      // Enable display, LoRa, serial communication with
antenna power boost using Indian band
6      Heltec.begin(true, true, true, true, BAND);
7    }
8    void loop() {
9      Serial.print("Sending packet: "); // print details on
serial terminal of arduino
10     Serial.println(Packet_Number);
11     //Send packet with maximum power of 20 dBm
12     LoRa.beginPacket(); // start packet
13     LoRa.setTxPower(20,RF_PACONFIG_PASELECT_PABOOST);
14     LoRa.print("Sent Packet: ");// include data into packet
15     LoRa.print(Packet_Number); // include data into packet
16     LoRa.endPacket();// ends the packet
17     Packet_Number++;
18   }
19
```

```
1    #include "heltec.h"
2    #define BAND    868E6    // Set LoRa band (here Indian)
3    void setup() {
4      // Enable display, LoRa, serial communication with
antenna power boost using Indian band
5      Heltec.begin(true, true, true, true, BAND);
6    }
7    void loop() {
8      // Receive packet
9      int pktSz = LoRa.parsePacket(); // parse the received
packet and calculate its packet size
10     if (pktSz) {
11       // If packet print "Received" and packet data
12       Serial.print("Received ");
13       // read packet
14       while (LoRa.available()) {
15         Serial.print((char)LoRa.read()); \\ print the
received data from LoRa on to the serial terminal of arduino.
16       }
17       Serial.print(" \n");
18     }}
19
```

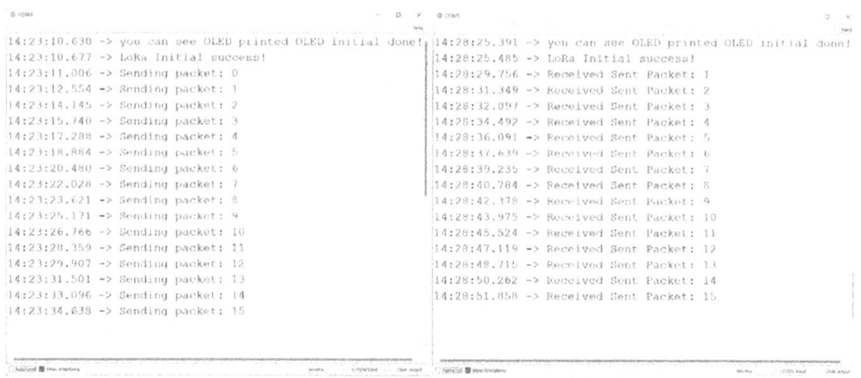

```
14:23:10.630 -> you can see OLED printed OLED initial done!
14:23:10.677 -> LoRa Initial success!
14:23:11.006 -> Sending packet: 0
14:23:12.554 -> Sending packet: 1
14:23:14.145 -> Sending packet: 2
14:23:15.740 -> Sending packet: 3
14:23:17.288 -> Sending packet: 4
14:23:18.884 -> Sending packet: 5
14:23:20.480 -> Sending packet: 6
14:23:22.028 -> Sending packet: 7
14:23:23.621 -> Sending packet: 8
14:23:25.171 -> Sending packet: 9
14:23:26.766 -> Sending packet: 10
14:23:28.359 -> Sending packet: 11
14:23:29.907 -> Sending packet: 12
14:23:31.501 -> Sending packet: 13
14:23:33.096 -> Sending packet: 14
14:23:34.638 -> Sending packet: 15
```

```
14:28:25.391 -> you can see OLED printed OLED initial done!
14:28:25.485 -> LoRa Initial success!
14:28:29.756 -> Received Sent Packet: 1
14:28:31.349 -> Received Sent Packet: 2
14:28:32.897 -> Received Sent Packet: 3
14:28:34.492 -> Received Sent Packet: 4
14:28:36.091 -> Received Sent Packet: 5
14:28:37.639 -> Received Sent Packet: 6
14:28:39.235 -> Received Sent Packet: 7
14:28:40.784 -> Received Sent Packet: 8
14:28:42.378 -> Received Sent Packet: 9
14:28:43.975 -> Received Sent Packet: 10
14:28:45.524 -> Received Sent Packet: 11
14:28:47.119 -> Received Sent Packet: 12
14:28:48.715 -> Received Sent Packet: 13
14:28:50.262 -> Received Sent Packet: 14
14:28:51.858 -> Received Sent Packet: 15
```

FIGURE 4.14 Serial monitor output of a simple packet transmitter-receiver.

- Basic serial port test (in baud rate 115200); support for the serial communication can be tested simply at serial terminal of an Arduino. The serial communication in the development board can be enabled using the Heltec.begin() function by making corresponding argument as 'true' as in line 5 of Listing.6. The data to be printed can be sent to serial port using Serial.printf() command as in line 7.

```
1   #include "heltec.h"
2   #define BAND    865E6  // Set LoRa band (here Indian)
3   void setup()
4   {
5     Heltec.begin(true, true, true, true, BAND); //
initialize the serial communication
6     delay(1000);
7     Serial.printf("Hello!"); // Print data serially.
8   }
9   void loop()
10  {
11  }
12
```

- Other Arduino basic functions such as timers, interrupts, etc., can be tested as per the requirement.

The complete and comprehensive code for the testing development board for its communication using Wi-Fi is available in Listing.5. The output for the same on the inbuilt LCD screen is shown in Figure 4.15.

The provided listing is the combination of listings 1 to 6, in which void WIFISetUp(void) function setup the Wi-Fi and display its connection status. Specifically, code from line 8 to 13 initialize the Wi-Fi. The connection status during the Wi-Fi connection is printed on the OLED display at lines 16 to 39. The function void WIFIScan (void) at lines 41 to 83 scans the available Wi-Fi in the area and display those Wi-Fi SSIDs on OLED display. Then in void setup(), we begin the heltec development board and call the above declared functions.

```
1
2 #include "heltec.h"
3 #include "WiFi.h"
4
5 void WIFISetUp(void)
6 {
7   // Setup a fresh WiFi station
8   WiFi.disconnect(true);
9   delay(2000);
10  WiFi.mode(WIFI_STA); // set wifi in station mode
11  WiFi.setAutoConnect(true);
12  WiFi.begin("SSID","Password");
13  delay(200);
14
15  byte count = 0;
16  while(WiFi.status() != WL_CONNECTED && count < 10)
17  {
18    count ++; // wait for 10 seconds for connection
19    delay(1000);
20    Heltec.display -> drawString(0, 0, "Connection in Progress..."); //
       display connection status on OLED display
21    Heltec.display -> display();
22  }
23  Heltec.display -> clear();
24  if(WiFi.status() == WL_CONNECTED) // if true then connection
      successful
25  {
26    Heltec.display -> drawString(0, 0, "Connected, congratulations!");
27    Heltec.display -> display();
28  }
29  else
30  {
31    Heltec.display -> clear();
32    Heltec.display -> drawString(0, 0, "Connection ");
33    Heltec.display -> display();
34 //    while(1);
35  }
36  Heltec.display -> drawString(0, 15, "WIFI Setup done");
37  Heltec.display -> display();
38  delay(500);
39 }
40
41 void WIFIScan(void)  // scans for the available wifi connections in the
       area
42 {
43  Heltec.display -> drawString(0, 15, "Started Scanning");
44  Heltec.display -> display();
45
46  int n = WiFi.scanNetworks();
47  Heltec.display -> drawString(0, 25, "Completed Scanning");
48  Heltec.display -> display();
49  delay(500);
50  Heltec.display -> clear();
51
52  if (n == 0)
53  {
54    Heltec.display -> clear();
```

```
      Heltec.display -> drawString(0, 0, "Unable to find connection");
      Heltec.display -> display();
      while(1);
    }
    else
    {
      Serial.print(n);
      Heltec.display -> drawString(0, 0, (String)n);
      Heltec.display -> drawString(14, 0, "networks found:");
      Heltec.display -> display();
      delay(500);

      for (int i = 0; i < n; ++i) {
      // Print the list of all networks available in the area
        Heltec.display -> drawString(0, (i+1)*9,(String)(i + 1));
        Heltec.display -> drawString(6, (i+1)*9, ":");
        Heltec.display -> drawString(12,(i+1)*9, (String)(WiFi.SSID(i)));
        Heltec.display -> drawString(90,(i+1)*9, " (");
        Heltec.display -> drawString(98,(i+1)*9, (String)(WiFi.RSSI(i)));
        Heltec.display -> drawString(114,(i+1)*9, ")");
        //            display.println((WiFi.encryptionType(i) ==
      WIFI_AUTH_OPEN)?" ":"*");
        delay(10);
      }
    }

    Heltec.display -> display();
    delay(800);
    Heltec.display -> clear();
  }

  void setup()
  {
    Heltec.begin(true, false, true);// Enable Display, disable LoRa
      Enable, and enable Serial communication.
    Heltec.display->clear();
    WIFISetUp();

    WiFi.disconnect(true);//WIFI
    delay(1000);
    WiFi.mode(WIFI_STA);
    WiFi.setAutoConnect(true);
  }

  void loop()
  {
    WIFIScan();
    delay(2000);
  }
```

FIGURE 4.15 Performing factory test on Heltec ESP32 LoRa module.

In the previous example, we tested the development board for its communication using Wi-Fi and related functionalities. In this example, we test the LoRa communication functionality by sending data packets and receiving data packets over the LoRa network. To realize data transmission over LoRa, we use 865 MHz channel, which we can set using 'LoRaWan Region "REGION_IN865' ".

The listing for the same is given you in Listing.8. Line 6, initializes the Display, LoRa, Serial-Communication, Antenna-Power-Boost, and selects the frequency BAND for the communication. User can defined display setting as per his/her choice in lines 9 to 14. In void loop() function which runs repetitively has two parts: (a) status display of the packet transmission which is optional and (b) data packet transmission at lines 30 to 34.

After uploading the firmware, the LCD screen on development board shows 'Heltec LoRa Initial Success!' indicating the LoRa functionalities are working without error, as shown in Figure 4.15. To test data transmission, we send some random data over the LoRa network as shown in Figures 4.16 and 4.17. Listing for testing LoRa by sending data packets:

```
1  #include "heltec.h"
2  #define BAND    865E6  //you can set band here directly,e.g. 868E6,915
     E6
3  unsigned int counter = 0;
4  void setup()
5  {
6    Heltec.begin(true, true, true, true, BAND); // Initialize display,
       LoRa, serial comm., and power boost
7
8   // following block of a code is related to setting display
9    Heltec.display->init();
10   Heltec.display->flipScreenVertically();
11   Heltec.display->setFont(ArialMT_Plain_10);
12   Heltec.display->clear();
13   Heltec.display->drawString(0, 0, "Display Settings Completed!");
14   Heltec.display->display();
15   delay(2000);
16 }
17
18 void loop()
19 {
20   // following block of a code is related to displaying packet
       information
21   Heltec.display->clear();
22   Heltec.display->setTextAlignment(TEXT_ALIGN_LEFT);
23   Heltec.display->setFont(ArialMT_Plain_10);
24
25   Heltec.display->drawString(0, 0, "Sending packet: ");
26   Heltec.display->drawString(90, 0, String(counter));
27   Heltec.display->display();
28
29   // send packet
30   LoRa.beginPacket(); // begin data packet here
31   LoRa.setTxPower(14,RF_PACONFIG_PASELECT_PABOOST); // set transmit
       power
32   LoRa.print("hello "); // add data to be send into the packet
33   LoRa.print(counter); // add data to be send into the packet
34   LoRa.endPacket(); // end data packer here
35
36   counter++; // data for the LoRa packet.
37 }
```

4.6 DISPLAY PROGRAMMING

The visual displays are not necessary for sensor nodes but are very handy and helpful for troubleshooting and feedback. The Heltec ESP32 LoRa development board comes with inbuilt 0.96-inch Blue OLED display and can be used in various development activities.

In this example, the rectangular graphics design is displayed to demonstrate the use of OLED display. Heltec Automations have developed various functions and display commands and are available at https://github.com/HelTecAutomation/Heltec_ESP32/blob/master/src/oled/OLEDDisplay.h.

These functions are compatible with Arduino IDE and can be swiftly used for development. We describe the various graphics functions for drawing lines, rectangle, circle, filling the rectangle, and custom function for printing the saved data into buffer.

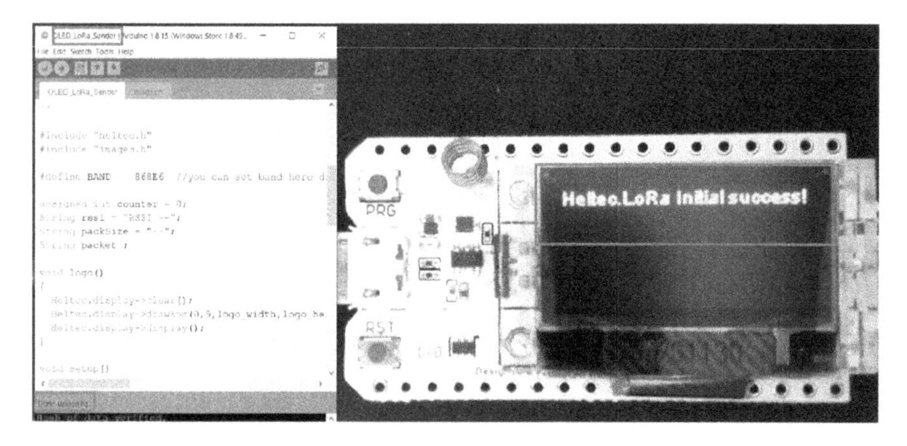

FIGURE 4.16 Testing LoRa functionalities.

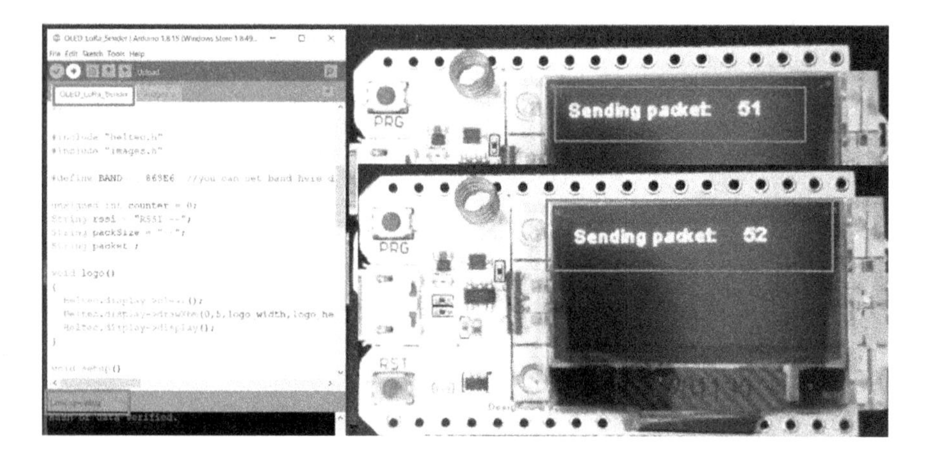

FIGURE 4.17 Transmitting LoRa packets.

- Listing for drawing rectangle on the OLED display.
  We use drawRect() function to draw rectangle, the function definition for the same is given below: The function, uses two arguments – DISPLAY_HEIGHT and DISPLAY_WIDTH – which are the height and width of the OLED display; instead user can use custom value for displaying the rectangle on the OLED display.

```
1       // setting done are considering the Adafruit_SSD1306 OLED
    display
2
3       void drawRect(void) {
4         for (int16_t i=0; i<DISPLAY_HEIGHT/2; i+=2) {//
    DISPLAY_HEIGHT and DISPLAY_WIDTH are the user arguments
5           Heltec.display->drawRect(i, i, DISPLAY_WIDTH-2*i,
    DISPLAY_HEIGHT-2*i);// drawRect function to draw rectangle.
6           Heltec.display->display();
7           delay(10);
8         }
9       }
10
```

- Listing for filling rectangle
  Same as the previous example, fillRect() accepts the arguments; DISPLAY_ HEIGHT and DISPLAY_WIDTH, and user can change the values according to the dimensions of the rectangle he/she wants to draw on the OLED display. The function definition for the same is given in Listing.10. The color to fill can be selected using setColour() command as shown in line 12; here we select 'white' color to fill the rectangle.

```
1       // setting done are considering the Adafruit_SSD1306 OLED
    display
2       void fillRect(void) {
3         uint8_t color = 1;
4         // DISPLAY_HEIGHT and DISPLAY_WIDTH are the user
    arguments
5         for (int16_t i=0; i<DISPLAY_HEIGHT/2; i+=3) {
6           Heltec.display->setColor((color % 2 == 0) ? BLACK :
    WHITE); // White color is used, alternate colors can be used
7           Heltec.display->fillRect(i, i, DISPLAY_WIDTH - i*2,
    DISPLAY_HEIGHT - i*2); \\ fillRect is the function to fill the
     rectangle
8           Heltec.display->display();
9           delay(10);
10          color++;
11        }
12        // Reset back to WHITE
13        Heltec.display->setColor(WHITE);
14      }
15
```

- Listing for drawing circle on OLED display
  To draw circles on the display we can use drawCircle() function as defined in Listing.11. The function itself has two parts: (a) In first part, line 3 to 9 uses OLED dimensions and draw circle at an increment of every 2 pixels as given in the for loop at line 3; (b) the second part considers the circle as a combination of

four segments drawn in four quadrants. The code in lines 17 to 19 defines code for the segment in the first quadrant labeled as 0b00000001 and uses function drawCircleQuads(). The same code is repeated to draw the rest of the segments to complete the circle as given in lines 20 to 27.

```
1     // setting done are considering the Adafruit_SSD1306 OLED
  display
2     void drawCircle(void) {
3       for (int16_t i=0; i<DISPLAY_HEIGHT; i+=2) { \\ draw
  circle for every 2 pixel increment
4         Heltec.display->drawCircle(DISPLAY_WIDTH/2,
  DISPLAY_HEIGHT/2, i);\\ drawCircle is the fucnction to draw
  circle
5       // DISPLAY_HEIGHT and DISPLAY_WIDTH are the user
  arguments
6         Heltec.display->display();
7         delay(10);
8       }
9       delay(1000);
10      Heltec.display->clear();
11
12      // This will draw the part of the circle in quadrant 1
13      // Quadrants are numbered like this:
14      //   0010 | 0001
15      //  ------|------
16      //   0100 | 1000
17      // Every quadrant draw one segment of the circle
18      Heltec.display->drawCircleQuads(DISPLAY_WIDTH/2,
  DISPLAY_HEIGHT/2, DISPLAY_HEIGHT/4, 0b00000001);
19      Heltec.display->display(); // draw one segment
20      delay(200);
21      Heltec.display->drawCircleQuads(DISPLAY_WIDTH/2,
  DISPLAY_HEIGHT/2, DISPLAY_HEIGHT/4, 0b00000011);
22      Heltec.display->display(); // draw second segment
23      delay(200);
24      Heltec.display->drawCircleQuads(DISPLAY_WIDTH/2,
  DISPLAY_HEIGHT/2, DISPLAY_HEIGHT/4, 0b00000111);
25      Heltec.display->display(); // draw third segment
26      delay(200);
27      Heltec.display->drawCircleQuads(DISPLAY_WIDTH/2,
  DISPLAY_HEIGHT/2, DISPLAY_HEIGHT/4, 0b00001111);
28      Heltec.display->display(); // draw forth segment
29    }
30
```

- Listing for drawing lines on the OLED display.
  Listing.12 presents the demo for graphics using lines drawn at different locations on the OLED display. drawLines() function executes this demo.
  To draw a line on the OLED display, drawLine() function provided by Heltec Automation is used. The function drawLine(0, 0, DISPLAY_WIDTH-1,i) takes four arguments; first two arguments define the start point of a line and last two define end point of a line. By varying starting and ending point of a line, simple graphics is developed. In the Listing.12, the function drawLine() is repeated multiple times as in lines 4, 9, 17, 22, 30, 35, 42, and 47.

```
1     // Adapted from Adafruit_SSD1306
2     void drawLines() {
3       for (int16_t i=0; i<DISPLAY_WIDTH; i+=4) {
4         Heltec.display->drawLine(0, 0, i, DISPLAY_HEIGHT-1);
   \\ drawLine is basic function for drawing line.
5     // takes four arguments: first two arguments define the
   start point of  a line and last two defines end point of a
   line.
6       Heltec.display->display();
7         delay(10);
8       } // This loop display line starting from origin i.e.
   left hand top corner of OLED and rotate line along the width
   of OLED
9       for (int16_t i=0; i<DISPLAY_HEIGHT; i+=4) {
10        Heltec.display->drawLine(0, 0, DISPLAY_WIDTH-1, i);
11        Heltec.display->display();
12        delay(10);
13      } // This loop display line starting from origin i.e.
   left hand top corner of OLED and rotate line along the height
   of OLED
14      delay(250);
15
16      Heltec.display->clear();
17      for (int16_t i=0; i<DISPLAY_WIDTH; i+=4) {
18        Heltec.display->drawLine(0, DISPLAY_HEIGHT-1, i, 0);
19        Heltec.display->display();
20        delay(10);
21      } // This loop display line starting from coordinate (0,
    DISPLAY_HEIGHT-1) and rotate line along the width of OLED
22      for (int16_t i=DISPLAY_HEIGHT-1; i>=0; i-=4) {
23        Heltec.display->drawLine(0, DISPLAY_HEIGHT-1,
   DISPLAY_WIDTH-1, i);
24        Heltec.display->display();
25        delay(10);
26      } // This loop display line starting from coordinate (0,
    DISPLAY_HEIGHT-1) and rotate line along the height of OLED
27      delay(250);
28
29      Heltec.display->clear();
30      for (int16_t i=DISPLAY_WIDTH-1; i>=0; i-=4) {
31        Heltec.display->drawLine(DISPLAY_WIDTH-1,
   DISPLAY_HEIGHT-1, i, 0);
32        Heltec.display->display();
33        delay(10);
34      } // This loop display line starting from coordinate (
   DISPLAY_WIDTH-1, DISPLAY_HEIGHT-1) and rotate line along the
   width of OLED
35      for (int16_t i=DISPLAY_HEIGHT-1; i>=0; i-=4) {
36        Heltec.display->drawLine(DISPLAY_WIDTH-1,
   DISPLAY_HEIGHT-1, 0, i);
37        Heltec.display->display();
38        delay(10);
39      }// This loop display line starting from coordinate (
   DISPLAY_WIDTH-1, DISPLAY_HEIGHT-1) and rotate line along the
   height of OLED
40      delay(250);
41      Heltec.display->clear();
```

```
42        for (int16_t i=0; i<DISPLAY_HEIGHT; i+=4) {
43          Heltec.display->drawLine(DISPLAY_WIDTH-1, 0, 0, i);
44          Heltec.display->display();
45          delay(10);
46        }// This loop display line starting from coordinate (
    DISPLAY_WIDTH-1,0) and rotate line along the height of OLED
47        for (int16_t i=0; i<DISPLAY_WIDTH; i+=4) {
48          Heltec.display->drawLine(DISPLAY_WIDTH-1, 0, i,
    DISPLAY_HEIGHT-1);
49          Heltec.display->display();
50          delay(10);
51        } // This loop display line starting from coordinate (
    DISPLAY_WIDTH-1,0) and rotate line along the width of OLED
52        delay(250);
53      }
54
```

- Listing to develop custom function for storing and displaying the data on OLED display.
  The buffer is used to store the data or characters to be displayed. To use this buffer, first we need to initialize it by defining its size in the form of number of lines and number of characters per line. For example in Listing.13, five lines and 30 characters per line are declared as in line 4. Once the size of buffer is declared, it can store the data inside it for printing and the same can be printed using command drawLogBuffer().

```
1       void printBuffer(void) {
2         // Initialize the log buffer
3         // allocate memory to store 8 lines of text and 30 chars
    per line.
4         Heltec.display->setLogBuffer(5, 30);
5
6         // Some test data
7         const char* test[] = {
8             "Hello",
9             "World" ,
10            "----",
11            "Show off",
12            "how",
13            "the log buffer",
14            "is",
15            "working.",
16            "Even",
17            "scrolling is",
18            "working"
19        };
20
21        for (uint8_t i = 0; i < 11; i++) {
22          Heltec.display->clear();
23          // Print to the screen
24          Heltec.display->println(test[i]);
25          // Draw it to the internal screen buffer
26          Heltec.display->drawLogBuffer(0, 0);
27          // Display it on the screen
28          Heltec.display->display();
29          delay(500);
30        }
31      }
32
```

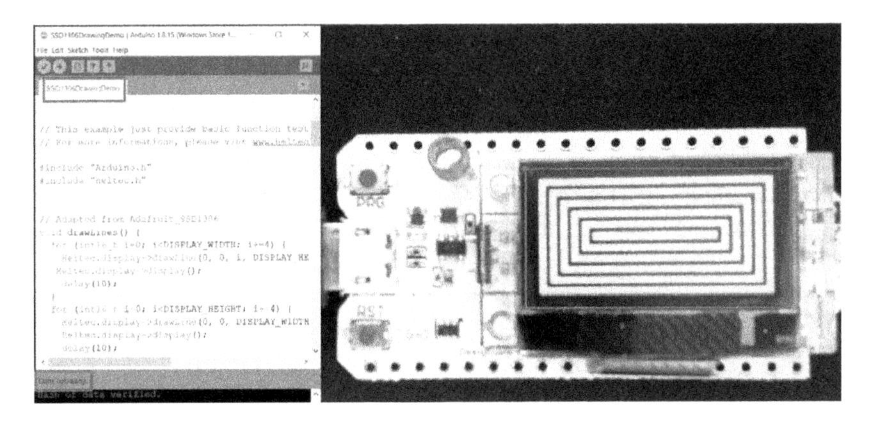

FIGURE 4.18 Displaying graphics on OLED display.

4.7 SENSOR INTERFACING (DATA ACQUISITION)

Data acquisition from the field is one of the important rather primary requirements of any IoT system. This is achieved by deploying sensor nodes at required points in the field. The sensor node is a hardware with sensors, a processor, battery, and communication technology either Wi-Fi, Bluetooth, LoRa, or RF. Following are the few characteristics of a typical sensor node in an IoT framework.

- It should support analog and digital sensors.
- Processor should support a low or ultra-low power configuration.
- Processor should be powerful enough to handle preprocessing on the sensor data if required.
- Communication technology should be power efficient.
- Sensor node should run sufficiently long enough on a single charge of a battery.

In this section, a sensor node using ESP32 LoRa development board is demonstrated, as it supports both analog and digital sensors, has ESP32 with ultra-low power core, and is enabled with long-range RF (LoRa) communication support and thus becomes an eligible option for using it as a sensor node in IoT framework.

We interface BMP280 sensor [7] from Bosch with heltec development board to measure temperature and pressure. We display these readings on OLED display and also transmit over the LoRa network.

The listing is performed in three stages: reading the sensor value, displaying it on the OLED display, and, lastly, transmitting these values over the LoRa network.

The function drawFrontFaceDemo() accept four arguments – temperature, pressure, altitude, and real altitude – and displays these values on the OLED display using function drawString(arg1, arg2, arg3) in which, arg1 is starting column, arg2 is starting line, and arg3 is the text itself.

In void loop(), at line 40, we check the sensor availability using !bmp.begin() and if the sensor is available then we read the values of temperature, pressure, altitude, and real altitude at lines, 44, 49, 54, and 65, respectively. The function to read the sensor values is provided by the library BMP180.h, which we included at the top of the listing.

Lastly, we want to send these sensor values over the LoRa network. We create a data packet for transmission at lines 75 to 82. LoRa.beginPacket() initiates the data packet and then LoRa.print() command adds sensor data into the packet. Finally, LoRa.endPacket() ends the packet and transmit it on the LoRa network.

Listing for testing LoRa by sending data packets:

```
1  #include "Arduino.h"
2  #include "heltec.h"
3  #include <Wire.h>
4  #include <BMP180.h>
5  #include "string.h"
6
7  #define BAND     865E6  // Set LoRa band (here Indian)
8
9  BMP085 bmp;
10
11 uint8_t T[20] = {"Temperature"};
12
13 void setup() {
14   pinMode(Vext,OUTPUT);
15   digitalWrite(Vext,HIGH);
16   Serial.begin(115200); // initialize serial comm. at  a baud rate of
       115200
17
18 }
19 void drawFontFaceDemo(double T,double P,double A,double R) {
20   //this block of code print the sensor values on the OLED display
21     Heltec.begin(true , false , true);
22 //   Heltec.display->flipScreenVertically();
23     Heltec.display->clear();
24     Heltec.display->setTextAlignment(TEXT_ALIGN_LEFT);
25     Heltec.display->setFont(ArialMT_Plain_10);
26     Heltec.display->drawString(0    , 0  ,    "Temperature =");
27     Heltec.display->drawString(76   , 0  ,    (String)T);
28     Heltec.display->drawString(106  , 0  ,    " *C");
29     Heltec.display->drawString(0    , 16 ,    "Pressure =");
30     Heltec.display->drawString(56   , 16 ,    (String)P);
31     Heltec.display->drawString(100  , 16 ,    "Pa");
32     Heltec.display->drawString(0    , 32 ,    "Altitude    =");
33     Heltec.display->drawString(56   , 32 ,    (String)A);
34     Heltec.display->drawString(86  , 32  ,   " m");
35     Heltec.display->drawString(0    , 48 ,    "Real altitude =");
36     Heltec.display->drawString(76   , 48 ,    (String)R);
37     Heltec.display->drawString(106  , 48 ,    " m");
38     Heltec.display->display();
39 }
40 void loop() {
41     if (!bmp.begin()) { // check the sensor availability
42   Serial.println("Could not find a valid BMP085 sensor, check wiring!")
     ;
43   while (1) {}
44   }
45     double T =bmp.readTemperature(); // read Temperature value from
       sensor
46     Serial.print("Temperature = ");
47     Serial.print(T);
48     Serial.println(" *C");
49
50     double P =  bmp.readPressure();// read Pressure value from sensor
51     Serial.print("Pressure = ");
52     Serial.print(bmp.readPressure());
53     Serial.println(" Pa");
54
```

```
55      double A = bmp.readAltitude(); // read Altitude value from sensor
56      // Standard Altitude calculation
57      Serial.print("Altitude = ");
58      Serial.print(bmp.readAltitude());
59      Serial.println(" meters");
60
61
62      Serial.print("Sea-level Pressure (calculated) = ");
63      Serial.print(bmp.readSealevelPressure()); // read sea level
        pressure from the sensor and display on serial terminal.
64      Serial.println(" Pa");
65
66      double R = bmp.readAltitude(101500); // read altitude at 101500 sea
          level pressure value from sensor
67    // Precise altitude calculation
68      Serial.print("Real altitude = ");
69      Serial.print(bmp.readAltitude(101500));
70      Serial.println(" meters");
71
72      Serial.println();
73      delay(500);
74    drawFontFaceDemo(T, P, A, R) ; // call function to display data on
        OLED
75      // send packet
76    LoRa.beginPacket(); // starts packet
77
78    LoRa.setTxPower(14,RF_PACONFIG_PASELECT_PABOOST);
79    LoRa.print(T); // add sensor value to data packet
80    LoRa.print(P);
81    LoRa.print(A);
82    LoRa.print(R);
83    LoRa.endPacket(); // ends packet and transmit
84    delay(5000);
85 }
```

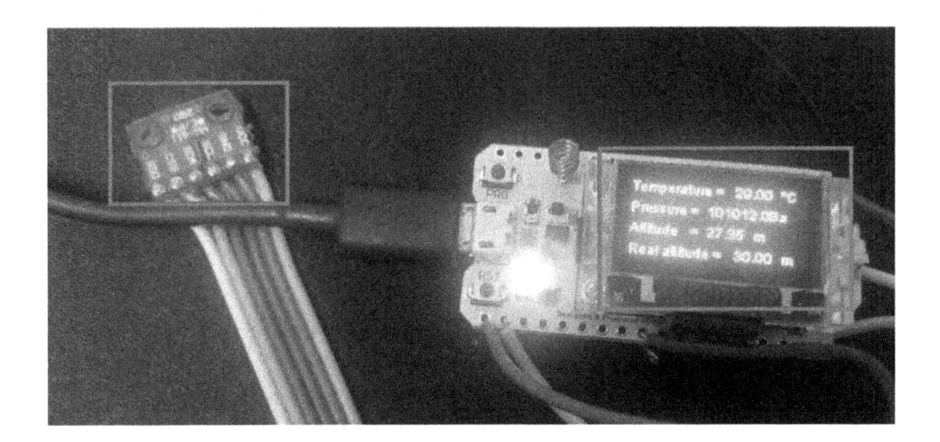

FIGURE 4.19 Displaying readings from BMP280.

4.8 APPLICATIONS IN HOME AUTOMATION, INDUSTRY, AND SMART AGRICULTURE

The reliable and long-distance connectivity is of highest concern in IoT networks, especially for Industry 4.0 and smart agriculture long-range wireless IoT connectivity is of prime interest due to large deployment area. LoRa and LoRa Wide Area Network (LoRaWAN) are ideal for such applications. They cover a long range at an expense of little power and low bandwidth; such networks are commonly known as Low-Power, Wide-Area Networks (LPWAN). LoRaWAN stands as number one in the non-cellular LPWA technologies and thus is a best fit for smart cities, environmental monitoring, smart agriculture, and far more, including Industry 4.0 [8], as illustrated in Figure 4.20.

4.8.1 Real life LoRa Application for Smart Agriculture

In the previous sections, we have seen all the elements for development of LoRa based IoT application and associated coding. This section presents end-to-end application for monitoring temperature, humidity, and pressure in the agriculture field. We deploy BME280 sensor [9] interfaced with Heltec ESP32 LoRa development board in the field to capture temperature, humidity, and pressure. The readings from the BME280 are transmitted over the 865 MHz LoRa channel. The receiver is the another Heltec ESP32 LoRa development board programmed to receive the readings and display on OLED display. Following is the complete code for transmitter and receiver.

Code for Weather Station Using ESP32 LoRa Heltec Development Board

Listing for transmitter:

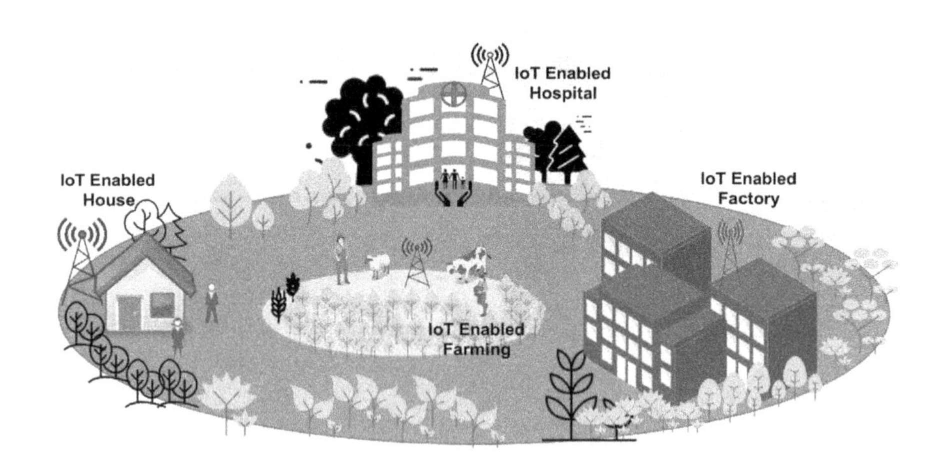

FIGURE 4.20 IoT applications and ecosystem.

To interface the BME280 sensor with heltec LoRa module, we include header files wire.h, Adafruit_Sensor.h, and Adafruit_BME280.h. In void setup() function, we enable OLED Display, LoRa, Serial-Communication, Antenna-Power-Boost, and select the frequency BAND for the communication using Heltec.begin() and then we define the settings for OLED display at lines 21 to 25. The function drawFrontFaceDemo() is defined to display sensor values on OLED display and then we read the sensor values and transmit them over the LoRa network as explained in the previous example.

```
1
2 #include "Arduino.h"
3 #include "heltec.h"
4 // header files corresponding to libraries of the sensor used
5 #include <Wire.h>
6 #include <Adafruit_Sensor.h>
7 #include <Adafruit_BME280.h>
8 #include "string.h"
9
10 #define SEALEVELPRESSURE_HPA (1013.25)
11 Adafruit_BME280 bme;
12
13
14 #define BAND    865E6 // Set LoRa band (here Indian)
15
16 //uint8_t T[20] = {"Temperature"};
17
18 void setup() {
19   pinMode(Vext,OUTPUT);
20   digitalWrite(Vext,LOW);
21   // following block of code heltec module and display
22   Heltec.begin(true, true, true, true, BAND);
23   Heltec.display->init();
24   Heltec.display->flipScreenVertically();
25   Heltec.display->setFont(ArialMT_Plain_10);
26   Heltec.display->drawString(0, 0, "LoRa Initial success!");
27
28   Serial.begin(115200); // begin serial comm.
29
30 }
31 // function to display sensor values on the OLED display
32 void drawFontFaceDemo(double T,double P,double H) {
33
34     Heltec.display->drawString(0    , 0  ,    "Temperature =");
35     Heltec.display->drawString(76   , 0  ,    (String)T);
36     Heltec.display->drawString(106  , 0  ,    " *C");
37     Heltec.display->drawString(0    , 16 ,    "Pressure =");
38     Heltec.display->drawString(56   , 16 ,    (String)P);
39     Heltec.display->drawString(100  , 16 ,     "Pa");
40     Heltec.display->drawString(0    , 32 ,    "Humidity =");
41     Heltec.display->drawString(56   , 32 ,    (String)H);
42     Heltec.display->drawString(86  , 32 ,    " %");;
43     Heltec.display->display();
44 }
45
46 void loop() {
47     if (!bme.begin()) { // check sensor availability
48   Serial.println("Could not find a valid BME280 sensor, check wiring!")
       ;
49   while (1) {}
50   }
51       // read the sensor values
52     double T =bme.readTemperature();
53     double P =  bme.readPressure();
54     double H =  bme.readHumidity();
55
56   drawFontFaceDemo(T, P, H) ; // call function to display data on OLED
```

```
57     // send packet
58     LoRa.beginPacket(); // beigns the data packet
59     LoRa.setTxPower(14,RF_PACONFIG_PASELECT_PABOOST);
60     LoRa.print("Weather Station");
61     LoRa.print(T); // add data values into data packet
62     LoRa.print(P);
63     LoRa.print(H);
64     LoRa.endPacket(); // ends the data packet and trasmit
65
66     delay(5000);
67
68 }
```

Listing for receiver:

The listing for receiver involves setting an OLED display to show received sensor reading in function LoRaData() and parse the data packet to extract packet size and data itself. In void setup(), we receive the LoRa data packet at line 29 using LoRa.receive(). At lines 33 and 34, we extract the packet size and call function void cbk() defined at line 13, which extracts the data and saves it into a variable ‘packet’ at line 15 and displays it by calling function LoRaData().

```
1  #include "heltec.h"
2  #define BAND    865E6  // Set LoRa band (here Indian)
3  String packet ;
4
5  void LoRaData(){  // setup of OLED
6    Heltec.display->clear();
7    Heltec.display->setTextAlignment(TEXT_ALIGN_LEFT);
8    Heltec.display->setFont(ArialMT_Plain_10);
9    Heltec.display->drawStringMaxWidth(0 , 26 , 128, packet);
10   Heltec.display->display();
11 }
12
13 void cbk(int packetSize) { // extract the data from the complete packet
14   packet =""; // hold the data
15   for (int i = 0; i < packetSize; i++) { packet += (char) LoRa.read();
     }
16   LoRaData();// print the data on the OLED
17 }
18
19 void setup() {
20   Heltec.begin(true, true, true, true, BAND); // begin heltec module
21
22   Heltec.display->init();
23   Heltec.display->flipScreenVertically();
24   Heltec.display->setFont(ArialMT_Plain_10);
25   Heltec.display->drawString(0, 0, "LoRa has been initialized.");
26   Heltec.display->drawString(0, 10, "Wait for incoming data...");
27   Heltec.display->display();
28   delay(1000);
29   LoRa.receive(); // receive the data from LoRa network
30 }
31
32 void loop() {
33   int packetSize = LoRa.parsePacket(); // parse the received packet
34   if (packetSize) { cbk(packetSize);  } // calculate packet size
35   delay(10);
36 }
```

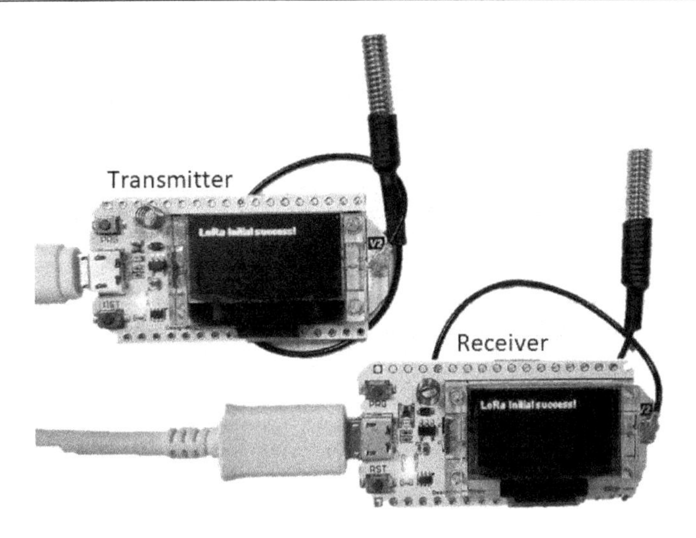

FIGURE 4.21 A simple LoRa transmitter-receiver model.

4.9 CONCLUSIONS

LoRa is a promising and futuristic technology for IoT-based applications with a wide range of real-world applications. This chapter provides a practical getting-started resource for LoRa communication. We began with the need for and salient features of LoRa followed by an overview of Heltec LoRa test platform. We explored the basic functionalities of the platform, a simple transmitter-receiver communication model and data acquisition modules. Further, the applications related to home automation and IoT smart agriculture are presented. LoRa can provide a good coverage of signals and is suitable for low-throughput, low-power, and long-range networks.

REFERENCES

[1] Semtech. Lora® and lorawan®: A technical overview. Technical report, Semtech Corporation, 2020. Accessed: 29–08–2021.
[2] LoRa Alliance. Rp002-1.0.2 lorawan® regional parameters. https://lora-alliance.org/wp-content/uploads/2020/11/RP_2-1.0.2.pdf, October 2020. Accessed: 29–08–2021.
[3] ESPRESSIF. *ESP32 Series Datasheet*, 4 2021. Rev. 3.7.
[4] Heltec. Wifi lora 32(v2.1) pinout diagram. https://resource.heltec.cn/ download/WiFi_LoRa_32/WIFI_LoRa_32_V2.pdf, January 2020. Accessed: 29–08–2021.
[5] Arduino. Guide. www.arduino.cc/, February 2018. Accessed: 30–08–2021.
[6] HelTecAutomations. Heltec docs. https://github.com/ HelTecAutomation/HeltecDocs/blob/master/en/source/esp32/quick_ start.md, May 2019. Accessed: 30–08–2021.

[7] BOSCH. Bmp280 datasheet. https://cdn-shop.adafruit.com/ datasheets/BST-BMP280-DS001–11.pdf, May 2015. Accessed: 31–08–2021.
[8] iSCOOP. Lora and lorawan: The technologies, ecosystems, use cases and market. www.i-scoop.eu/internet-of-things-iot/lpwan/ iot-network-lora-lorawan/, February 2020. Accessed: 30–08–2021.
[9] Adafruit. Bme280 datasheet. https://cdn-learn.adafruit.com/downloads/pdf/adafruit-bme280-humidity-barometric-pressure-temperature-sensor-breakout. pdf, 2021. Accessed: 31–08–2021.

Multi-Sensor System and Internet of Things (IoT) Technologies for Air Pollution Monitoring

5

Ghizlane Fattah[1], Jamal Mabrouki[2], Fouzia Ghrissi[1], Mourade Azrour[3], and Younes Abrouki[2]

[1]*Water treatment and reuse structure, civil hydraulic and environmental engineering laboratory, Mohammadia School of Engineers, Mohammed V University in Rabat,Morocco*

[2]*Laboratory of Spectroscopy, Molecular Modelling, Materials, Nanomaterial, Water and Environment, CERNE2D, Mohammed V University in Rabat, Morocco*

[3]*Moulay Ismail University, Faculty of Sciences and Techniques, Department of Computer Science, IDMS Team, Errachidia, Morocco*

Contents

DOI: 10.1201/9781003244714-5

5.1 INTRODUCTION

At the end of the 18th century, due to the problems caused by the burning of coal in large cities [1], it became necessary to set up an air quality management system. At the time, due to technological limitations and the lack of suitable instrumentation for measuring this pollution, there were no regulations. Since then, between exponential industrialization and a drastic increase in human activity, the problem of air pollution has continued to grow. With technological advances in the measurement of gaseous species in the atmosphere, with detection limits (LODs) reaching a few particles per trillion (1012) (ppt) [2], dozens of standards and directives have been developed to regulate many pollutants in the atmosphere [3]. Some of these species can affect our natural environment, such as acid rain, linked to sulfur oxide emissions, which destroys vegetation, or have a negative impact on human health, ranging from simple irritation [4] to allergies such as asthma, or trigger serious diseases with lethal consequences [5].

The study of air pollution aims to understand its particularities in order to limit its effects as much as possible [6]. Many efforts have been efforts have been made to control and reduce emissions of pollutants from transport, industry, and agriculture and this has had a remarkable impact on the levels of sulfur dioxide and carbon monoxide in the atmosphere: these, after having shown increasing amounts during the first half of the 20th century, have been reduced by almost a third over the last two decades. For example, sulfur dioxide emissions in France have fallen from 1326 Gg in 1990 to 303 Gg in 2009 [7]. Over the same period, carbon monoxide emissions have decreased from 10,890 Gg to 3950 Gg.

On the other hand, the situation is worsening for other pollutants, such as nitrogen oxides and ozone, whose levels often exceed environmental alert thresholds in large cities. Ozone, which had average background levels of between 10 and 15 ppb at the beginning of the century, is now reaching 30 to 40 ppb in many places around the world [8]. In order to avoid the constraint of the obligation of geographical mobility to measure the quality of the air in situ [9], now, there is a tendency towards the recourse method for monitoring air pollution remotely by using multi-gas sensor systems [10] based on chemical and/or semiconductor gas sensors [11]. These sensor arrays aim to complete the range of instrumentation currently available by offering the possibility of rapidly tracing a pollution signature without necessarily carrying out an exhaustive study by standard physico-chemical analysis of the composition of the atmosphere that constitutes it. Spatial and temporal monitoring of air pollution is a field where the use of this type of multi-sensor system is a complementary tool to heavy analysis techniques, due to their low cost and small size [12]. In particular, they make it possible to develop simpler measurement networks that are easy to implement and mobile measurement networks to geographically map an area affected by a specific pollution. Usually, this type of sensor network is dedicated to the qualification of olfactory nuisances, as shown by the recent developments of 'electronic noses'. We propose to use a multi-sensor system to measure atmospheric pollution in the broad sense, a subject rarely addressed by this type of device. The objective of this thesis is therefore to design an autonomous portable system based commercial gas sensors, for the detection and classification of different types and levels of air pollution that may occur in an urban or photochemical pollution background [13]. The proposed study is definitely different from the work carried out in the laboratory on multi-sensor systems by our desire to use commercially available sensors in order to better control the problems of reproducibility and sensor drift, as well as to prefigure possible technology transfers.

We propose to familiarize the reader with the problem of air pollution and the various methods and regulations associated with IoT. We continue this bibliographical study with a description of the advantages and disadvantages of using gas sensors to measure air pollution [14]. The different physical elements and mathematical models associated with systems based on sensor network, together with an overview of the problems usually encountered in real field qualifications of air pollution are presented next.

5.2 LITERATURE REVIEW

Air pollution can be defined as the presence of pollutants (gaseous or particulate) in the atmosphere that can cause harmful effects on the environment and human health [15]. The sources of this pollution can be either natural (forest fire, volcanic eruption, etc.) or anthropogenic, i.e., linked to human activity [16]. In the latter case, pollution is often the direct result of industrial progress in recent centuries, such as the continuous and sometimes careless emission of pollutants associated with combustion processes (motor vehicles, industrial plants, energy production by industrial installations, energy production by combustion . . .). For the past few decades, studies have shown a link between the degradation of the environment and human health, and the presence of these pollutants in the atmosphere [17–18]. Thus, air pollution is considered to be responsible for approximately 800,000 premature deaths each year worldwide [19]. Figure 5.1 shows the

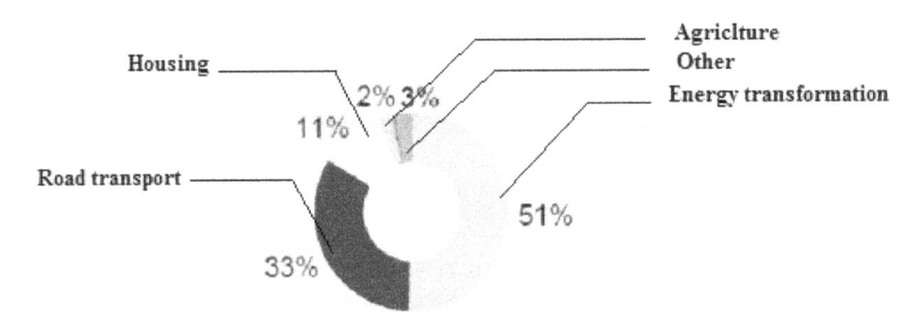

FIGURE 5.1 SO_2 emission sources during 2015.

projected evolution of premature deaths related to excess ozone between 2000 and 2030. In all geographical areas, the situation is already very worrying, with nearly 8 to 15 premature deaths per million inhabitants per year, but forecasts for the next 20 years are very pessimistic, with in particular an increase of a factor of 10 in Asia.

The accumulation of these pollutants is also at the origin of the acid rain phenomena, which has a very negative impact on vegetation, and of global warming [20]. In order to clearly define the context of this work and the specifications of the diagnostic tool that we propose to develop, we have listed in the following paragraph the primary or secondary atmospheric pollutants, their sources, and the associated regulations at the European level where applicable.

Pollutants can be classified as primary or secondary. By definition, primary pollutants are those that come directly from an identified pollutant source, such as carbon monoxide from combustion processes in exhaust gases from cars, and sulfur dioxide from industry [21]. Secondary pollutants are not emitted directly from known and identified sources. Instead, they are formed in the air as a result of reactions involving the compounds emitted by primary sources, especially during particular meteorological episodes. A major example of a secondary pollutant is ground-level ozone, one of many secondary pollutants that form photochemical smog, but other examples include NPA (peroxyacetyl nitrates) or VOCs (volatile organic compounds), known as 'secondary VOCs'.

Carbon monoxide (CO): It results from the incomplete combustion of fossil fuels (coal, oil, etc.). This gas is toxic (for living beings) and can cause fatal asphyxiation in high concentrations. It is considered to be one of the main compounds in the indoor air quality issue; the AFSSET4 has introduced a guide value of 100 mg/m^3 for a 15-minute exposure in CO as a limit value for this species in the sense of indoor air quality. For outdoor air, the European Directive 2008/50/EC indicates a limit value of 10 mg/m^3 for a maximum daily maximum over eight hours [22]. This gas is found in abundance in cities and on motorways because of the combustion engine vehicles. This makes it one of the tracers of traffic-type air pollution.

Sulphur dioxide (SO_2): In general, this gas is produced naturally by volcanoes. It can also be produced by various industrial processes such as metallurgical industries and oil refineries. This gas, which is toxic to all living things, helps to create particles (sulfuric aerosols) that prevent the sun's rays from reaching the earth, so it plays a role in cooling the planet [23]. However and in accordance with European directives, during the 1980s, monitoring of the main industrial installations was introduced and the implementation of source reduction measures led to a drastic reduction in SO_2 into the atmosphere (Figures 5.1 and 5.2 show the different sources of SO_2 emissions recorded during 2010). We notice that 84% of the emissions are produced by energy production and the industrial sector. Thus, taking into account all the regulations put in place and the possibility of detecting the concentration of this gas near certain industrial sites, the concentration of this gas remains relatively low. Concentration of this gas can be detected near certain industrial sites. (Graphs 5.1, 5.2, and 5.3 show the evolution of the average annual SO_2 in France in different urban and industrial sites.) We notice on the one hand, that ambient concentrations have been steadily decreasing since the 2000s, and, on the other hand, that the average SO_2 concentrations are always higher in industrial sites than in urban sites. This gas can thus be considered as a potential tracer of emissions from industrial sources.

Nitrogen oxides (NOx): These include nitrogen monoxide (NO) and nitrogen dioxide (NO_2). These are highly toxic gases that result from the oxidation of nitrogen from the air by oxygen at high temperatures, a phenomenon that usually occurs during combustion processes, particularly in internal combustion engines and power plants [24]. It is in areas with heavy traffic that we observe high concentrations of NO and NO_2. However, the NOx emission concentration records for the year 2010 showed that the main sources of NOx emissions recorded in 2010, we find that 55% of NOx emissions, come from road transport. NOx could therefore be a tracer of urban pollution on a global scale and, more specifically, of major combustion sourccs such as urban combustion plants and road traffic.

Volatile organic compounds (VOCs): Their definition is specified in the European Directive 2008/50/EC. They are 'organic compounds from anthropogenic and biogenic sources, other than methane, capable of producing oxidants. and biogenic sources, other than methane, capable of producing photochemical oxidants by reaction with nitrogen oxides under the influence of solar radiation'. Another definition was introduced by Directive 1999/13/EC:

> any organic compound containing at least the element carbon and one or more of the following elements: hydrogen, halogen, oxygen, sulphur, phosphorus, silicon or nitrogen with the exception of carbon oxides and inorganic carbonates and bicarbonates, having a vapour pressure of 0.01 kPa or more at a temperature of 293.5°C at a temperature of 293.15 K or having a corresponding volatility under the particular conditions of use. conditions of use.

Anthropogenic emission of VOCs to the atmosphere comes from sources such as incomplete combustion in engines or power plants, evaporation of refined products, the use of solvents in industry, paints, etc.

Suspended particulate matter: These particles (known as 'PM') are generally fine solid or liquid particles in suspension in the air. These particles are defined in Directive 1999/30/EC as 'particles passing through a calibrated inlet with a separation efficiency of 50% for an aerodynamic diameter of aerodynamic diameter of 10 μm (PM10) or 2.5 μm (PM2.5)'. These particles come from natural sources such as volcanic eruptions, vegetation (pollen, etc.), forest fires, etc., or from anthropogenic sources such as industrial emissions, combustion of fossil fuels, etc. There are four types of particles [25]:

- **PM10:** airborne particles with an aerodynamic diameter of less than 10 μm
- **PM2.5:** with a diameter of less than 2.5 μm, known as 'fine particles'
- **PM1:** with a diameter of less than 1.0 μm, known as 'very fine particles'
- **PM0.1:** with a diameter of less than 0.1 μm, known as 'ultrafine particles' or nanoparticles

All these particles are dangerous to human health and vegetation. However, only PM10 and PM2.5 are regulated and targeted by the European Directives. Increasing concentrations of particulate matter in the air are mainly of industrial and residential origin [26].

Methane: This gas is not regulated as a pollutant in the atmosphere by European Directives, but can be a good tracer of certain anthropogenic pollution. It is found in its

TABLE 5.1 Main Air Pollutants Measured, Their Sources and Effects

POLLUTANTS	*MAIN SOURCES*	*HEALTH EFFECTS*
SO_2	Industries (thermal plants, refineries, etc.)	Highly soluble, rapidly absorbed by moist surfaces in the mouth and nose. Respiratory irritant, contributes to the exacerbation of bronchial disorders
NOx (NO et NO_2)	Combustion industries (transport, thermal installations, etc.)	Irritation of the respiratory system, asthma attacks, and bronchiolitis
CO	Installations de combustion, transport, domestique	Damage to the central nervous system and sensory organs
O_3	Secondary pollutant formed as a result of chemical reactions between VOCs and NOx in the presence of ultraviolet radiation	Eye, throat, and lung irritant; may impair respiratory function and resistance to infections
COV-NM	Combustion, use and evaporation of solvents and industrial fuels, etc.	Irritating to eyes and lungs can lead to bronchitis through chronic intoxication. In the long term, they would be responsible for cancers.
PM10 et PM2.5	Vehicles (especially diesel) and combustion in some industries	Very active irritant by alteration of the respiratory functions, asthma attacks with long term chronic bronchitis

natural state, produced by living organisms. It constitutes the bulk of natural gas that is used as a fossil fuel. It is one of the main greenhouse gases and plays an important role in global warming. Since the industrial era, the concentration of methane in the atmosphere has been increasing.

Ozone: In the lower atmosphere, particularly in urban areas, precursor emissions are significant due to the presence of large quantities of hydrocarbon emissions and nitrogen oxides in this type of area. Because of this, a large amount of ozone will form but will react very quickly with nitric oxide leaving a small amount of ozone which in turn is consumed in human respiration.

5.3 AIR QUALITY SYSTEM

Standardized methods for the measurement of regulated gases require the use of analyzers incorporating complex components to ensure that measurements meet the quality requirements of the directives and standards (UV or IR radiation source, filter, oven, etc.)

[27–28], which makes these analyzers quite complex and expensive. These methods generally ensure selective measurements with a high degree of accuracy in a relatively short response time [29]. Their major negative points are their non-portability and complexity of use. These tools need regular calibration, and maintenance can be costly.

Air quality monitoring and air pollution measurement networks are spread across France and in other countries. Each network is responsible for monitoring the air in its region, and sets up different measuring stations, each containing one or more air physico-chemical analyzers, based on the standardized measurement methods listed earlier. These stations are placed at different fixed points in the main agglomerations to meet the requirements of the Directives. The measurements of the concentration of each pollutant in the air are measured continuously for 24 hours in the case of automatic measurements or are carried out at regular intervals [30].

These types of measuring stations require a team of qualified personnel (technicians, computer specialists, engineers, etc.) to ensure the proper functioning of the equipment, its calibration, and the recovery and processing of data. These teams must also have skills in the electro technical and electronic maintenance of the equipment. In addition to the cost in human resources, the construction and equipment budget for a single station exceeds several tens of thousands of euros. This severely limits the number of possible installations, which is generally limited to the requirements of the European Directives or for monitoring specific points such as around industrial sites. Several specific parameters and criteria are attributed to pollution measurement stations by the European Directives [31]. The choice and classification of the stations correspond to the air pollution problems encountered in the geographical area where they are located (urban area, industrial area . . .) [32–33]. The location itself meets specific classification criteria. These criteria depend on the number of inhabitants in an urban area, the position of the station, the daily car traffic, the ratio between the different pollutants encountered, and finally the height of the sampling point sampling point [34]. Two groups of stations are distinguished: background stations and proximity stations [35]. The background stations monitor the average exposure of people and the average exposure of people and the environment to air pollution phenomena. This is the case of the so-called ‘urban’ and ‘peri-urban’ stations. The proximity stations, such as the so-called ‘industrial’ and ‘traffic’ stations, provide information on the maximum levels of pollutant concentrations measured in their representative areas, as well as the exposure levels of the population located near the sources. The different criteria for classifying the types of stations are as follows the following:

- A ratio of the annual average concentration ‘R’ of nitrogen monoxide to that of nitrogen dioxide is the marker for the definition of a traffic station site [36]:

$$R = \frac{[NO]}{[NO_2]}$$

- For an R-value below 1.5 the station is classified as urban or suburban; for a value of R greater than 1.5 the station is classified as traffic. For the location of traffic stations, the maximum distance to a busy road should not exceed 10 meters to be as close as possible to the source.

Urban and suburban stations and peri-urban stations are generally under the influence of strong photochemical pollution in the summer period. However, at traffic sites, the NO from traffic can react with the O_3 present in the air, transforming it into NO_2, thus creating a NO sink. Therefore, even in periods of photochemical pollution, traffic sites can have a very low O_3 concentration. Measuring ozone at traffic stations is therefore not relevant [36] and none of the traffic stations have an O_3 analyzer, thus creating a significant difference in the available data sets.

5.4 MATERIALS AND METHODS

5.4.1 The Multi-Sensor

In general, the sensors used in these systems are solid state sensors, either commercially available or developed by their designers for a specific application [37–38]. These systems are operated using pattern recognition and numerical analysis methods [39–40]. By using a group of sensors, whose individual response is different depending on the nature of the gases likely to be present, a fingerprint is formed that can be unique and specific to each gas mixture, provided that appropriate signal processing is applied. The search for this fingerprint is the multi-variable approach. This aspect is strongly exploited in all studies on sensor networks and e-nose [41]. Figure 5.2 shows the basic architecture of a multi-sensor system with signal processing. The signals of several gas sensors exposed together to the same gas environment are exploited in order to obtain adequate information on the gas species.

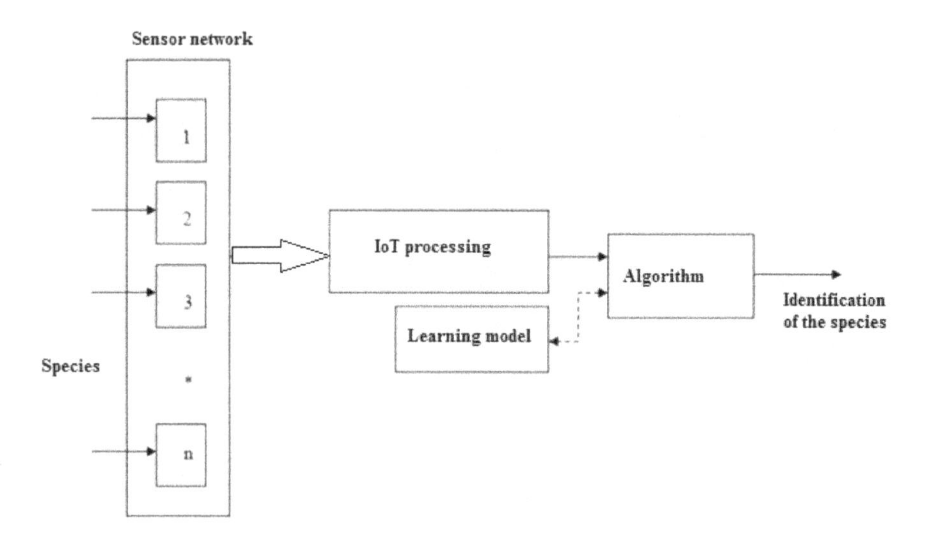

FIGURE 5.2 Basic signal processing diagram of a sensor network.

The use of a multi-variable approach is a necessity in a network of non-selective sensors [42–43]. The methods mainly used as pattern recognition methods [44] are aimed at finding a correlation between the output signals of these sensors and the chemical composition of their gaseous environment [45–46]. They can be divided into two groups: supervised and unsupervised methods. The supervised technique consists of identifying gas species by comparison with known species during the calibration of the system. In this case, an approach to calibrate the sensors under different gas species at the sensors under different gas species to be measured will be necessary. The more these species are the more numerous these species are, the longer and more complex the calibration process will be. This calibration will make it possible to build a behavioral model that will be used by a designated mathematical algorithm. This is described in the section called 'learning model'. In contrast, the objective of the unsupervised technique is to detect and discriminate unknown species by amplifying the differences between their associated input vectors [47–48]. These methods, which are also part of multivariate analysis, will be developed in the section on neural networks, but by default the methods we present are supervised. Multi-sensor systems couple partially selective chemical sensors (especially solid state sensors) with a pattern recognition algorithm based on multi-variable based pattern recognition algorithm for 'chemical fingerprinting' [49–50]. The choice of algorithm is an important element of these systems, so researchers need to develop pattern recognition methods that are appropriate for the intended application [51].

5.4.2 Multi-Sensor Systems and Air Quality

Several studies have been conducted with multi-sensor systems applied to air quality [52–53]. In general, these studies can be divided into two categories. First, there are those carried out in the laboratory. They concern the monitoring of one or more toxic gas species in a synthetic laboratory atmosphere, such as the measurement of H_2S and NO_2 concentrations in a gas mixture [54–55] or the detection of VOC mixtures using solid-state sensors [56]. In addition, other studies are carried out directly in the field, in polluted sites, where one or more sensors are placed in specific sites to detect the presence of one or more gaseous species in the atmosphere. Other studies have been conducted on the stability of signals [57–58] and drift of sensor responses, seeking analytical solutions to these problems, at the risk of making the designed systems more complex [59]. The quality specialists proposed a self-calibration method to solve the sensor drift problem during their work on monitoring NO_2 concentration in the atmosphere [60–61]. Thus, their system was installed in air quality monitoring stations and their self-calibration method is based on the comparison of NO_2 levels measured by the analyzers of the air monitoring stations and the concentrations calculated from the sensor signals. This self-calibration has managed to reduce errors caused by sensor drift, but it requires access to information on NO_2 concentrations at the monitored sites. In addition, A.C. Romain and J. Nicolas studied the drift of TGS-type sensors for seven years. They showed the importance of developing a system capable of compensating for this drift on the one hand and of taking into account the non-reproducibility of the manufacture of different sensors of the same type [62]. Even if one sensor is replaced by another, its behavior, and therefore its drift, will never be strictly identical to that of the original sensor of the same type and from the same supplier.

It is therefore necessary to balance the system each time a sensor is exchanged [63]. In their work, they succeeded in developing a correction model that estimates the slope of a sensor's drift, and calculates a multiplicative factor for each sensor. Even with such a correction, a quarterly calibration under a standard gas is necessary to maintain efficiency of the model.

5.5 RESULTS AND DISCUSSION

After validating our system in the laboratory, the next step is to carry out measurements on polluted sites. We decided to study the pollution during two different seasons, one in summer to be able to encounter photochemical pollution episodes and the other in autumn. For each campaign we chose to install our cells containing our multi-sensor systems in several measuring stations representative of different types of air pollution monitoring sites. In order to make a relevant correlation between the sensors' responses and the physic-chemical data of the air pollutants from the measurement stations, an appropriate study of the different events and characteristics of the air pollution at the chosen measurement sites became essential. For this purpose, we worked on a one-year history (2018) of data on the concentrations of pollutants in the atmosphere from each measuring station studied. The gases selected for this study are NO, NO_2 and O_3, as their data are complete for all stations.

5.5.1 Analysis of Pollutant Concentrations by Station Type

The analysis of this history has been directed toward finding repeatable patterns and characteristics that can define and distinguish one type of pollution from another. We have studied the daily, weekly, and seasonal patterns of these data. A statistical study allowed us to show the repeatability of the inter-station data. The results are presented in Table 5.2.

Table 5.2 shows the properties of the pollutants NO, NO_2, and O_3 over a one-year period. We note that the urban and peri-urban sites studied have similar O_3 values, which is logical given the regional nature of the photochemical pollution episodes that cause O_3 concentrations. We can observe this similarity of behavior between the different types of sites via the graphs of the annual averages of the pollutants presented in Figure 5.3. We also show in Figure 5.4 the correlation between the NO and NO_2 concentrations in a

TABLE 5.2 Result of the Measurement Carried Out on the Stations

URBAN STATION	*NO (UG/M³)*	*NO_2 (UG/MG)*	*O_3 (UG/M³)*
Site 1	3	21	40
Site 2	2	23	38
Site 3	2	18	43

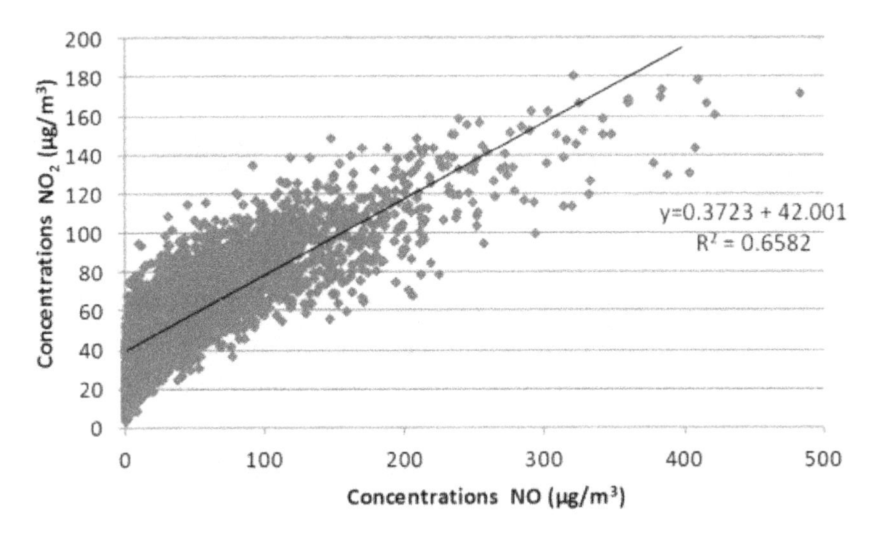

FIGURE 5.3 Correlation between NO and NO_2 concentrations at site 1 during one year, y and R2 established for NO>20µg/m³.

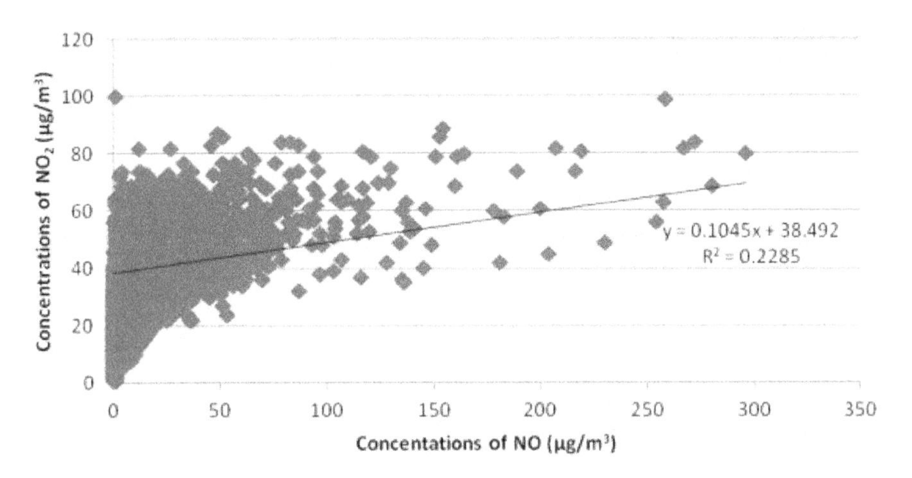

FIGURE 5.4 Correlation between NO and NO_2 concentrations at an urban site 2 during one year, y and R2 established for NO>20µg/m³.

traffic site (Roubaix). In order to differentiate the sites and to identify their characteristics, Figures 5.3 and 5.4 show the correlations between NO and NO_2 concentrations at different study sites.

Thus, for the traffic sites, we can observe that a high NO concentration is always accompanied by a high NO_2 concentration of which a finding generally observed at peak traffic times. Thus it appears that when the concentration exceeds 20 µg/m³ the relationship between NO and NO_2 is well correlated (for example R^2=0.65 for the Roubaix site). On the other hand, this relationship is more noticeable in urban or peri-urban site as shown

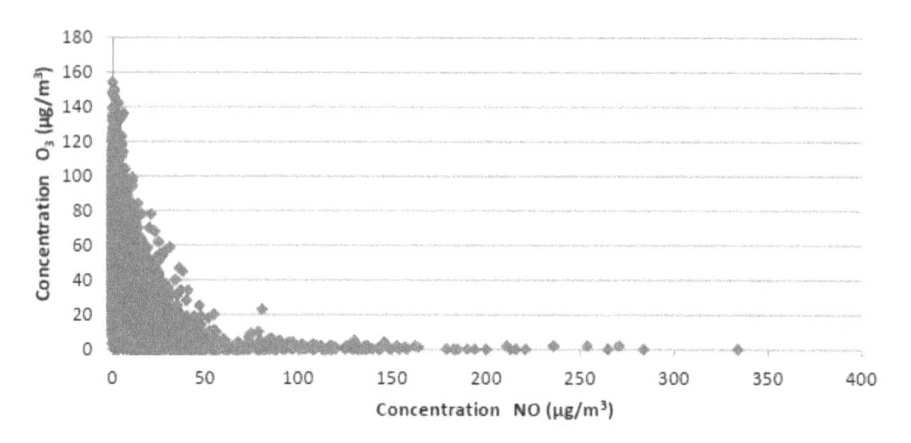

FIGURE 5.5 Correlation between O_3 and NO concentrations at site 1.

in Figure 5.4: the presence of a high NO_2 concentration is independent of the presence of NO in the atmosphere.

Traffic sources are less important in these areas, which is why NO concentrations are relatively low during the year. This is reflected in Table 5.2, where the average NO and NO_2 concentrations at urban and peri-urban sites are lower than those at traffic sites. Figure 5.5 shows the inversely proportional relationship between NO and O_3 at the urban site of Douai: a high O3 concentration is always accompanied by a low NO concentration, and vice versa.

The relationships shown in Figures 5.3, 5.4, and 5.5 are observed at all stations. NO and NO_2 are strongly correlated at urban 1 sites, and O_3 and NO are anti-correlated at urban 2 and peri-urban sites.

5.6 CONCLUSION

In conclusion, we have shown in this work the feasibility of a non-regulatory but nevertheless relatively efficient monitoring of air pollution by multi-sensor measurement systems based on semi-conductor gas sensors thanks to the use of multi-variable methods such as neural networks. We have already taken a step forward with the use of commercial sensors compared to systems integrating 'laboratory' type sensors. However, future technology transfer requires further improvement of sensor performance in terms of both reproducibility and drift. Our study shows that such multi-sensor systems can certainly be used in the very near future for monitoring outdoor air quality and probably indoor air as well. These systems could also be used for the identification of industrial sources in 'receptor' sites where such identification is sometimes difficult. The use of such multi-sensor systems integrating sensors could identify specific signatures of the different industrial sources potentially impacting a site, thus allowing the identification of the industrial sources. The use of such multi-sensor systems integrating sensors could identify specific signatures of the different industrial sources potentially impacting a site, thus allowing the attribution of a pollution index to these sources.

REFERENCES

[1] Lin, Y. C., Zhang, Y. L., Song, W., Yang, X., & Fan, M. Y. (2020). Specific sources of health risks caused by size-resolved PM-bound metals in a typical coal-burning city of northern China during the winter haze event. *Science of the Total Environment*, 734, 138651.

[2] Heland, J., Kleffmann, J., Kurtenbach, R., & Wiesen, P. (2001). A new instrument to measure gaseous nitrous acid (HONO) in the atmosphere. *Environmental Science & Technology*, 35(15), 3207–3212.

[3] Hewitt, C. N., Harrison, R. M., & Radojevic, M. (1986). The determination of individual gaseous ionic alkyllead species in the atmosphere. *Analytica Chimica Acta*, 188, 229–238.

[4] Schiavon, G., Zotti, G., Toniolo, R., & Bontempelli, G. (1995). Electrochemical detection of trace hydrogen sulfide in gaseous samples by porous silver electrodes supported on ion-exchange membranes (solid polymer electrolytes). *Analytical Chemistry*, 67(2), 318–323.

[5] Hewitt, C. N., & Metcalfe, P. J. (1989). Sampling of gaseous alkyllead compounds using cryo-trapping: Validation and field results. *Science of the Total Environment*, 84, 211–221.

[6] Allegrini, I., De Santis, F., Di Palo, V., Febo, A., Perrino, C., Possanzini, M., & Liberti, A. (1987). Annular denuder method for sampling reactive gases and aerosols in the atmosphere. *Science of the Total Environment*, 67(1), 1–16.

[7] Allegrini, I., De Santis, F., Di Palo, V., Febo, A., Perrino, C., Possanzini, M., & Liberti, A. (1987). Annular denuder method for sampling reactive gases and aerosols in the atmosphere. *Science of the Total Environment*, 67(1), 1–16.

[8] Knake, R., Jacquinot, P., & Hauser, P. C. (2001). Amperometric detection of gaseous formaldehydein the ppb range. *Electroanalysis*, 13(8–9), 631–634.

[9] Kumwenda, B., Cleland, J. A., Prescott, G. J., Walker, K. A., & Johnston, P. W. (2018). Geographical mobility of UK trainee doctors, from family home to first job: A national cohort study. *BMC Medical Education*, 18(1), 1–10.

[10] Shaban, K. B., Kadri, A., & Rezk, E. (2016). Urban air pollution monitoring system with forecasting models. *IEEE Sensors Journal*, 16(8), 2598–2606.

[11] Lewis, A. C., Lee, J. D., Edwards, P. M., Shaw, M. D., Evans, M. J., Moller, S. J., . . . White, A. (2016). Evaluating the performance of low cost chemical sensors for air pollution research. *Faraday Discussions*, 189, 85–103.

[12] Adedeji, O. H., Oluwafunmilayo, O., & Oluwaseun, T. A. O. (2016). Mapping of traffic-related air pollution using GIS techniques in Ijebu-Ode, Nigeria. *The Indonesian Journal of Geography*, 48(1), 73.

[13] Hu, K., Sivaraman, V., Luxan, B. G., & Rahman, A. (2015). Design and evaluation of a metropolitan air pollution sensing system. *IEEE Sensors Journal*, 16(5), 1448–1459.

[14] Ghizlane, F., Mabrouki, J., Ghrissi, F., & Azrour, M. (2022). Proposal for a high-resolution particulate matter (PM10 and PM2. 5) capture system, comparable with hybrid system-based internet of things: Case of quarries in the western rif, Morocco. *Pollution*, 8(1), 169–180.

[15] Seyyednejad, S. M., Niknejad, M., & Koochak, H. (2011). A review of some different effects of air pollution on plants. *Research Journal of Environmental Sciences*, 5(4), 302.

[16] Makri, A., & Stilianakis, N. I. (2008). Vulnerability to air pollution health effects. *International Journal of Hygiene and Environmental Health*, 211(3–4), 326–336.

[17] Thu, M. Y., Htun, W., Aung, Y. L., Shwe, P. E. E., & Tun, N. M. (2018, November). Smart air quality monitoring system with LoRaWAN. In *2018 IEEE International Conference on Internet of Things and Intelligence System (IOTAIS)* (pp. 10–15). IEEE.

[18] Hu, K., Rahman, A., Bhrugubanda, H., & Sivaraman, V. (2017). HazeEst: Machine learning based metropolitan air pollution estimation from fixed and mobile sensors. *IEEE Sensors Journal*, 17(11), 3517–3525.

[19] Hu, K., Sivaraman, V., Bhrugubanda, H., Kang, S., & Rahman, A. (2016, October). SVR based dense air pollution estimation model using static and wireless sensor network. In *2016 IEEE SENSORS* (pp. 1–3). IEEE.

[20] Genner, A., Martín-Mateos, P., Moser, H., & Lendl, B. (2020). A quantum cascade laser-based multi-gas sensor for ambient air monitoring. *Sensors*, 20(7), 1850.
[21] Tran, T. V., Dang, N. T., & Chung, W. Y. (2017). Battery-free smart-sensor system for real-time indoor air quality monitoring. *Sensors and Actuators B: Chemical*, 248, 930–939.
[22] Choi, Y., Kanaya, Y., Park, S. M., Matsuki, A., Sadanaga, Y., Kim, S. W., . . . Jung, D. H. (2020). Regional variability in black carbon and carbon monoxide ratio from long-term observations over East Asia: Assessment of representativeness for black carbon (BC) and carbon monoxide (CO) emission inventories. *Atmospheric Chemistry and Physics*, 20(1), 83–98.
[23] Kumar, V., Ranjan, D., & Verma, K. (2021). Global climate change: The loop between cause and impact. In *Global Climate Change* (pp. 187–211). Elsevier.
[24] Vilas Boas, F. M., Borges-da-Silva, L. E., Villa-Nova, H. F., Bonaldi, E. L., Oliveira, L. E. L., Lambert-Torres, G., . . . da Silva, E. G. (2021). Condition monitoring of internal combustion engines in thermal power plants based on control charts and adapted nelson rules. *Energies*, 14(16), 4924.
[25] Sharma, P., Yadav, P., Ghosh, C., & Singh, B. (2020). Heavy metal capture from the suspended particulate matter by Morus alba and evidence of foliar uptake and translocation of PM associated zinc using radiotracer (65Zn). *Chemosphere*, 254, 126863.
[26] Kończak, B., Cempa, M., & Deska, M. (2021). Assessment of the ability of roadside vegetation to remove particulate matter from the urban air. *Environmental Pollution*, 268, 115465.
[27] Schymanski, D., Oßmann, B. E., Benismail, N., Boukerma, K., Dallmann, G., Von der Esch, E., . . . Ivleva, N. P. (2021). Analysis of microplastics in drinking water and other clean water samples with micro-Raman and micro-infrared spectroscopy: Minimum requirements and best practice guidelines. *Analytical and Bioanalytical Chemistry*, 1–26.
[28] Yadav, M., Gupta, R., & Sharma, R. K. (2019). Green and sustainable pathways for wastewater purification. In *Advances in Water Purification Techniques* (pp. 355–383). Elsevier.
[29] Aasen, H., Honkavaara, E., Lucieer, A., & Zarco-Tejada, P. J. (2018). Quantitative remote sensing at ultra-high resolution with UAV spectroscopy: A review of sensor technology, measurement procedures, and data correction workflows. *Remote Sensing*, 10(7), 1091.
[30] Cai, P., Nie, W., Chen, D., Yang, S., & Liu, Z. (2019). Effect of air flowrate on pollutant dispersion pattern of coal dust particles at fully mechanized mining face based on numerical simulation. *Fuel*, 239, 623–635.
[31] Mabrouki, J., Azrour, M., Dhiba, D., Farhaoui, Y., & El Hajjaji, S. (2021). IoT-based data logger for weather monitoring using arduino-based wireless sensor networks with remote graphical application and alerts. *Big Data Mining and Analytics*, 4(1), 25–32.
[32] Collins, J. P., Kinzig, A., Grimm, N. B., Fagan, W. F., Hope, D., Wu, J., & Borer, E. T. (2000). A new urban ecology: Modeling human communities as integral parts of ecosystems poses special problems for the development and testing of ecological theory. *American Scientist*, 88(5), 416–425.
[33] Fattah, G., Ghrissi, F., Mabrouki, J., & Kabriti, M. (2021). Control of physicochemical parameters of spring waters near quarries exploiting limestone rock. In *E3S Web of Conferences* (Vol. 234, p. 00018). EDP Sciences.
[34] Oliver, J. E. (2000). City size and civic involvement in metropolitan America. *American Political Science Review*, 94(2), 361–373.
[35] Borsdorf, A., & Haller, A. (2020). Urban montology: Mountain cities as transdisciplinary research focus. In *The Elgar Companion to Geography, Transdisciplinarity and Sustainability*. Edward Elgar Publishing.
[36] Xiao, Q., Geng, G., Liang, F., Wang, X., Lv, Z., Lei, Y., . . . He, K. (2020). Changes in spatial patterns of PM2.5 pollution in China 2000–2018: Impact of clean air policies. *Environment International*, 141, 105776.
[37] Wang, Y., Xiang, H., Zhao, R., & Huang, C. (2018). A renewable test strip combined with solid-state ratiometric fluorescence emission spectra for the highly selective and fast determination of hydrazine gas. *Analyst*, 143(16), 3900–3906.

[38] Rodrigues, R., Du, Y., Antoniazzi, A., & Cairoli, P. (2020). A review of solid-state circuit breakers. *IEEE Transactions on Power Electronics*, 36(1), 364–377.
[39] Zhang, Z., & Sun, C. (2021). Structural damage identification via physics-guided machine learning: A methodology integrating pattern recognition with finite element model updating. *Structural Health Monitoring*, 20(4), 1675–1688.
[40] Sun, G., Guan, X., Yi, X., & Zhou, Z. (2018). Grey relational analysis between hesitant fuzzy sets with applications to pattern recognition. *Expert Systems with Applications*, 92, 521–532.
[41] Salgueiro, P. A. M. D. S. (2020). Network analysis of connectivity thresholds in fragmented landscapes. A multi-species approach using birds in pine and oak forests. Doctoral Thesis. University of Évora, Évora, Portugal, 139–148.
[42] Dasari, V. S., Pouryazdan, M., & Kantarci, B. (2018, May). Selective versus non-selective acquisition of crowdIoT data and its dependability. In *2018 IEEE International Conference on Communications Workshops (ICC Workshops)* (pp. 1–6). IEEE.
[43] Panicker, S., Gostar, A. K., Bab-Hadiashar, A., & Hoseinnezhad, R. (2018, October). Accelerated multi-sensor control for selective multi-object tracking. In *2018 International Conference on Control, Automation and Information Sciences (ICCAIS)* (pp. 183–188). IEEE.
[44] Güntner, A. T., Abegg, S., Wegner, K., & Pratsinis, S. E. (2018). Zeolite membranes for highly selective formaldehyde sensors. *Sensors and Actuators B: Chemical*, 257, 916–923. Are aimed at finding a correlation between the.
[45] Wu, C., Liu, J., Zhao, P., Li, W., Yan, L., Piringer, M., & Schauberger, G. (2017). Evaluation of the chemical composition and correlation between the calculated and measured odour concentration of odorous gases from a landfill in Beijing, China. *Atmospheric Environment*, 164, 337–347.
[46] Machala, Z., Tarabová, B., Sersenová, D., Janda, M., & Hensel, K. (2018). Chemical and antibacterial effects of plasma activated water: Correlation with gaseous and aqueous reactive oxygen and nitrogen species, plasma sources and air flow conditions. *Journal of Physics D: Applied Physics*, 52(3), 034002.
[47] Zhao, L., Wang, L., Tan, J., Duan, J., Ma, X., Zhang, C., . . . Xu, R. (2019). Changes of chemical composition and source apportionment of PM2.5 during 2013–2017 in urban Handan, China. *Atmospheric Environment*, 206, 119–131.
[48] Tao, J., Zhang, L., Cao, J., & Zhang, R. (2017). A review of current knowledge concerning PM 2.5 chemical composition, aerosol optical properties and their relationships across China. *Atmospheric Chemistry and Physics*, 17(15), 9485–9518.
[49] Reichenbach, S. E., Zini, C. A., Nicolli, K. P., Welke, J. E., Cordero, C., & Tao, Q. (2019). Benchmarking machine learning methods for comprehensive chemical fingerprinting and pattern recognition. *Journal of Chromatography A*, 1595, 158–167.
[50] Aykas, D. P., Shotts, M. L., & Rodriguez-Saona, L. E. (2020). Authentication of commercial honeys based on Raman fingerprinting and pattern recognition analysis. *Food Control*, 117, 107346.
[51] Tulukcu, E., Cebi, N., & Sagdic, O. (2019). Chemical fingerprinting of seeds of some salvia species in Turkey by using GC-MS and FTIR. *Foods*, 8(4), 118.
[52] Gerboles, M., Spinelle, L., & Signorini, M. (2015). AirSensEUR: An open data/software/hardware multi-sensor platform for air quality monitoring. Part A: Sensor shield. Available online: www. ama-science. org/proceedings/details/2118 (accessed on 30 October 2015).
[53] Doni, A., Murthy, C., & Kurian, M. Z. (2018). Survey on multi sensor based air and water quality monitoring using IoT. *Indian Journal of Scientific Research*, 17(2), 147–153.
[54] Di Natale, C., D'Amico, A., Davide, F. A., Faglia, G., Nelli, P., & Sberveglieri, G. (1994). Performance evaluation of an SnO2-based sensor array for the quantitative measurement of mixtures of H2S and NO2. *Sensors and Actuators B: Chemical*, 20(2–3), 217–224.
[55] Yang, B., Carotta, M. C., Faglia, G., Ferroni, M., Guidi, V., Martinelli, G., & Sberveglieri, G. (1997). Quantification of H2S and NO2 using gas sensor arrays and an artificial neural network. *Sensors and Actuators B: Chemical*, 43(1–3), 235–238.

[56] Capone, S., Forleo, A., Francioso, L., Rella, R., Siciliano, P., Spadavecchia, J., . . . Taurino, A. M. (2003). Solid state gas sensors: State of the art and future activities. *Journal of Optoelectronics and Advanced Materials*, 5(5), 1335–1348.
[57] Ritter, T., Hagen, G., Lattus, J., & Moos, R. (2018). Solid state mixed-potential sensors as direct conversion sensors for automotive catalysts. *Sensors and Actuators B: Chemical*, 255, 3025–3032.
[58] Dunning, S. G., Nuñez, A. J., Moore, M. D., Steiner, A., Lynch, V. M., Sessler, J. L., . . . Humphrey, S. M. (2017). A sensor for trace H2O detection in D2O. *Chem*, 2(4), 579–589.
[59] Chen, K., Juhasz, M., Gularek, F., Weinhold, E., Tian, Y., Keyser, U. F., & Bell, N. A. (2017). Ionic current-based mapping of short sequence motifs in single DNA molecules using solid-state nanopores. *Nano Letters*, 17(9), 5199–5205.
[60] Ogen, Y. (2020). Assessing nitrogen dioxide (NO2) levels as a contributing factor to coronavirus (COVID-19) fatality. *Science of the Total Environment*, 726, 138605.
[61] Rahman, A., Luo, C., Khan, M. H. R., Ke, J., Thilakanayaka, V., & Kumar, S. (2019). Influence of atmospheric PM2.5, PM10, O3, CO, NO2, SO2, and meteorological factors on the concentration of airborne pollen in Guangzhou, China. *Atmospheric Environment*, 212, 290–304.
[62] Zoran, M. A., Savastru, R. S., Savastru, D. M., & Tautan, M. N. (2020). Assessing the relationship between ground levels of ozone (O3) and nitrogen dioxide (NO2) with coronavirus (COVID-19) in Milan, Italy. *Science of The Total Environment*, 740, 140005.
[63] Zheng, C., Zhao, C., Li, Y., Wu, X., Zhang, K., Gao, J., . . . Chai, F. (2018). Spatial and temporal distribution of NO2 and SO2 in inner Mongolia urban agglomeration obtained from satellite remote sensing and ground observations. *Atmospheric Environment*, 188, 50–59.

Semantic Web Technology–Based Secure System for IoT-Enabled E-Healthcare Services

6

Nikita Malik[1] and Dr. Sanjay Kumar Malik[2]

[1]Research Scholar, USIC&T, GGSIP University, Delhi; Assistant Professor, MSI, GGSIP University, Delhi, India

[2]Professor, USIC&T, GGSIP University, Delhi, India

Contents

DOI: 10.1201/9781003244714-6

6.1 INTRODUCTION

6.1.1 IoT

Introduced by Kevin Ashton, Internet of Things (IoT) refers to widely distributed real-world devices, appliances, sensors, and 'things' in general, having storage and processing capability, interconnected through a dynamic and collaborative network infrastructure. The idea is to access real-time information and services anytime and from anywhere within the Internet's infrastructure, broadening the scope of data sharing and remote-control ability beyond the machine-to-machine (M2M) scenarios (Szilagyi & Wira, 2016). IoT environment broadly comprises of the domains of devices – the sensors, actuators, processors; connectivity – wired or wireless network; and platform – data generation and analysis technologies (Choi & Choi, 2019). The things in IoT devices offer sensing, measuring, understanding, and modifying abilities in the collaborative environment, but their limited processing and storage capacities raise concerns regarding their performance, reliability, privacy, and security (Srinivasa, Siddesh, & Hanumantha, 2017).

6.1.2 SWT

The Semantic Web is Sir Tim Berners Lee's vision of a highly intelligent or meaningful web system that aims at associating meaning with the data for the machines to be able to understand and process it globally. This web of linked data provides a better representation of knowledge and serves in decision making, scheduling, and other tasks efficiently by requiring minimum human involvement (Berners Lee, Hendler, & Lassila, 2001). Semantic Web Technologies (SWTs) are the web technologies supporting semantic web, and form a part of its layered architecture or stack. The SWTs contextualize and give meaning to the data, enabling its linking, automation, sharing, reuse and integration across various applications. RDF (Resource Description Framework) and Ontologies are the two most prominently used SWTs for graph-like knowledge representation and common understanding, SWRL (Semantic Web Rule Language) for rules to reason over the knowledge base, and SPARQL (SPARQL Protocol and RDF Query Language) for querying and accessing this shared data (Dragoni, Solanki, & Blomqvist, 2017).

6.1.3 E-Healthcare

The digitization in healthcare sector is aimed towards adoption of IT technologies to develop e-health (electronic health) and m-health (mobile health) services and applications for maintaining medical records, on-the-go patient health monitoring through devices, sharing of health status and receiving doctor's prescription remotely, etc. Instruments of data acquisition, that is the personal health monitoring and other medical devices, are the smart things that are interconnected in the IoT environment to effectively facilitate e-healthcare services. This web-based system of healthcare is both time and cost efficient in providing access to medical care anytime and anywhere. It also opens up opportunities for healthcare to reach remote areas lacking infrastructure and medical professionals, as well as provide for the currently demanded personalized healthcare services (Malik & Malik, Using IoT and Semantic Web Technologies for Healthcare and Medical Sector, 2020) (Tiwari, Jain, Abraham, & Shandilya, 2018).

6.1.4 Security Concerns

IoT involves a large number of devices interconnected and exchanging information with one another. This calls for a need of an adjustable infrastructure that can handle the scalable, dynamic environment for all kinds of security and privacy threats (Mishra, Jain, Rai, & Gandhi, Security Challenges in Semantic Web of Things, 2018). Privacy here refers to the ability of determining that what self-information goes where, and security provides the ability that these decisions are adopted and respected. Their basic goals are generally the same, but may differ depending on the service being offered (Shafik & Matinkhah, 2019).

Protecting the confidential medical data collected from various devices requires imposing strict security policies that restrict sharing among only authorized personnel, and detection and prevention of any vulnerability, breaches in IoT enabled e-healthcare (Malik & Malik, Using IoT and Semantic Web Technologies for Healthcare and Medical Sector, 2020). The primary security requirements that should be met for an e-healthcare application are the security of data storage, data flow, web services, and resilience to agreeable user interferences (Konev et al., 2019).

6.2 SECURITY CHALLENGES AND REQUIREMENTS OF IOT-BASED E-HEALTHCARE

IoT devices are developed with the necessary network connectivity capabilities but often do not implement strong network security. Network security is a critical factor when deploying IoT devices. Measures must be taken to safeguard the authenticity, integrity, and security of the data, the path from the sensor to the collector, and the connectivity to the device. Factors that impact network security in the IoT:

- **Increased count of devices:** The quantity or number of interconnecting sensors and other smart devices is growing at an alarming rate, raising the opportunity for attacks. Sensors and smart devices tend to be small devices, with varying OS, CPU, and memory capabilities. These entities are mostly expected to be inexpensive and single purpose devices with simple, basic network connectivity.

- **Unconventional location of devices:** Some IoT devices are connected in a manner that they are able to interact with the physical world, such as those located in appliances, automobiles, on or in our bodies and in our homes. Sensors may gather data from the refrigerator or the heating system. They could also be located in city lampposts or attached to tree trunks. These nontraditional locations make it difficult or impossible to achieve physical security of the devices. The devices should be manufactured to be resistant to tampering, and they should be placed so that they are not obvious and are very difficult to access.
- **Lack of upgradeability:** IoT smart devices may be remotely located where they are inaccessible and where human intervention or configuration is almost impossible. The devices are often designed to stay in service for years much longer than what is typical for conventional high-tech equipment. Some IoT devices are designed without the upgrading ability intentionally, or they might be deployed in situations that make it nearly impossible or difficult to update or reconfigure. New vulnerabilities keep getting uncovered over time, but if a device is non-upgradeable, then the vulnerability will exist for the rest of its lifetime. If a device is upgradeable, the typical consumer may not have a technology background; therefore, the upgrade process should perform automatically or be easy enough to be performed by a layperson.

IBM's 2015 Point of View on IoT security presented a review of threats and protections applicable at the different layers of the IoT ecosystem, as described next and shown in Figure 6.1 (Hahn et al., 2015).

1. **Sensing layer:** At this layer, the sensors or devices are susceptible to attack by hackers wherein they can do damage by gaining access to things like pacemakers, baby monitors, etc. Possible threats include unauthorized data access, malware on intruded device sending wrong information or to the wrong party, data leakage, information gathering for conducting planned attacks, DoS (Denial of Service). It is therefore very crucial to protect these sensors and monitor them to raise alerts in case of any intrusion attempts (Moh & Raju, 2019).
2. **Network layer:** For operation of the IoT, availability, scalability, and manageability are very critical. Applications monitoring IoT devices must be able to get the data timely to be considered useful. So, to cripple the smart systems' effectiveness, intruders most often target networks. Loads of data is sent on the network by the attackers in order to congest it and commonly lead to a DoS attack (Moh & Raju, 2019).
3. **Service layer:** This layer provides the bridging between the topmost interface layer and the bottommost hardware layer. This implies that an attack here would hamper the servicing to end users by impacting important functions of information and device management. Therefore, essential aspects for securing service layer include privacy and access control, authentication, confidentiality, and integrity protection (Moh & Raju, 2019).
4. **Interface layer:** This layer in the IoT ecosystem is considered the most vulnerable as it lies on the top and serves as a gateway to the layers below, such that a compromise in the mechanisms of authentication and authorization here would

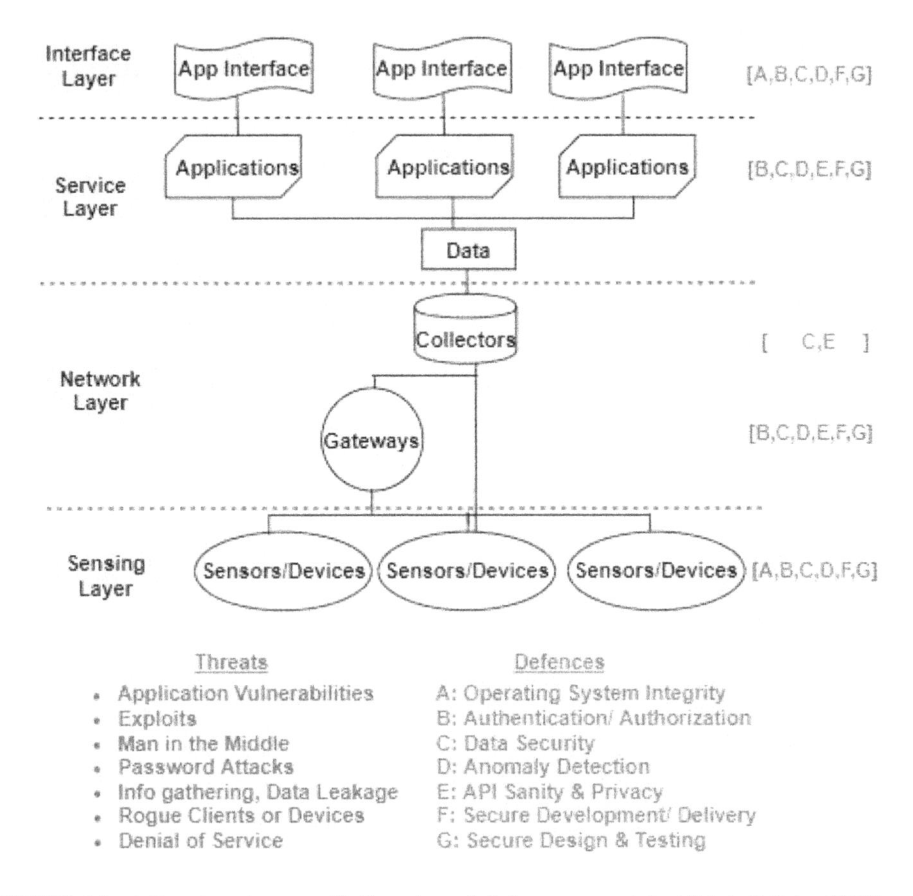

FIGURE 6.1 IoT system layers with threats and defenses annotated (Moh & Raju, 2019).

ripple down its effects to the edge. Possible attack mechanisms include phishing, SQL injection, recovery of insecure passwords, cross site scripting, exploiting known default credentials, etc. (Moh & Raju, 2019).

IoT devices have constrained resources such as power, processing capacity and that can limit their performance. They can be combined with cloud computing technologies for improved healthcare services, forming an IoT-cloud environment, which offers better device and protocol availability and efficiency, but issues around security and reliability remain a challenge. The different security threats occurring in IoT-cloud ecosystem include: physical change or damage to smart IoT devices, DoS attack against availability of the devices having limited resources, exposure of personal data from sensors and other medical devices, modifying critical data like health status or prescriptions, injecting malicious codes or forged information for misleading or causing malfunctioning (Choi & Choi, 2019).

The OWASP (Open Web Application Security Project) identifies and summarizes the top ten vulnerabilities and security challenges in IoT (OWASP-IoT-Top-10–2018-final,

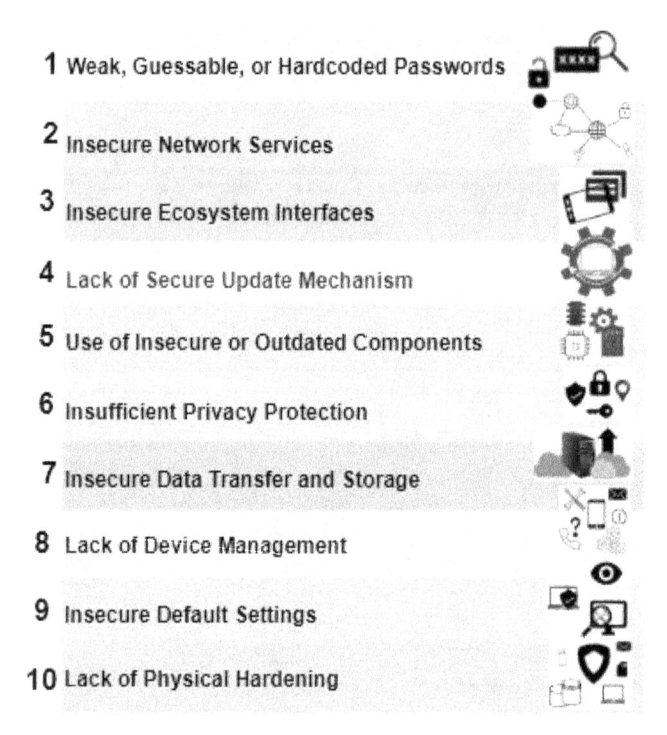

1 Weak, Guessable, or Hardcoded Passwords
2 Insecure Network Services
3 Insecure Ecosystem Interfaces
4 Lack of Secure Update Mechanism
5 Use of Insecure or Outdated Components
6 Insufficient Privacy Protection
7 Insecure Data Transfer and Storage
8 Lack of Device Management
9 Insecure Default Settings
10 Lack of Physical Hardening

FIGURE 6.2 OWASP's IoT top 10 security issues (OWASP-IoT-Top-10–2018-final, 2018).

2018), as shown in Figure 6.2, which can be used as a handy reference. It also provides various countermeasures as recommendations for improving security in IoT environments (Khan & Salah, 2018).

An IoT-enabled healthcare system does not include security features on its own, such as a CT scans radiation can be altered to make it life threatening for the patient. This unencrypted and unauthorized access is major vulnerabilities to the remote health devices. Other vulnerabilities than an attacker can exploit in IoT healthcare are (Tiwari, Jain, Abraham, & Shandilya, 2018):

- Including spoofed or deceptive devices to the system
- Injecting inaccurate information for bypassing access to the system
- Modifying and corrupting sensor data being sent over
- Stealthily monitoring communication channels between smart devices and server to leak confidential personal data to unauthenticated users
- Stealing private data records like health records of patients, from storage
- Tampering with critical information like doctor prescriptions for patients (Tiwari, Jain, Abraham, & Shandilya, 2018)

Most IoT security challenges are in continuation of the network and routing security issues. Privacy in IoT refers to the awareness of the risks levied by the surrounding smart devices and the services, control over the collecting and processing of personal data by the smart things and, regulates subsequent use and distribution of that information to outside entities (Ziegeldorf, Morchon, & Wehrle, 2014). Threats to data and device privacy in IoT-based healthcare systems, are categorized as shown in Table 6.1.

As these IoT devices are accessed on the web to form WoT (Web of Things), new security challenges spring up. The devices, with their constrained resources, are not able to support HTTPS (Hypertext Transfer Protocol-Secure) security scheme that employs intensive algorithms of cryptography such as AES (Advanced Encryption Standard), SHA (Secure Hash Algorithm) etc. Also, OAuth (Open Authorization) (Moh & Raju, 2019) like web authentication schemes, which are critical for sensitive and personal data's privacy,

TABLE 6.1 Privacy Threats in IoT Applications

Identification	It refers to linking a name or an address with an individual as an identifier. This may further augment other threats such as profiling or tracking people.
Profiling	It refers to using data from IoT devices for classification of individuals into groups. Profiling is useful for providing optimized personalization in e-commerce through targeted advertisements, recommender systems, etc., however, may lead to privacy violations like social engineering, selling profile data, unsolicited content, etc.
Localization and Tracking	Ascertaining, fetching, monitoring, and recording the subject's location through time and space, although essential in some IoT applications, can also lead to disclosure of private information.
Life Cycle Transitions	It refers to configuration updating and data backup and restoring. This may cause a threat if some data is incorrectly made available on a wrong device.
Inventory Attack	Smart things are accessible on the web and can be queried for data by attackers to compile and create an inventory of various things and their locations.
Linkage	It refers to collecting and combining data about a subject in different contexts and from various sources and gathering insights about it. The insights drawn can be inaccurate or users may not have the access permissions to that data.
Privacy-violating Interaction & Presentation	It refers to communication of confidential information in a way that it unnecessarily gets disclosed to outsiders.

Source: Ziegeldorf, Morchon, & Wehrle (2014)

are insufficient in WoT devices. Current researches in the field however are lacking in exploring encryption at application level, ownership transfer and overall WoT security frameworks (Xie et al., 2016).

The basic major requirements for security in IoT-based healthcare communication systems include (Yeh, 2016):

- **Agreement on generation and usage of a session key for securing communication and authenticating entities:** Usage of SSL (Secure Sockets Layer) protocol or other techniques is not sufficient for providing user anonymity along with secure communication channel between the authenticated entities, the way an agreed upon session key does, and is therefore an essential property.
- **Careful usage of bitwise exclusive-or (EX-OR) module:** The bitwise exclusive-or operations utilized during protocol runs can be on attacker's target list instead of the one-way hash functions, and therefore should be embedded within the hash function computations.
- **Immunity of identification information towards spoofing attack:** Besides maintaining the privacy of the personal health-related data, the accuracy of the sensing device and communication network must also be ensured. This includes authenticating the user's identity without breaching its untraceability and anonymity. For these requirements, GPS (global positioning system) information of individual's location provides unique and legitimate identification verification, and should guarantee resistance against spoofing attacks.
- **Resistance towards MITM (Man-in-the-Middle) attack:** The messages sent over for authentication can be intercepted by a malicious entity and counterfeited authentication messages may be inserted in their place, making the two legal communicating parties believe that it is a legitimate connection, while it actually isn't, and the session has been hijacked by an attacker in the middle of the communication. This can be avoided by encrypting the identities of legitimate communicating parties within the protocol message used for authentication of entities.
- **Guaranteeing multiple security policies simultaneously:** It is essential for IoT-based healthcare systems to embed all the privacy and security requirements-mutual authentication among communicating parties to protect against spoofing and illegal data access, maintaining user anonymity and untraceability of the devices having sensitive personal data, immunity from forgery or replay attacks during operations (Yeh, 2016).

6.3 STATE OF THE ART MEASURES FOR IOT SECURITY

As more and more smart sensors are installed, the potential for security issues is increased. Sensors are often connected to the same network as our home or small business devices, which causes the breach of one device to radiate outwards to affect all connected devices. For securing the IoT network, identification of devices and their authentication is an integral part of maintaining the integrity of the system and protecting it from interventions from third parties (Konev et al., 2019). With the given new, innovative, and dynamic nature of IoT technologies, designing tailor-made security solutions based on the IoT objects and targeted services is expected now. The traditional mechanisms of ensuring protection of devices and information seem insufficient for the smart devices and therefore need to be refined to fit the specific security requirements, presenting a promising solution towards securing IoT enabled applications such as for e-healthcare (Yeh, 2016).

However, this still involves challenges as the large heterogeneous private data needs to be converted to higher levels of abstraction which requires high processing capabilities that are missing in the resource constrained IoT devices. Also, there is a need for associating descriptions to devices and captured information to be able to perform common understandable reasoning over them to achieve results, along with scalable ways to access these semantic services (Tiwari, Jain, Abraham, & Shandilya, 2018).

With the alarming rate at which attacks on IoT devices are growing and impacting the economies globally, it is imminent that the IoT cybersecurity be made effective, for which, a legislation was introduced by the US Senate in August 2017, called as the IoT Cybersecurity Improvement Act, which focused on setting recommendations for secure development, configuration management, vulnerability patching, etc. (Warner, Gardner, Wyden, & Daines, 2017).

Different kinds of protection for various company-side operations can be provided by security services by deciding at design time what novel services will be implemented to address specific business processes or how existing services will be adapted. This requires adaptations to meet the device constraints also, like implementing compressed IPSec (Internet Protocol Security) protocol in place of link layer for IoT end-to-end security (Mozzaquatro et al., 2018).

For automatic and contactless information retrieval for on-demand patient care, an authentication mechanism is proposed (Yeh, 2016) where all the security parameters are agreed upon in the initialization phase and shared among the body sensors, processing units and body sensor network, after which, a key exchange and agreement among communicating parties ensures authentication as well as secured data exchanges between them (Yeh, 2016).

The principal strategy to securing IoT devices is twofold: reducing the number of vulnerable smart things that can be exploited, and influencing the potential attackers, as well as setting up global scheme for punishing the guilty. In wake of the latest IoT attacks that were launched, the top-most recommended actions include (Moh & Raju, 2019):

1. Ensuring that all default credentials be changed and passwords be made strong.
2. Updating the devices whenever the security patches are available.
3. Unless necessary, disabling the UPnP (Universal Plug and Play) setting on routers.
4. Purchasing IoT devices from reputed companies that offer secure devices (Moh & Raju, 2019).

IoT devices' constrained resources and capabilities make service performance challenging, and as the need for customized security and data anonymization becomes essential, security in IoT environments can shift to the level of networks – the edge. The advantages

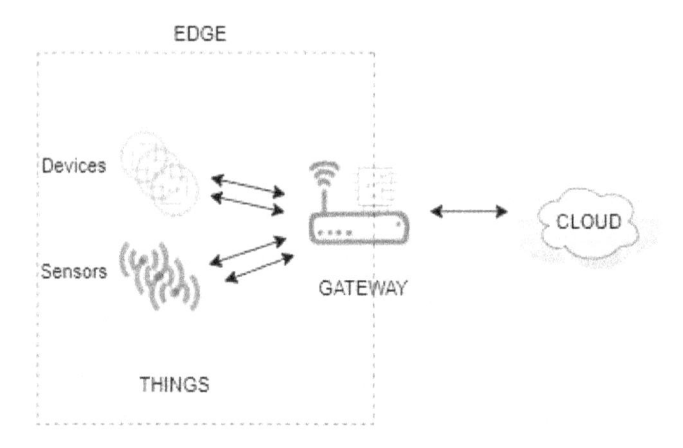

FIGURE 6.3 IoT and edge computing.

of edge computing and IoT can be used to integrate more resources and provide better added services (Wang et al., 2018). Figure 6.3 shows the relationship between the two technologies.

Few existing commercial edge layer solutions include Amazon's AWS Greengrass and Microsoft Azure IoT Edge. Edge-based security solutions are suitable to be implemented for security in IoT-based systems because:

- As compared to the things' layer, edge layer has more resources. So, some security operations can be outsourced to the edge layer.
- Edge layer is placed close to the things layer. So, it can satisfy real time requirements in security design.
- Edge layer has additional information than things to make security decisions.
- Relatively more stable relation between the edge layer and the things helps building trust.
- It is easy for the edge layer to get help from the cloud when needed, because it has high-speed connection to the cloud.
- The edge layer can be powerful or reasonably powerful depending on the needs of security operations, i.e. its flexible enough to deploy various security services.

Gateway devices, as compared to edge devices, have more computing power and resources, and may also be used to enforce security in IoT environments at a larger impact as compared to focusing on individual edge devices (sensors). Machine learning (ML) techniques like artificial neural networks (ANN) may also be used in gateways for monitoring the subsystem components, or at the application layer for monitoring the entire system's state (Canedo & Skjellum, 2016).

IoT security also includes significant tasks of pattern discovery in data, detecting any outliers, predicting values, extracting features critical to security purposes, etc. Outlier detection task can further be divided into the use cases of malware detection, intrusion detection, and data anomaly detection. ML algorithms and techniques can be used for these use cases, as shown in Figure 6.4 (Moh & Raju, 2019).

Use Case	*ML Algorithm*
Malware Detection	SVM
	Random Forest
Intrusion Detection	PCA
	Naïve Bayes
	KNN
Anomaly Detection	Naïve Bayes
	ANN

FIGURE 6.4 Categorization of ML solutions for outlier detection (Moh & Raju, 2019).

Detecting anomalies is essentially a problem of classification, and the most commonly used ML algorithms of Naïve Bayes, support vector machine (SVM), ANN etc. can be applied for that. For malware detection, ANNs aren't used generally as they take longer time in training (Moh & Raju, 2019). Modern day IoT networks like Nest use open authentication protocols, but are susceptible to forgery. A framework based on deep neural network, which uses the wireless transmitters' radio frequency properties and in-situ ML algorithms at the receiver to provide real-time authentication to wireless nodes in the IoT environment, serves as another ML based solution towards IoT security as discussed in Chatterjee, Das, Maity, and Sen (2018).

6.4 SECURITY WITH SEMANTIC WEB IN IOT

IoT devices are heterogeneous and not always interoperable with one another in that they don't utilize commonly agreed upon vocabulary for expressing their data. For enriching this data in order to develop interoperable applications, that are smarter, is the need of the hour and SWTs are utilized to this effect (Tiwari, Jain, Abraham, & Shandilya, 2018). IoT environment, in order to offer context reasoning, needs to be annotated, providing descriptions (semantics) for common understanding of the devices and the data. Reasoning based on semantic ontologies can help draw inferences to recognize, predict, and therefore defend against security breaches in the IoT-enabled system (Choi & Choi, 2019). Several researchers have worked towards building ontologies for health and medical care such as in (Gonzalez-Gil, Martinez, & Skarmeta, 2020) and (Rhayem, Mhiri, Salah, & Gargouri, 2017). Certain situations require for rules and alerts to be described for the collected data from devices to be able to query, store and analyze it. The ontologies can be integrated with the IoT-enabled healthcare system, and after authentication, the system can be exploited to answer any user queries by reasoning on this knowledge base of ontologies and rules (Rhayem, Mhiri, Salah, & Gargouri, 2017). Significant aspects in this include data sharing, interoperability, and maintaining consistency. For this, SWTs and principles are adopted which led to the development of Semantic Web of Things (SWoT) (Tiwari, Jain, Abraham, & Shandilya, 2018).

IoT and WoT devices aren't capable of appropriately representing the human and machine relationship and fail in semantic collaboration (Tiwari, Jain, Abraham, & Shandilya, 2018). Inclusion of the concept of semantics brings about a clear understanding of the data being shared, enriching it with structure and meaning. The smart devices' data becomes unambiguous and interpretable, making them completely interoperable. SWoT for e-healthcare proves to be pivotal in discovery of smart medical wearables, and their monitoring for accessing information (Malik & Malik, Using IoT and Semantic Web Technologies for Healthcare and Medical Sector, 2020). Authors in (Rhayem, Mhiri, & Gargouri, Semantic Web Technologies for the Internet of Things: Systematic Literature Review, 2020) present a detailed, systematic review of the different works in the direction of integration of SWTs with IoT for various applications such as e-healthcare.

Security threats to information in any system are atypical, but these threats pose a greater danger in semantic systems because:

- The data is structured, which allows the attackers who obtain this data to comprehend its meaning and perform complex operations on it, to be able to use it for their own purposes.
- A semantic graph offers a complete knowledge base of a domain, and is not just a database which means that the attacker is not just able to infer knowledge from it using appropriate means, but also exchange of confidential information among semantic web agents is possible to be carried out.
- The data is open, as the communication is with LOD (Linked Open Data), which acts as an open information leakage channel. However, limited set of data can be configured for open access (Konev et al., 2019).

Figure 6.5 shows the different categories for security related incidents and various types of such incidents that take place in a SWoT environment.

The major challenge faced in ensuring information security in SWoT applications is the enforcement of access control, i.e. forbidding any unauthorized access to data or devices. Besides this, the detection of any kind of malware or infection analysis is also considered, which is a rather knowledge-intensive process. The listed security issues include checking for any fraud, spamming or phishing schemes for collecting data, or

Incident Category	*Incident Type*
Information Security	Unauthorized Deletion/ Access/ Updation
Fraud	Wastage of Resources
Information Collection	Phishing, Sniffing
Malware	Infection, Undetermined
Availability	Dos/ DDoS

FIGURE 6.5 Listing of SWoT security incidents (Mishra, Jain, Rai, & Gandhi, Security Challenges in Semantic Web of Things, 2018).

making the entire system go down or unavailable through launching of distributed denial of service like attacks (Mishra, Jain, Rai, & Gandhi, Security Challenges in Semantic Web of Things, 2018). Other aspects of security critical for designing reliable and dependable SWoT applications are presented in Table 6.2.

TABLE 6.2 SWoT Security Aspects

Confidentiality	Sensitive data of SWoT services needs to be kept confidential. Although there already exist several symmetric and asymmetric techniques of encryption for protecting the data, choosing a single scheme would be specific to a certain kind of device or application, which needs to be overcome.
Integrity	Crucial data is interchanged among SWoT services and with other third parties also which demand for protection against any malicious or accidental damage to the critical data during sensing, transferring or storage. MAC (Message Authentication Codes) applying one-way hash functions are used to ensure the integrity of data, but the choice is again specific to device or dependent on application, which needs to be addressed.
Availability	Services hosted in the SWoT environment may be sensor hosted and should be available at any time and from everywhere, producing semantic based information whenever needed. No secure protocol exists for this and practical measures for assuring availability of SWoT applications and services may be accepted.
Authentication	SWoT applications require mutual authentication i.e. identity verification for exchanging data which is then used for actuation and decision making. The user needs to be ensured that the source of service that he is accessing is authentic and the service provider is assured that an authentic service is attempting to approach. An authentication technique would require to register the users' identities, which poses a challenge in SWoT scenario due to resource limitations of the sensor nodes.
Authorization	It refers to describing policies or the access rules which provide certain privileges for some roles. Easy-to-use, dynamic and re-usable policies need to be described in SWoT environment to facilitate the authorization mechanism, and externalizing this policy definition and enforcement process is important for authorizing SWoT services.
Access Control	This enforcement process follows the access controls' outcomes and allows only authorized users the permission to access the resources. Controlling access and revealing critical user data to just the authorized parties is of high significance in SWoT scenarios.
Trustworthiness	SWoT applications dealing with safety critical services or services that are personalized and delicate in nature require analyzing the various entities involved for their trustworthiness. This is of high significance because un-trusted or malicious sensor nodes or data can cause a disaster in critical services because of intentional misbehavior or unintentional errors. Incorporating some kind of trustworthiness analysis is of importance to assure trustworthiness of SWoT.

Source: Mishra, Jain, Rai, & Gandhi, Security Challenges in Semantic Web of Things (2018)

Technologies, such as the IoT, that deal with information distribution and/or manipulation on the web raise the issues of data privacy and security policies, for which semantic web and its technologies, not only help in streamlining the information sharing among the web users, but also additionally contribute towards flexible and smarter handling of privacy and security issues by integrating information (Ngoc & Van, 2020). Each layer of the semantic web stack, as explained in (Malik & Malik, Security in Web Semantics: A Revisit, 2018), includes some security measures to realize the overall security of the semantic web (Malik & Malik, Security in Web Semantics: A Revisit, 2018).

Further, ontologies serve in the detection and analysis of issues in security, and are applied as an annotation vocabulary or knowledge schema or representation schema generally. SWTs can assist in controlling access to RDF data through different access control frameworks that are built on the semantic web languages of OWL, SPARQL, etc., which prevent any unauthorized access to the resources. These SWTs support designing of various access control and security enforcement policies. Also, for maintaining the confidentiality and integrity of shared data, authorized lightweight encryption mechanisms which work fast on SWoT devices are being promoted (Mishra, Jain, Rai, & Gandhi, Security Challenges in Semantic Web of Things, 2018). Figure 6.6 summarizes some of the discussed security solutions of SWoT.

The component, to which data is transferred to, once the authorization process is completed, can be responsible for semantic data's security using an ontological security model which restricts user access to inaccessible resources and to the system generated knowledge, as explained by (Konev et al., 2019).

With the continuing growth in malwares and revolutionized hacking techniques, intelligent security systems (ISS) that employ artificial intelligence (AI) methods for building improved defenses against dangerous threats in SWoT applications are much needed and are being popularized. The ISS function beyond the general use cases such as home security, aiming to provide better security and protection of endpoints in SWoT (Shafik & Matinkhah, 2019).

The smart systems' privacy, protected through access control mechanisms, is generally coarse-grained i.e. limited to controlling access to the data store as a whole. Existing

SWoT Security Solutions

Employing SWTs in IoT for:

- Authorized lightweight Encryption for confidentiality and integrity
- Detection and analysis of security issues through annotation vocabularies
- Ensuring privacy and Controlled access to knowledge graph elements
- Designing role/ context based Access control policies
- Secure Capability profiling of devices
- Interpretation and combining of Security policies
- Specification and interoperability of secure Web services
- Artificial intelligence based methods for intelligent smart systems

FIGURE 6.6 Summary of a few SWoT security solutions.

Sesame or Jena RDF repositories employing such mechanisms need to introduce triple level access controls for fine grained privacy. Semantic technology-based models using role or context-centric policies, in which the access control rules are associated with the context or the situation of an element instead of with the subjects, offer a good privacy solution in smart information systems. The current information about the entities' environments, or the context, is matched with the requester's required context in order to allow access and permit performing an action. Context based access control prove suitable for critical situations or when context defines the user's role or affects the execution of the rules (Hosseinzadeh, Virtanen, Diaz-Rodriguez, & Lilius, 2016).

With the focus on SWTs' representation capabilities aiding in the reasoning, monitoring, and interpretation of security policies on access control, privacy, and management of intellectual property, a lot of research work on their implications in aiding of understanding and combining various security policies and interpreting them appropriately for the data consumers as well as producers is also under progress (Ngoc & Van, 2020).

In view of SWoT interoperability, a solution in the form of a SWoTPAD framework, and a semantic language-SWoTPADL, developed by the PAD-LSI/USP (Parallel and Distributed Computing Group of the Integrated System Laboratory of the University of Sao Paulo) has been established, which provides a strong basis for developing interoperable SWoT services, with extensible components to include automatic service composition and specifications for their security requirements (Silva, Perez-Alcazar, & Kofuji, 2019).

Nowadays, IoT devices are mostly unified technically, having implementation stacks that are similar to each other. Analyzing them by grouping based on some criteria like portability or application domains or common components' presence is therefore useful. This profiling of IoT devices based on their capabilities or features can be secured effectively using SoWTs. SWoT-enabled security solutions identify the technical information of IoT devices without needing to access them physically, simply by analyzing their discovered information semantically, and can profile the existing as well as future IoT products' functionality that is sensitive to security and privacy. This can help in discovering, categorizing, and comparing the SWoT devices' security-sensitive capabilities at low cost and at early stages (Bytes, Adepu, & Zhou, 2019).

6.5 MODEL FOR SWOT-BASED SECURE SYSTEM FOR E-HEALTHCARE

6.5.1 Models' Comparison

Multiple models for SWoT-based healthcare systems have been proposed as well as implemented over the years. Table 6.3 highlights few of those models, drawing a comparison between them on the basis of different parameters.

The criteria used for evaluation of SWoT models include use of an ontology editor, providing proper semantic representation and semantic annotation, offer visualization, taxonomy modeling ability, can deal with complex relationships, offers security and

TABLE 6.3 Comparison of a Few SWoT Models

PARAMETER	*SWOT4CPS (WU ET AL., 2017)*	*IOT-BASED HCIS (SEZER, BURSA, CAN, & UNALIR, 2016)*	*HIERARCHICAL MHS (SINGH, TRIPATHI, & ALBERTI, 2017)*	*HCIOT (RHAYEM, MHIRI, SALAH, & GARGOURI, 2017)*	*OB-CPS (MOZZAQUATRO ET AL., 2018)*	*S3HC (TIWARI, JAIN, ABRAHAM, & SHANDILYA, 2018)*	*SEMANTIC IOT ARCHITECTURE FOR HC-PH COLLABORATION (LIM & RAHMANI, 2020)*	*SDN/NFV AWARE IOT SYSTEM (ZARCA, BAGAA, BERNABE, TALEB, & SKARMETA, 2020)*
Ontology Editor	✓		×	×	✓	✓	✓	×
Semantic Annotation & Representation	✓		✓	×	×	✓	✓	✓
Visualization	✓		✓	×	✓	✓	✓	✓
Taxonomy Modeling	✓		×	✓	×	×	×	✓
Complex Relationships			×	✓	✓	✓	✓	✓
Security & Privacy	×		×	×	✓	✓		✓
Integration & Fusion	✓		✓	✓	✓	✓	✓	✓
Analysis & Reasoning	✓		✓	✓	✓	✓	✓	✓

privacy mechanisms, facilitates integration and fusion and offers suitable analysis and reasoning. A comparison of the various SWoT models for healthcare domain on the basis of these parameters reflects the need for a generic model which is adaptable and can conform to all the criteria.

6.5.2 Proposed Generic SWoT Model

The IoT devices communicating with each other can lead to breaches in the system channels if the communication links are insecure, leading to exposure of private personal data to manipulation or theft. The semantic models used for interoperability are also required to be properly authenticated for further enhancing the security of devices in heterogeneous environment. To monitor the end users and maintain the CIA (confidentiality, integrity, and authentication) goals of security, a model for secure e-healthcare services has been designed.

Based on how SWTs are used in IoT enabled healthcare applications as discussed in (Malik & Malik, Using IoT and Semantic Web Technologies for Healthcare and Medical Sector, 2020), and using the layers for security semantic web as discussed in (Malik & Malik, Security in Web Semantics: A Revisit, 2018). Figure 6.7 shows a generic model drafted for a secure SWoT to represent the IoT devices and data and to reason within healthcare systems.

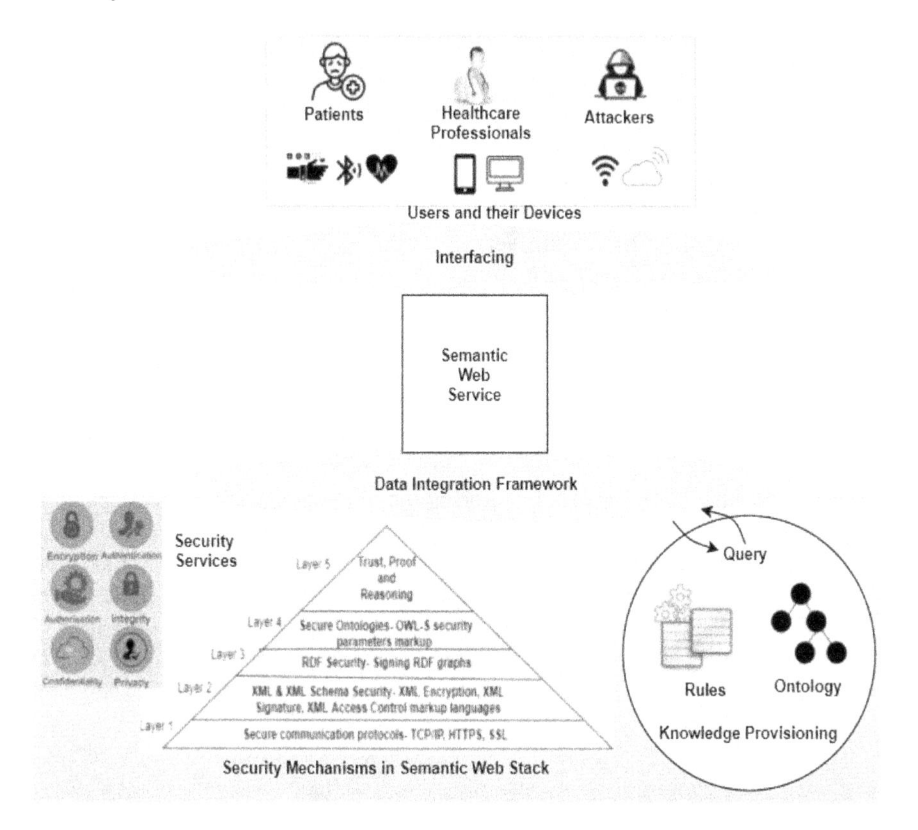

FIGURE 6.7 SWoT-based secure model for IoT-enabled healthcare.

The model shown comprises of six components broadly, namely Knowledge Provisioning unit, Secure Semantic Stack/Layers, Data Integration Framework, Semantic Web Service, Interfacing and, Users with their devices.

- **Knowledge provisioning unit:** Conceptually, a knowledge base is represented as a collection of all the concepts or terminologies (TBox – Terminology box) and their instances or assertions (ABox – Assertions box). These contents form an ontology 'explicit specification of a conceptualization', which forms the base for extracting facts from available knowledge and applying rules for reasoning. All confidential health records of patients, from monitoring devices and by doctors, is stored in the ontology form for knowledge provisioning. Rules and queries are required for end users to interact with the knowledge base and are interpreted for the actual meaning of the request by an inference engine or a reasoner, which then logically checks and returns the expected results (Tiwari, Jain, Abraham, & Shandilya, 2018).
- **Secure semantic layers:** Along with all the knowledge, security aspects are equally important. Starting from the bottom most layer of the semantic web stack, there's security mechanisms for all the layers – security protocols for data transmission, secure standard representation language, security of knowledge graphs, secure interoperability markups, application trust and proof. The security services or goals remain the same as for any information system, which are confidentiality, integrity, authentication, authorization, access control, and privacy (Malik & Malik, Security in Web Semantics: A Revisit, 2018).
- **Data integration framework:** Knowledge from the different data sources needs to be unified as a database in the knowledge provisioning unit, which is facilitated through an integration framework. This acts as layer providing information from services, and which when correlated with reasoning on ontologies, offers cybersecurity solutions, thereby integrating all aspects for a secure, semantic e-healthcare service (Mozzaquatro, Agostinho, Goncalves, Martins, & Jardim-Goncalves, 2018).
- **Semantic web service:** The heart of the whole e-healthcare system is the semantic web services responsible for all the semantic data annotation, processing, actuating, and querying of the knowledge base at run-time. All the data sent over by the users' smart healthcare devices such as heart rate sensor or posture monitor is enriched by adding semantic annotations to it, and these descriptions to the data are then stored in the form of ontologies and rules in the knowledge base. Besides monitoring the sent over data such as the patients' vital health status, the service also decides the actions to be performed (actuation), like changing the monitoring device's state, based on the doctor's recommendation. The service provides the inference engine needed for querying and reasoning over the knowledge to gain insights. It can also responsibly identify known threats based off of the knowledge and present the appropriate security service to ward off any more similar occurrences in the future (Tiwari, Jain, Abraham, & Shandilya, 2018) (Mozzaquatro et al., 2018).
- **Interfacing:** The end users access the web services through diverse medical devices and communication networks. The interface layer allows all kinds of users and devices to retrieve or send information through the service towards a knowledge database.

- **Users:** Patients, doctors, and attackers are the kind of end users that participate in the web-based healthcare service. Patients can share their medical situation or health status through smart devices and the doctors can monitor for daily activities of the patient and prescribe treatment, while the attackers may try and identify vulnerabilities in the devices or service design and make malicious attempts at accessing, stealing, modifying, or damaging sensitive information (Malik & Malik, Using IoT and Semantic Web Technologies for Healthcare and Medical Sector, 2020).

6.6 CONCLUSION

Healthcare is a huge industry, moving rapidly towards IT adoption for meeting the current demands of personalized healthcare services around the clock. IoT plays a significant role in the implementation of e-healthcare devices and SWT framework for the s further manage the structuring of data and its interoperability in IoT enabled healthcare services. As Internet is a hostile environment and IoT and SWoT are both based off of the web, security of the data forms an important aspect in e-healthcare applications. Through this chapter, various security issues in IoT and SWT based systems were identified, along with the state-of-the-art solutions applied in different works. A generic model for secure design of SWT based IoT enabled e-healthcare system was then discussed, having components of knowledge provisioning, semantic secure layers, data processing and integration and, end user devices. Using machine learning algorithms and data mining, with cloud computing know-how in SWoT networks etc., the model can be extended further by integrating techniques for minimizing vulnerabilities, controlled authorized access and enforced strong encryption and security policies.

REFERENCES

Berners Lee, T., Hendler, J., & Lassila, O. (2001). The semantic web. *Scientific American, 284*(5), 28–37.

Bytes, A., Adepu, S., & Zhou, J. (2019). Towards semantic sensitive feature profiling of IoT devices. *IEEE Internet of Things*, 8056–8064.

Canedo, J., & Skjellum, A. (2016). Using machine learning to secure IoT systems. In *14th Annual Conference on Privacy, Security and Trust (PST)* (pp. 219–222). IEEE.

Chatterjee, B., Das, D., Maity, S., & Sen, S. (2018). RF-PUF: Enhancing IoT security through authentication of wireless nodes using in-situ machine learning. *IEEE Internet of Things Journal, 6*(1), 388–398.

Choi, C., & Choi, J. (2019). Ontology-based security context reasoning for power IoT-cloud security service. *IEEE Access, 7*, 110510–110517.

Dragoni, M., Solanki, M., & Blomqvist, E. (2017). Semantic web challenges. In *Proceedings of 4th Semantic Web Evaluation Challenge at ESWC*. Springer.

Hahn, T., Matthews, S., Wood, L., Cohn, J., Regev, S., Fletcher, J., et al. (2015). *IBM Point of View: Internet of Things Security*. White Paper.

Hosseinzadeh, S., Virtanen, S., Diaz-Rodriguez, N., & Lilius, J. (2016). A semantic security framework and context-aware role-based access control ontology for smart spaces. In *Proceedings of the International Workshop on Semantic Big Data* (pp. 1–6). ACM Publications.

Khan, M., & Salah, K. (2018). IoT security: Review, blockchain solutions, and open challenges. *Future Generation Computer Systems*, *82*, 395–411.

Konev, A., Khaydarova, R., Lapaev, M., Feng, L., Hu, L., Chen, M., et al. (2019). CHPC: A complex semantic-based secured approach to heritage preservation and secure IoT-based museum processes. *Computer Communications*, 240–249.

Lim, S., & Rahmani, R. (2020). Toward semantic IoT load inference attention management for facilitating healthcare and public health collaboration: A survey. In *The 10th International Conference on Current and Future Trends of Information and Communication Technologies in Healthcare (ICTH). 177* (pp. 371–378). Procedia Computer Science.

Malik, N., & Malik, S. K. (2018). Security in web semantics: A revisit. In *5th International Conference on Computing for Sustainable Global Development -INDIACom* (pp. 678–683). IEEE.

Malik, N., & Malik, S. K. (2020). Using IoT and semantic web technologies for healthcare and medical sector. In V. Jain, R. Wason, J. M. Chatterjee, & D.-N. Le (Eds.), *Ontology-Based Information Retrieval for Healthcare Systems* (pp. 91–115). Wiley-Scrivener.

Mishra, S., Jain, S., Rai, C., & Gandhi, N. (2018). Security challenges in semantic web of things. In *International Conference on Innovations in Bio-Inspired Computing and Applications* (pp. 162–169). Springer.

Moh, M., & Raju, R. (2019). Using machine learning for protecting the security and privacy of internet of things (IoT) systems. *Fog and Edge Computing: Principles and Paradigms*, 223–257.

Mozzaquatro, B., Agostinho, C., Goncalves, D., Martins, J., & Jardim-Goncalves, R. (2018). An ontology-based cybersecurity framework for the internet of things. *Sensors*, *18*(9), 3053.

Ngoc, B., & Van, B. (2020). Privacy, security, and policies: A review of problems and solutions with semantic web technologies. *Frontiers in Intelligent Computing: Theory and Applications*, 1–10.

OWASP-IoT-Top-10–2018-final. (2018). Retrieved from owasp.org: https://owasp.org/www-pdf-archive/OWASP-IoT-Top-10–2018-final.pdf.

Rhayem, A., Mhiri, M., & Gargouri, F. (2020). Semantic web technologies for the internet of things: Systematic literature review. *Internet of Things*, 100206.

Rhayem, A., Mhiri, M., Salah, M., & Gargouri, F. (2017). Ontology-based system for patient monitoring with connected objects. *Procedia Computer Science*, *112*, 683–692.

Sezer, E., Bursa, O., Can, O., & Unalir, M. O. (2016). Semantic web technologies for IoT-based health care information systems. In Ö. Can & F. Scioscia (Eds.), *SEMAPRO: The Tenth International Conference on Advances in Semantic Processing* (pp. 45–48). IARIA.

Shafik, W., & Matinkhah, M. (2019). Privacy issues in social web of things. In *5th International Conference on Web Research (ICWR)* (pp. 208–214). IEEE.

Silva, A., Perez-Alcazar, J., & Kofuji, S. (2019). Interoperability in semantic web of things: Design issues and solutions. *International Journal of Communication Systems*, *32*(6), e3911.

Singh, D., Tripathi, G., & Alberti, A. M. (2017). Semantic edge computing and IoT architecture for military health services in battlefield. In *14th IEEE Annual Consumer Communications & Networking Conference* (pp. 185–190). IEEE.

Srinivasa, K., Siddesh, G., & Hanumantha, R. R. (2017). *Internet of Things*. Cengage Publications.

Szilagyi, I., & Wira, P. (2016). Ontologies and semantic web for the internet of things-a survey. In *IECON* (pp. 6949–6954). IEEE.

Tiwari, S. M., Jain, S., Abraham, A., & Shandilya, S. (2018). Secure semantic smart healthcare (S3HC). *Journal of Web Engineering*, *17*(8), 617–646.

Wang, T., Zhang, G., Liu, A., Bhuiyan, M., & Jin, Q. (2018). A secure IoT service architecture with an efficient balance dynamic based on cloud and edge computing. *IEEE Internet of Things Journal*, *6*(3), 4831–4843.

Warner, M., Gardner, C., Wyden, R., & Daines, S. (2017). Internet of things cybersecurity improvement act of 2017. In *Proceedings of 115th US Congress* (p. 1691). Congressional Record.

Wu, Z., Xu, Y., Yang, Y., Zhang, C., Zhu, X., & Ji, Y. (2017). Towards a semantic web of things: A hybrid semantic annotation, extraction, and reasoning framework for cyber-physical system. *Sensors*, *17*(2), 403.

Xie, W., Tang, Y., Chen, S., Zhang, Y., & Gao, Y. (2016). Security of web of things: A survey (short paper). In *International Workshop on Security* (pp. 61–70). Springer.

Yeh, K.-H. (2016). A secure IoT-based healthcare system with body sensor networks. *IEEE Access*, *4*, 10288–10299.

Zarca, A. M., Bagaa, M., Bernabe, J. B., Taleb, T., & Skarmeta, A. F. (2020). Semantic-aware security orchestration in SDN/NFV-enabled IoT systems. *Sensors*, *20*(13), 3622.

Ziegeldorf, J., Morchon, O., & Wehrle, K. (2014). Privacy in the internet of things: Threats and challenges. *Security and Communication Networks*, *7*(12), 2728–2742.

Fuzzy Weighted Nuclear Norm Based Two-Dimensional Linear Discriminant Features for Fruit Grade Classification

7

G. Yogeswararao, R. Malmathanraj, and P. Palanisamy

Research scholar

Assistant professor

Professor

Department of Electronics and Communication Engineering

National Institute of Technology, Tiruchirappalli, India

Contents

DOI: 10.1201/9781003244714-7

7.1 INTRODUCTION

The best quality vegetables and fruits are rich in minerals, fiber, antioxidants, and vitamins. Especially specific types of tropical fruits and vegetables processed into syrups and juices provide nutrients to the human body. The traditional grading of quality fruits requires more skilled manpower since it involves fatigue factor and also it is time-consuming. The computer algorithm-based solution to the sorting and grading problem have been investigated [1]. In the past few decades, various computer vision and machine learning algorithms for image feature extraction have been developed to address the postharvest horticulture problems [2–4]. The automatic banana grading system has been developed using gray level co-occurrence matrix, neural network arbitrations and support vector machine classifiers [5]. The image processing and k-means clustering methods have been used to classify the grade of nuts [6]. The shape features along with artificial neural networks have been used for cucumber fruit classification based on image processing techniques [7]. The fruit skin damage detection has been investigated using fast discrete curvelet transform and color texture features [8]. The nutrient deficiency of apple fruit has been analyzed and recognized using a multiclass deep neural network model [9]. The size and mass of the pistachio has been predicted using a random forest learning algorithm [10].

Among the available dimensional reduction algorithms, linear discriminant analysis (LDA) has been the best dimensional reduction technique to get the optimum features [11]. The main aim of LDA is to maximize the ratio of the between-class (SB) to within-class scatter (SW) matrices and map the lower-dimensional space. LDA agonizes with two problems, (1) small sample size (SSS) problem and (2) linearity problem. The kernel functions as a probable solution to the aforementioned problems have been listed [12]. The polynomial kernel discriminant analysis for classification has been developed [13]. An excellent alternative solution is the two-dimensional feature extraction approach, i.e., two-dimensional principal component analysis (2DPCA) [14] and two-dimensional linear discriminant analysis (2DLDA) [15]. The Frobenius-norm (F-norm) based two-directional 2DPCA [(2D) 2PCA] have been developed for face recognition [16] and two-directional 2DLDA [(2D) 2LDA] have been developed for pattern recognition with fewer coefficients than PCA and 2DLDA [17]. The two-dimensional joint local and nonlocal discriminant analysis for deep learning model has been developed for face image recognition [18]. Using fuzzy k-nearest neighbor (FKNN) membership, the fuzzy two-dimensional LDA (F2DLDA) has been developed [19] and also fuzzy two-dimensional discriminant analysis using local graph embedding have been developed for face and palm biometrics [20]. The fuzzy logic and regression models for image feature extraction and crop yield forecasting has also been investigated [21]. Recently the fuzzy fractional weight incorporated in F2DLDA named as Fractional Fuzzy 2DLDA (FF2DLDA) approach for pomegranate fruit grade classification has been proposed [22]. Similarly, L1 norm based feature extraction techniques has been developed to improve the robustness of LDA [23–25]. The F-norm is sensitive to outliers in data and the noise, due to which recently the nuclear norm-based minimization (NNM) technique has received much attention in the fields of image classification and pattern recognition. Based on the nuclear norm, 2DPCA (N-2DPCA) has been proposed [26], which could get global information of data effectively. Recently horizontal, vertical and bilateral 2DLDA with the nuclear norm for effective image representation using various data sets has been proposed [27].

7.2 LITERATURE SURVEY

The training process of traditional 2DLDA uses binary class labeling to grade and classify the fruits and vegetables. The change in environmental conditions significantly leads to change in color and maturity of the fruits. The binary class labeling is not suitable in real time fruit grade assignment problems. therefore, it is always advantageous to have the fuzzy membership value assigned to the fruit image instead of the binary value. In order to get the best discriminant features, the fuzzy k-nearest neighbor (FKNN) based membership values has been incorporated in SB and SW matrices in 2DLDA called fuzzy 2DLDA (F2DLDA) has been proposed [19]. The pomegranate fruit grading has been investigated using two feature extraction adaptive mathematical principles of research on 2DLDA and combination of 2DLDA and LDA, which is 2DLDA-LDA, with SVM. In both approaches, the SVM with polynomial training, radial basis function, multilayer perception classifiers, quadratic programming, and liner classifier kernel functions have been used to identify the retrieved features [28]. The gray level co-occurrence matrix (GLCM) and pixel run length matrix (PRLM) features, as well as two distinct classifiers, have been used to analyze semi-ripened, ripened, and over-ripened pomegranate MRI image samples. Based on the type I, type II discriminant analysis contribution model, the optimal amount of features has been chosen to achieve the best classification accuracy [29].

A unique texture and color gradient features-based methodology for automatic pomegranate and mango fruit grade identification has been recently developed. The local binary pattern (LBP), GLCM, and PRLM have been used to determine the texture of the fruits, while average color gradients, variances, and color coordinates of the three primary colors red, green, and blue have been used to represent the fruits' color gradients (CG). The KSVM has been used to grade and classify the retrieved features using texture and color gradient approaches. The gradient features such as 59 structural LBP features, 11 statistical GLCM and 7 PRLM features, and nine CG features have been used in the proposed automatic grade analysis system to help enhance all performance metrics [30]. The color, shape, and spot features have been used to examine the quality and grade classification of unripe, semi-ripe, and ripe mango fruits. The median filtering and Otsu threshold based segmentation have been done first. The lengths of the major and minor axes of mango fruits have also been determined. Finally, the strip histogram was used to determine the mango surface sports [31]. SqueezeNet, a deep learning model based on pre-trained transfer learning, has been proposed for grading the mango quality [32]. By combining shape, size, and color features in digital image analysis, an automated image processing method for identifying the defect and maturity of mango fruits has been established [33]. The skin color of the Langdo type mangos does not change over time. Normal imaging will not be able to forecast its maturity in this scenario. For maturity prediction, infrared, x-ray, or thermal imaging can be used. Mango grading has been done based on maturity and weight, as well as eccentricity and area-based size features along with a fuzzy classifier has been used to predict size features [34]. Similarly, the mangos have been graded using pre-trained convolutional neural network (CNN) features and an SVM classifier [35]. The main contribution of the chapter as follows:

- The present work identifies and recognizes the healthy pomegranate and mango fruits, in the open cofilab data set and self-built mango digital database.
- In this chapter, a novel non-destructive feature extraction technique called fuzzy weighted nuclear norm-based 2DLDA is proposed for unilateral (row and column) and bidirectional procedures.
- The four existing feature extraction based pattern recognition approach including (1) 2DLDA, (2) F2DLDA, (3) Nuclear norm based 2DLDA (NN-2DLDA), and (4) fuzzy weighted nuclear norm-based 2DLDA (FNN-2DLDA) are analyzed.
- The support vector machine (SVM) algorithm together with four feature extraction techniques and the proposed technique are implemented and compared for accurate grade assignment and classification of pomegranate and mango fruits.

The rest of the chapter is organized as follows: the proposed fuzzy weighted nuclear norm based 2DLDA technique and its algorithm are described in materials and methods Section 7.3; the average classification accuracy, statistical performance measurements, and hypothesis test results and discussions are reported in Section 7.4. and, finally, Section 7.5 concludes the chapter.

7.3 MATERIALS AND METHODS

The problem with traditional 2DLDA based fruit grading is its binary nature of grade assignment of fruit. The change in environmental conditions significantly leads to change in color and maturity of the fruits. Since it is always advantageous to have the fuzzy membership value assigned to the fruit image instead of the binary value.

7.3.1 Fuzzy Membership Value for Proposed FNN-2DLDA

Assume that there are N training samples with L known pattern classes $c_1, c_2, \ldots, c_L$, $I = \{I_j^i\}\ (i = 1,2,\ldots,L, j = 1,2,\ldots,t_i)$, where '$i$' and t_i are the class count and particular class training samples which satisfies $\sum_{i=1}^{L} t_i = N$. Then the fuzzy membership value for each sample and each class mean matrix $\tilde{I}_i$ are calculated as given below.

$$\tilde{I}_i = \frac{\sum_{j=1}^{N} \mu_{ij}^{p} I_j}{\sum_{j=1}^{N} \mu_{ij}^{p}} \tag{7.1}$$

where μ_{ij} is fuzzy membership value corresponding to j^{th} sample in i^{th} class, 'p' is a constant, $\tilde{I}_i$ is i^{th} class mean and $\tilde{I}$ is centroid of all classes. The fuzzy membership value calculation using FKNN is shown below.

$$\mu_{ij} = \begin{cases} 0.51 + 0.49 \times (\eta_{ij} / k) & \textit{if } i = \textit{label of the jth vector} \\ 0.49 \times (\eta_{ij} / k) & \textit{if } i \neq \textit{label of the jth vector} \end{cases} \tag{7.2}$$

where η_{ij} is neighbors ('j^{th}' vector belongs to 'i^{th}' class) count and k is the number of neighboring vectors in the vicinity of the vector. The final fuzzy membership matrix 'U' can be obtained

$$U = \left[\mu_{ij}\right], i = 1,2,\ldots,L,\ j = 1,2,\ldots,N. \tag{7.3}$$

7.3.2 Unilateral FNN-2DLDA

In recent years, the nuclear norm minimization (NNM) concept has been attracting abundant attention in convex relaxation of the low-rank matrix factorization problems [36]. The NNM regularizes each singular value equally, composing an efficiently calculated convex norm.

7.3.2.1 Row Information FNN-2DLDA

In this chapter, the training samples row numerical information has been induced using nuclear norm (NN) of a matrix. The NN is optimize the quantitative relation of the fuzzy SB (FSB) and fuzzy SW (FSW). In this research work, the row FNN-2DLDA (RFNN-2DLDA) is proposed. The objective function of RFNN-2DLDA is outlined as

$$W_{opt}^{row} = \arg\max_{W} \frac{\sum_{i=1}^{L} N_i \left\| (\tilde{I}_i - \tilde{I})W \right\|_*}{\sum_{i=1}^{L}\sum_{j=1}^{t_i} \left\| (\tilde{I}_j^i - \tilde{I}_i)W \right\|_*} \tag{7.4}$$

The conversion of the nuclear norm optimization problem can be implemented using F-norm optimization problem. The conversion matrix Y provides the following expression

$$\left\| Y \right\|_* = \left\| (YY^T)^{-1/4} Y \right\|_F^2 \tag{7.5}$$

using the expression (7.5), the nuclear-norm is used to transform into the F-norm. Thus, the expression (7.6) can be obtained from equation (7.4):

$$W_{opt}^{row} = \arg\max_{W} \frac{\sum_{i=1}^{L} N_i \left\| (\tilde{I}_i - \tilde{I})W \right\|_*}{\sum_{i=1}^{L}\sum_{j=1}^{t_i} \left\| (\tilde{I}_j^i - \tilde{I}_i)W \right\|_*} == \arg\max_{W} \frac{tr(W^T S_{NFB}^{row} W)}{tr(W^T S_{NFW}^{row} W)} \tag{7.6}$$

$$S_{NFB}^{row} = \sum_{i=1}^{L}\sum_{j=1}^{t_i} \mu_{ij} N_i (G_b \tilde{I}_i - G_b \tilde{I})^T (G_b \tilde{I}_i - G_b \tilde{I}) \tag{7.7}$$

$$S_{NFW}^{row} = \sum_{i=1}^{L}\sum_{j=1}^{t_i} \mu_{ij} (G_w I_j^i - G_w \tilde{I}_i)^T (G_w I_j^i - G_w \tilde{I}_i) \tag{7.8}$$

$$G_b = \left[(\tilde{I}_i - \tilde{I}) W W^T (\tilde{I}_i - \tilde{I})^T \right]^{-1/4} \tag{7.9}$$

$$G_w = \left[(I_j^i - \tilde{I}_i) W W^T (I_j^i - \tilde{I}_i)^T \right]^{-1/4} \tag{7.10}$$

where fuzzy weight matrix G_b is the distance from i^{th} class mean to the centroid point and G_w is the distance of samples in the same class to mean of particular class for unilateral row information. The S_{NFB}^{row} is the row nuclear norm based FSB (NN-FSB) calculation during which the centroid and every class mean is pre multiplied with G_b. The S_{NFW}^{row} is the row nuclear norm based FSW (NN-FSW) calculation during which the samples in i^{th} class and corresponding class mean is pre multiplied with G_w.

7.3.2.2 Column Information FNN-2DLDA

Additionally, in this chapter, unilateral FNN-2DLDA is abbreviated as CFNN-2DLDA, introduces the column numerical information to search the discriminant features to find the optimum projection direction. The column information of NN-FSB 'S_{NFB}^{col}', NN-FSW 'S_{NFW}^{col}' and the objective function for CFNN-2DLDA as given below:

$$W_{opt}^{col} = \arg\max_{W} \frac{\sum_{i=1}^{L} N_i \left\| (\tilde{I}_i - \tilde{I})^T W \right\|_*}{\sum_{i=1}^{L}\sum_{j=1}^{t_i} \left\| (\tilde{I}_j^i - \tilde{I}_i)^T W \right\|_*} == \arg\max_{W} \frac{tr(W^T S_{NFB}^{col} W)}{tr(W^T S_{NFW}^{col} W)} \tag{7.11}$$

$$S_{NFB}^{col} = \sum_{i=1}^{L}\sum_{j=1}^{t_i} \mu_{ij} N_i (H_b \tilde{I}_i - H_b \tilde{I})(H_b \tilde{I}_i - H_b \tilde{I})^T \tag{7.12}$$

$$S_{NFW}^{col} = \sum_{i=1}^{L}\sum_{j=1}^{t_i} \mu_{ij} (H_w I_j^i - H_w \tilde{I}_i)(H_w I_j^i - H_w \tilde{I}_i)^T \tag{7.13}$$

$$H_b = \left[(\tilde{I}_i - \tilde{I})^T W W^T (\tilde{I}_i - \tilde{I}) \right]^{-1/4} \tag{7.14}$$

$$H_w = \left[(I_j^i - \tilde{I}_i)^T W W^T (I_j^i - \tilde{I}_i) \right]^{-1/4} \tag{7.15}$$

where fuzzy weight matrices H_b is the distance from i^{th} class mean to the centroid point and H_w is the distance of samples in the same class to mean of particular class for unilateral column information. The S_{NFB}^{col} is the column nuclear norm based FSB (NN-FSB) calculation during which the centroid and every class mean is pre multiplied with H_b. The S_{NFW}^{col} is the column nuclear norm based FSW (NN-FSW) calculation during which the samples in i^{th} class and corresponding class mean is pre multiplied with H_w.

Using eigen analysis, determine eigen vectors $W_{opt}^{Nf\,row} = \{v_1^{Nf(row)}, v_2^{Nf(row)}, \ldots, v_k^{Nf(row)}\}$ and $W_{opt}^{Nf\,row} = \{v_1^{Nf(row)}, v_2^{Nf(row)}, \ldots, v_k^{Nf(row)}\}$ corresponding to k largest eigen values of the matrices $(S_{NFW}^{row})^{-1} S_{NFB}^{row}$ and $(S_{NFW}^{col})^{-1} S_{NFB}^{col}$ respectively. Then each training image sample I_j^i is projected onto $W_{opt}^{Nf\,row}$ and $W_{opt}^{Nf\,col}$ for row and column feature space representation of RFNN-2DLDA and CFNN-2DLDA. The $X_j^{i\,row} = I_j^i \times W_{opt}^{Nf\,row}$ is the new optimal projection direction of RFNN-2DLDA and $X_j^{i\,col} = (W_{opt}^{Nf\,col})^T \times I_j^i$ is for CFNN-2DLDA.

7.3.3 Bilateral FNN-2DLDA

The RFNN-2DLDA and CFNN-2DLDA operates along the row and column numerical information of image respectively. The RFNN-2DLDA and CFNN-2DLDA learns the optimal separation projection matrices $W_{opt}^{Nf\,row}$ and $W_{opt}^{Nf\,row}$ which projects a test image I_j^i onto new projection space i.e., $X_j^i = I_j^i \times W_{opt}^{Nf\,row}$ and $X_j^i = (W_{opt}^{Nf\,col})^T \times I_j^i$. After obtaining row and column projection matrices $W_{opt}^{Nf\,row}$ and $W_{opt}^{Nf\,col}$ from RFNN-2DLDA and CFNN-2DLDA, the bilateral FNN-2DLDA (BFNN-2DLDA) optimal separation projection direction solution is to combine each unilateral FNN-2DLDA projection directions. The $X_j^i = (W_{opt}^{Nf\,col})^T \times I_j^i \times W_{opt}^{Nf\,row}$ is the new optimal projection direction of BFNN-2DLDA.

7.3.4 Work Flow Graph of Research Work Involved

The detailed work flow graph of the techniques involved in this chapter are shown in Figure 7.1. The step by step algorithm of proposed FNN-2DLDA technique is explained below.

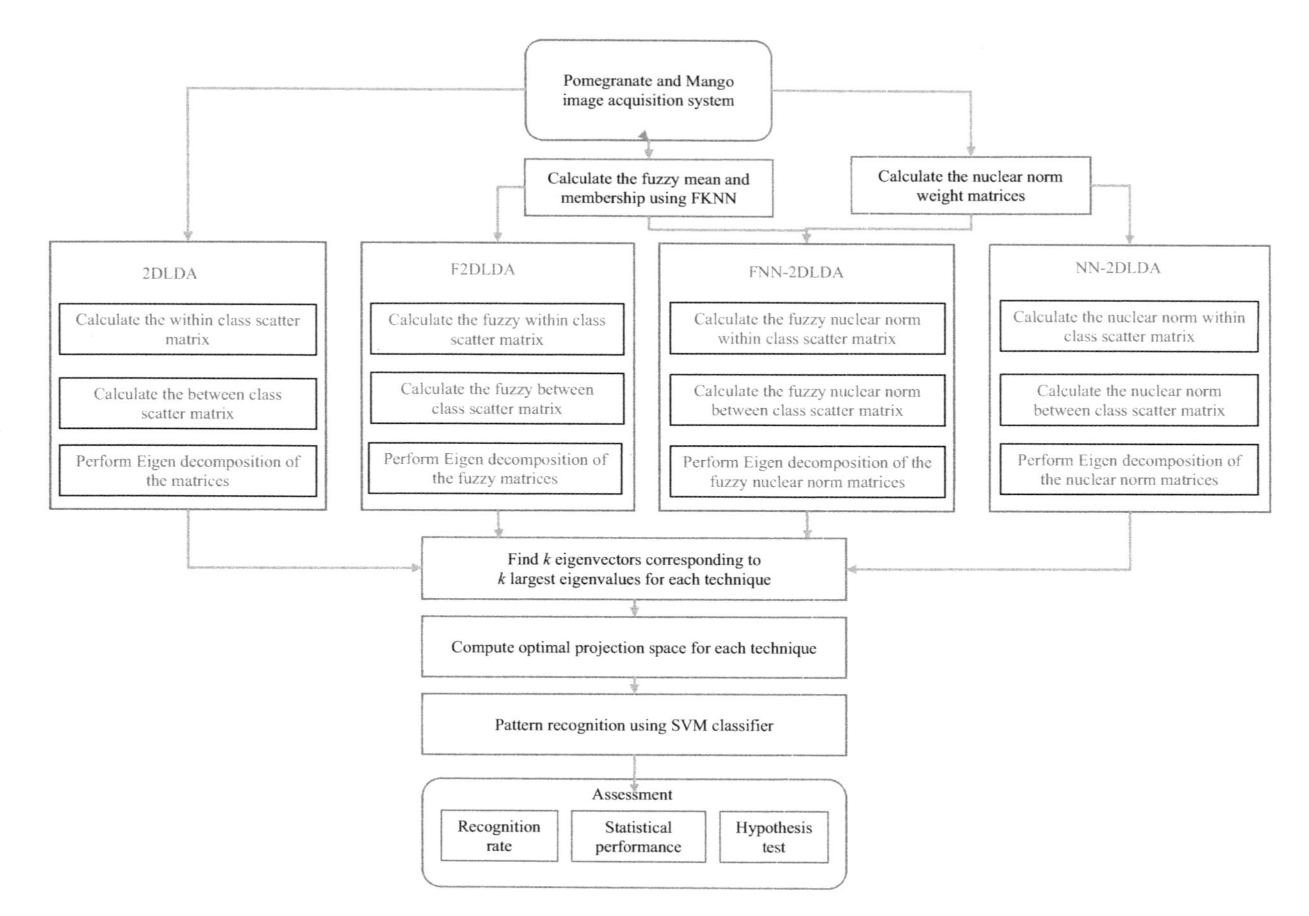

FIGURE 7.1 Work flow graph of research work involved.

7.3.5 Algorithm of FNN-2DLDA

1. A set of N training samples with $m \times n$ dimension, using fuzzy weighted NN features, each sample is now denoted by $m \times k$, $k \times n$ and $k \times k$ for unilateral (row and column information) and bilateral information, where $k < (m,n)$.
2. Compute the ith class mean $\tilde{I}_i$, centroid of samples $\tilde{I}$ and the degree of fuzzy membership U using FKNN algorithm.
3. Compute the weight matrices G_b, G_w (for row information) and H_b, H_w (for column information) using equations (7.9), (7.10) and (7.14), (7.15) respectively.
4. Generate the nuclear norm fuzzy between class scatter matrix S_{NFB}^{row}, S_{NFB}^{col} using equations (7.7), (7.12) and nuclear norm fuzzy within class scatter matrix S_{NFW}^{row}, S_{NFW}^{col} using equations (7.8), (7.13).
5. Usingeigenanalysis,determineeigenvectors $W_{opt}^{Nf\,row} = \{v_1^{Nf(row)}, v_2^{Nf(row)}, \ldots, v_k^{Nf(row)}\}$ and $W_{opt}^{Nf\,row} = \{v_1^{Nf(row)}, v_2^{Nf(row)}, \ldots, v_k^{Nf(row)}\}$ corresponding to k largest eigenvalues of the matrices $(S_{NFW}^{row})^{-1} S_{NFB}^{row}$ and $(S_{NFW}^{col})^{-1} S_{NFB}^{col}$ for RFNN-2DLDA and CFNN-2DLDA.

7.4 EXPERIMENTAL RESULTS AND DISCUSSION

7.4.1 Pomegranate Fruit Database

The experiment is conducted using a data set consisting of 328 (pixel sized 768×1024) pomegranate fruit images. The region of interest in original pixel area images (386×512) is used for experimental purposes to include the individual pomegranate image. Figure 7.2 shows the data samples of pomegranate images.

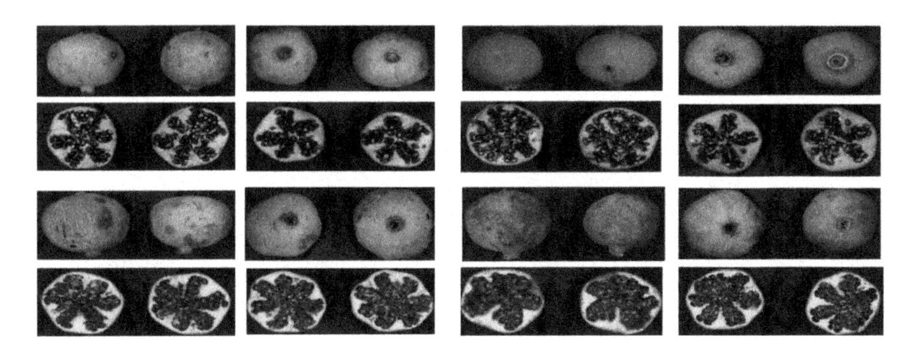

FIGURE 7.2 Pomegranate images for different classes.

7.4.1.1 Classification Accuracies of Unilateral and Bilateral Techniques for Pomegranate Database

The classification accuracy is used as one of the parameter for evaluating the performance of the proposed approach. Initially the experiment is performed on pomegranate database using unilateral approaches. The pomegranate training and validation samples are arbitrarily elite from the cofilab database. This procedure results in cross validation of the training and testing pomegranate samples. The F-norm and NN based 2DLDA, fuzzy 2DLDA techniques are adapted to classify the test samples using SVM. The average classification accuracy of unilateral 2DLDA, fuzzy 2DLDA and nuclear norm-based 2DLDA and proposed technique are depicted in Figure 7.3 and Figure 7.4.

The performance of the proposed approach is appraised using the recognition rate of bilateral approaches on pomegranate database. The average classification accuracy of bilateral F-norm based 2DLDA, fuzzy 2DLDA, nuclear norm-based 2DLDA and proposed technique are depicted in Figure 7.5. From Figures 7.3, 7.4, and 7.5, it is shown that the proposed FNN-2DLDA techniques outperform the other 2D feature extraction techniques. This is often because of the subsequently following factors:

- Considering the fuzzy membership value to each training image sample using FKNN algorithm, instead of binary value (traditional 2DLDA).
- Using fuzzy set theory, redefining the scatter matrices which incorporate the overlapped samples distribution in scatter matrices.
- The nuclear norm mainly provides global information of the data (image samples) which will help to get the best discriminant features in the projection.
- As compared to F-norm, nuclear norm is more robust while using low rank learning.

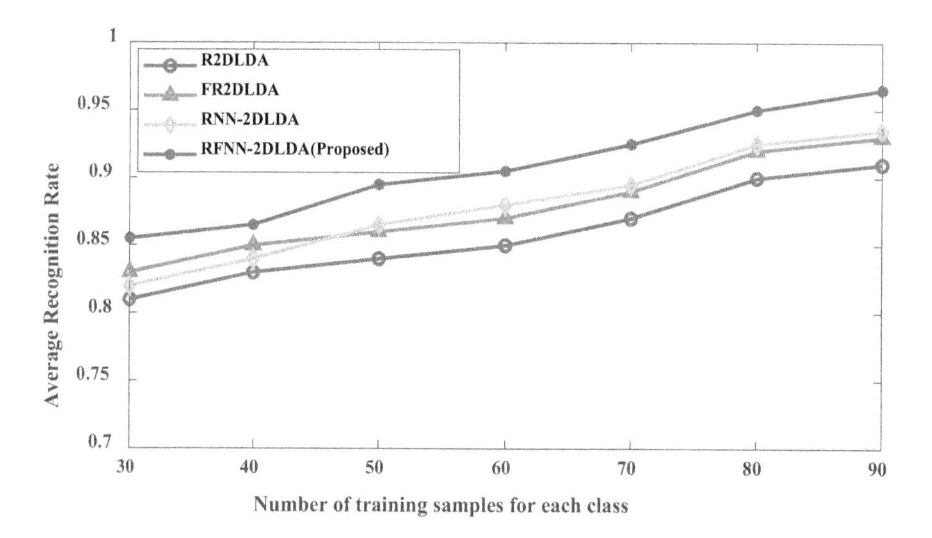

FIGURE 7.3 The recognition rates of row information 2D feature extraction technique for pomegranate database.

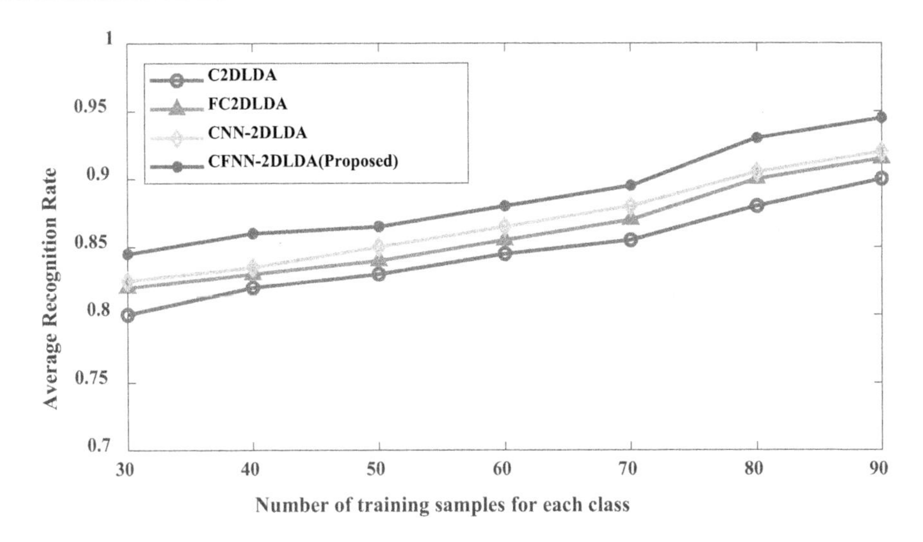

FIGURE 7.4 The recognition rates of column information 2D feature extraction technique for pomegranate database.

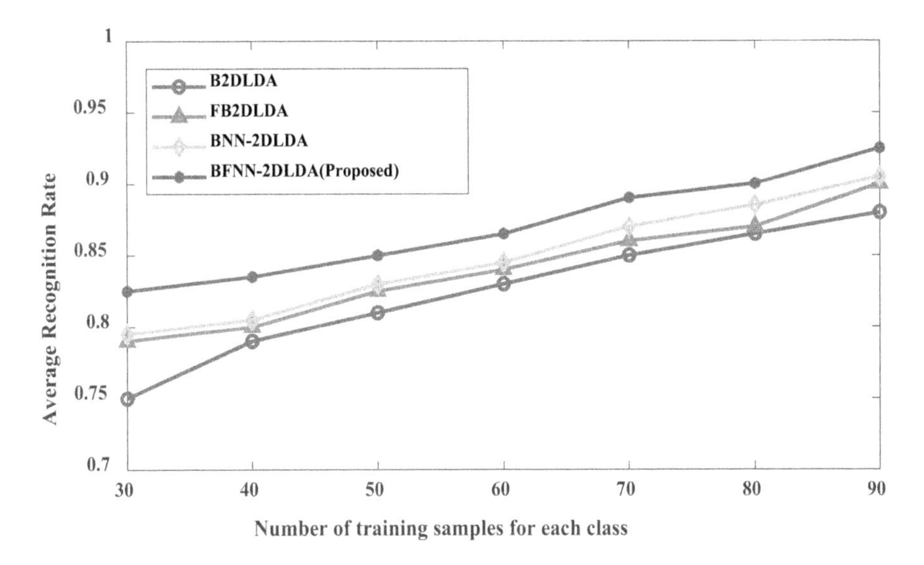

FIGURE 7.5 The recognition rates of bilateral 2D feature extraction technique for pomegranate database.

- The combination of fuzzy membership and nuclear norm provides more structure, global and discriminant information of the image samples.

The proposed unilateral and bilateral FNN-2DLDA offers sensible recognition rate compared with standard, fuzzy unilateral and fuzzy bilateral 2DLDA and NN-2DLDA techniques and same is plotted in Figures 7.3, 7.4, and 7.5.

7.4.1.2 Statistical Analysis of Unilateral and Bilateral Techniques with Pomegranate Fruit Database

The statistical performance measurements are used as the second parameter for appraise the effectiveness of the proposed technique. The tested statistical performance metrics are accuracy (ACC), specificity (SPE), sensitivity (SEN), precision (PRE), F-score, False Positive Rate (FPR), Negative Predictive Value (NPV), False Discovery Rate (FDR), False Negative Rate (FNR) and Matthew's Correlation Coefficient (MCC).

The ACC, SEN, SPE, PRE, F-score, NPV, FPR, FNR, FDR, and MCC of F-norm and nuclear norm-based unilateral and bilateral features of 2DLDA and F2DLDA techniques are displayed in Table 7.1. The statistical performance metrics indicate that the proposed FNN-2DLDA technique is giving the effective results when compared with prevailing techniques, since the fuzzy membership and nuclear norm are enclosed with in the FNN-2DLDA.

7.4.2 Mango Fruit Database

The experiment is conducted using a data set consisting of 200 (pixel sized 1024 × 1024) mango fruit images. The region of interest in original pixel area images (512 × 512) is used for experimental purposes to include the individual mango. The Figure 7.6 illustrates the mango fruit images of different classes. The image acquisition setup used to capture the mango images consists of selfie-stick mounted on fixed basement and a mobile camera (13 mega pixel). In order to have uniform illumination, all mango fruit images were captured in the outdoor environment on a sunny day.

TABLE 7.1 Statistical Analysis of F-Norm and Nuclear Norm-Based 2D Feature Extraction Techniques with Pomegranate Database (a), (b), and (c)

STATISTICAL PERFORMANCE MEASUREMENTS	*R2DLDA*	*FR2DLDA*	*RNN-2DLDA*	*RFNN-2DLDA (PROPOSED)*
ACC	0.9	0.925	0.925	0.95
SEN	0.8	1	0.85	1
SPE	1	0.85	1	0.9
PRE	1	0.8696	1	0.9091
F-score	0.8889	0.9302	0.9189	0.9524
NPV	0.8333	1	0.8696	1
FPR	0	0.15	0	0.1
FNR	0.2	0	0.15	0
FDR	0	0.1304	0	0.0909
MCC	0.8165	0.8597	0.8597	0.9045

(a)

STATISTICAL PERFORMANCE MEASUREMENTS	*C2DLDA*	*FC2DLDA*	*CNN-2DLDA*	*CFNN-2DLDA (PROPOSED)*
ACC	0.875	0.925	0.9	0.95
SEN	1	0.9	0.95	0.9
SPE	0.75	0.95	0.85	1
PRE	0.8	0.9474	0.8636	1
F-score	0.8889	0.9231	0.9048	0.9474
NPV	1	0.9048	0.9444	0.9091
FPR	0.25	0.05	0.15	0
FNR	0	0.1	0.05	0.1
FDR	0.2	0.526	0.1364	0
MCC	0.7746	0.8511	0.804	0.9045

(b)

STATISTICAL PERFORMANCE MEASUREMENTS	*B2DLDA*	*FB2DLDA*	*BNN-2DLDA*	*BFNN-2DLDA (PROPOSED)*
ACC	0.85	0.9	0.875	0.925
SEN	1	1	0.95	0.95
SPE	0.7	0.8	0.8	0.9
PRE	0.7692	0.8333	0.8261	0.9048
F-score	0.8696	0.9091	0.8867	0.9268
NPV	1	1	0.9412	0.9474
FPR	0.3	0.2	0.2	0.1
FNR	0	0	0.05	0.05
FDR	0.2308	0.1667	0.1739	0.0952
MCC	0.7338	0.8165	0.7586	0.8511

(c)

FIGURE 7.6 Semi-ripened and ripened mango sample images for different classes.

7.4.2.1 Classification Accuracies of Unilateral and Bilateral Techniques for Mango Database

The recognition rate is used as one of the parameters to appraise the performance of the proposed approach. Initially, the experiment is performed on mango database using unilateral approaches. The mango samples are arbitrarily elite for training from the self-built mango database resulting in cross validation process. The F-norm based 2DLDA, fuzzy 2DLDA, nuclear norm-based 2DLDA and proposed FNN-2DLDA are accustomed to classify the mango test samples with SVM. The mango sample average classification accuracy of unilateral 2DLDA, fuzzy 2DLDA and nuclear norm-based 2DLDA and proposed technique are depicted in Figures 7.7 and 7.8.

The performance of the proposed approach is appraised using the recognition rate of bilateral approaches on mango database. The average classification accuracy of bilateral 2DLDA, fuzzy 2DLDA and nuclear norm-based 2DLDA and proposed technique are depicted in Figure 7.9. The proposed unilateral and bilateral FNN-2DLDA gives best recognition rate compared with traditional, fuzzy unilateral, bilateral 2DLDA and NN-2DLDA techniques using mango database and same is plotted in Figures 7.7, 7.8 and 7.9.

7.4.2.2 Statistical Analysis of Unilateral and Bilateral Techniques with Mango Database

The statistical performance measurements are used as second parameter to apprise the effectiveness of the proposed technique, the tested statistical performance metrics are ACC, SEN, SPE, PRE, F-score, NPV, FPR, FNR, FDR and MCC using mango database.

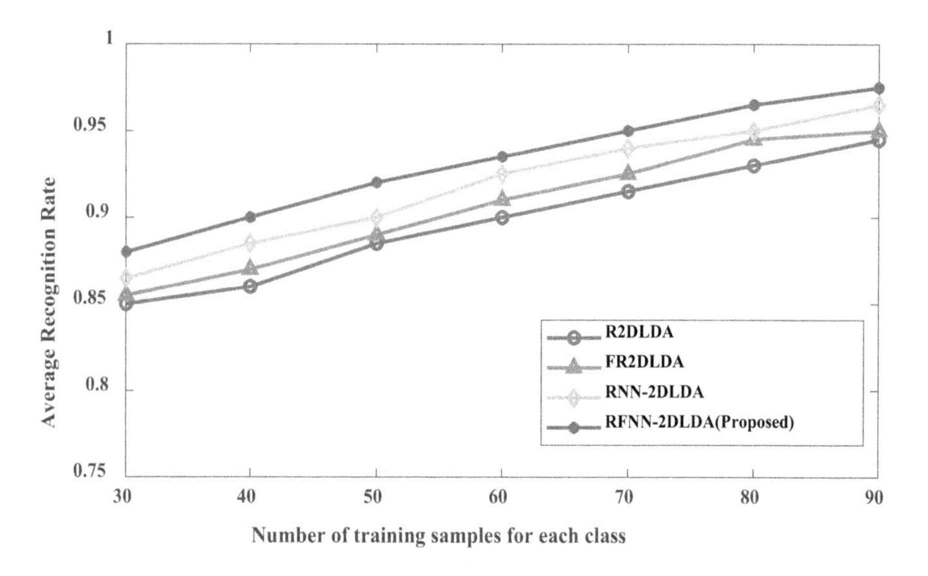

FIGURE 7.7 The recognition rates of row information 2D feature extraction technique for mango database.

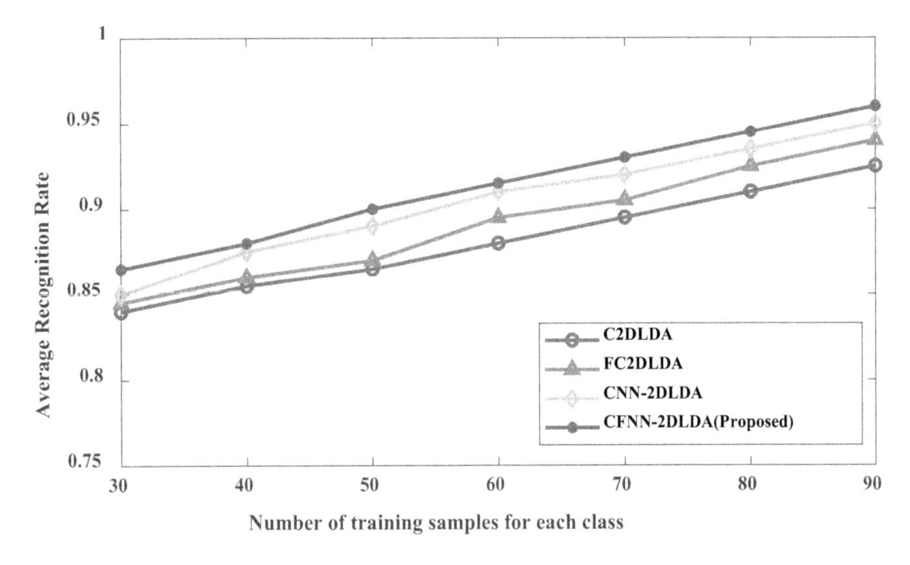

FIGURE 7.8 The recognition rates of column information 2D feature extraction technique for mango database.

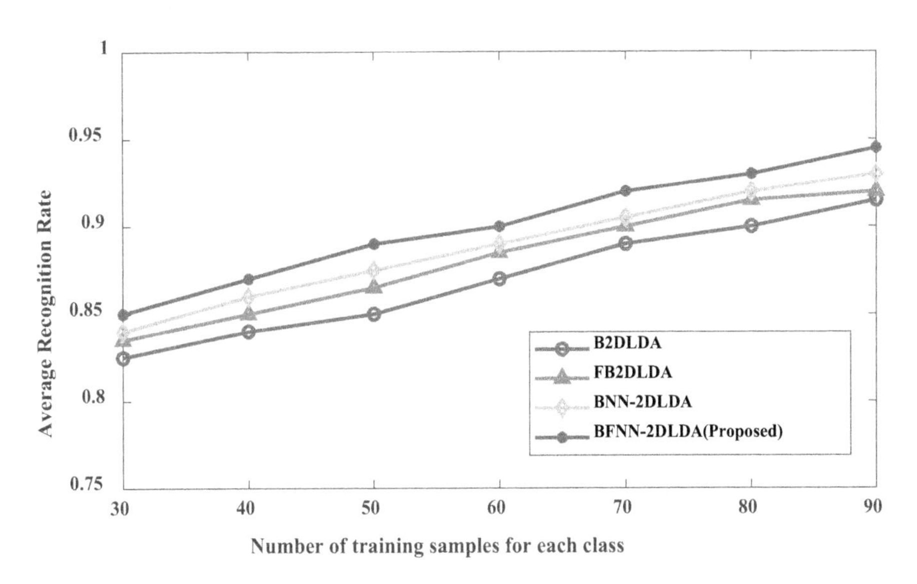

FIGURE 7.9 The recognition rates of bilateral 2D feature extraction technique for mango database.

The ACC, SEN, SPE, PRE, F-score, NPV, FPR, FNR, FDR, and MCC of unilateral and bilateral 2DLDA and F2DLDA supported F-norm and nuclear norm techniques using mango database are shown in Table 7.2. From Table 7.2, the statistical performance metrics inform that the proposed technique is notably giving the most effective results when

TABLE 7.2 Statistical Analysis of F-Norm and Nuclear Norm-Based 2D Feature Extraction Techniques with Mango Database (a), (b), and (c)

STATISTICAL PERFORMANCE MEASUREMENTS	*R2DLDA*	*FR2DLDA*	*RNN-2DLDA*	*RFNN-2DLDA (PROPOSED)*
ACC	0.875	0.9	0.9	0.925
SEN	0.9	0.85	1	0.95
SPE	0.85	0.95	0.8	0.9
PRE	0.8571	0.9444	0.8333	0.9048
F-score	0.878	0.8947	0.9091	0.9268
NPV	0.8947	0.8636	1	0.9474
FPR	0.15	0.05	0.2	0.1
FNR	0.1	0.15	0	0.05
FDR	0.1429	0.0556	0.1667	0.0952
MCC	0.7509	0.804	0.8165	0.8511

(a)

STATISTICAL PERFORMANCE MEASUREMENTS	*C2DLDA*	*FC2DLDA*	*CNN-2DLDA*	*CFNN-2DLDA (PROPOSED)*
ACC	0.85	0.875	0.9	0.925
SEN	0.95	0.85	0.95	0.9
SPE	0.75	0.9	0.85	0.95
PRE	0.7912	0.8947	0.8636	0.9474
F-score	0.8636	0.8718	0.9148	0.9231
NPV	0.9375	0.8571	0.9444	0.9048
FPR	0.25	0.1	0.15	0.05
FNR	0.5	0.15	0.05	0.1
FDR	0.2083	0.1053	0.1364	0.526
MCC	0.7144	0.7509	0.804	0.8511

(b)

STATISTICAL PERFORMANCE MEASUREMENTS	*B2DLDA*	*FB2DLDA*	*BNN-2DLDA*	*BFNN-2DLDA (PROPOSED)*
ACC	0.85	0.85	0.875	0.9
SEN	0.8	0.7	0.75	0.85
SPE	0.9	1	1	0.95
PRE	0.8889	1	1	0.9444
F-score	0.8421	0.8235	0.8571	0.8947
NPV	0.8182	0.7692	0.8	0.8636
FPR	0.1	0	0	0.05
FNR	0.2	0.3	0.25	0.15
FDR	0.111	0	0	0.0556
MCC	0.7035	0.7338	0.7746	0.804

(c)

compared with prevailing techniques, since the degree of fuzzy membership and nuclear norm are enclosed within the FNN-2DLDA. Mainly the MCC value is high when compared to that of the existing 2D feature extraction techniques which shows the proposed FNN-2DLDA technique is effective in predicting the individual grading of both pomegranate and mango fruits compared with the prevailing techniques.

7.4.2.3 Hypothesis Test Analysis for Pomegranate and Mango Fruit Database

In order to evaluate the performance of the proposed technique, 50 different attempts are made for proposed and existing techniques in both unilateral and bilateral approaches. The student's paired t-hypothesis test is performed in order to check the superiority in performance of the proposed technique over the existing techniques. The null hypothesis is considered as the proposed technique sample mean is lower than the existing techniques sample mean and the alternative hypothesis is considered as the proposed technique sample mean is higher than the existing techniques sample mean.

It can be seen from the Tables 7.3 and 7.4, the P_{value} within the hypothesis testing of unilateral and bilateral approaches for each pomegranate and mango databases are too small ($P_{value} < 0.05$). As P_{value} less than 0.05 in all the cases, which indicates that the performance of proposed technique is better than existing techniques with 95% confidence. Hence forth it can be understood that the performance of proposed FNN-2DLDA technique is also statistically better than that of the existing techniques in unilateral and bilateral procedures.

TABLE 7.3 T-Test Analysis for Pomegranate Fruit Database

TECHNIQUE	*P_{VALUE}*		
	UNILATERAL		*BILATERAL*
	ROW INFORMATION	*COLUMN INFORMATION*	
2DLDA	0.0006	0.0010	0.0011
F2DLDA	0.0011	0.0023	0.0050
NN-2DLDA	0.0160	0.0402	0.0256

TABLE 7.4 T-Test Analysis for Mango Fruit Database

TECHNIQUE	*P_{VALUE}*		
	UNILATERAL		*BILATERAL*
	ROW INFORMATION	*COLUMN INFORMATION*	
2DLDA	0.0132	0.0016	0.0155
F2DLDA	0.0246	0.0064	0.0187
NN-2DLDA	0.0334	0.0339	0.0449

7.5 CONCLUSION

This chapter presents pomegranate and mango fruit grading procedure supported by two-dimensional fruit images. During this approach, four non-destructive algorithms (F-norm based 2DLDA, F2DLDA, and nuclear norm-based 2DLDA and FNN-2DLDA) are applied to extract the features in unilaterally and bilaterally. Then the SVM is applied so as to spot that which feature extraction technique may provide the most effective grade classification for pomegranate and mango fruit. Out of four algorithms, the proposed FNN-2DLDA provides the most effective grade classification for pomegranate and mango fruit, because the fuzzy membership and nuclear norm approach included in traditional 2DLDA technique. The combination of fuzzy membership and nuclear norm provides more structure, global and discriminant information of the image samples in the proposed FNN-2DLDA technique. Considering the above reasons, the classification accuracy, statistical performance measurements and hypothesis test values of the proposed FNN-2DLDA giving the best fruit grade classification compared with prevailing techniques. The proposed work can be further extended as: (1) Preforming the classification by fuzzy C Means, K-Nearest Neighbor, neural networks instead of SVM and also the weights and other parameters of these techniques optimized by heuristic techniques. (2) The different norm and fuzzy weight-based unilateral and bilateral image feature extraction techniques. (3) Multidimensional features to 2D plot representation for best visualization of the results.

7.6 ACKNOWLEDGMENT

Authors wish to acknowledge Cofilab for giving access to use digital pomegranate images for grade analysis and simulations purpose.

7.7 CONFLICT OF INTEREST

The authors declare that they have no conflict of interest.

REFERENCES

[1] Scarlett Liu and Mark Whitty (2015) Automatic grape bunch detection in vineyards with an SVM classifier. *J. Appl. Log*, vol. 13, no. 4, Part 3, 643–653.

[2] Ferhat Kurtulmu, Sencer Öztüfekçi, and İsmail Kavdır (2018) Classification of chestnuts according to moisture levels using impact sound analysis and machine learning. *Journal of Food Measurement and Characterization*, vol. 12, pp. 2819–2834. https://doi.org/10.1007/s11694-018-9897-y.

[3] M.S. Iraji (2019) Comparison between soft computing methods for tomato quality grading using machine vision. *Food Measure*, vol. 13, pp. 1–15. https://doi.org/10.1007/s11694-018-9913-2.
[4] J.J. Lopez, M. Cobos, and E. Aguilera (2011) Computer-based detection and classification of flaws in citrus fruits. *Neural Comput & Applic,* vol. 20, pp. 975–981. https://doi.org/10.1007/s00521-010-0396-2.
[5] E.O. Olaniyi, A. Adekunle, T. Odekuoye, and A. Khashman (2017). Automatic system for grading banana using GLCM texture feature extraction and neural network arbitrations. *J. Food Process Eng,* vol. 40, p. e12575. https://doi.org/10.1111/jfpe.12575.
[6] S. Solak and T. Altinişik (2018) U. A new method for classifying nuts using image processing and k-means++ clustering. *J. Food Process Eng*, vol. 41, p. e12859. https://doi.org/10.1111/jfpe.12859.
[7] K. Kheiralipour and A. Pormah (2017) Introducing new shape features for classification of cucumber fruit based on image processing technique and artificial neural networks. *J. Food Process Eng,* vol. 40, p. e12558. https://doi.org/10.1111/jfpe.12558.
[8] S. Khoje and S. Bodhe (2015) Comparative performance evaluation of fast discrete curvelet transform and colour texture moments as texture features for fruit skin damage detection. *J. Food Sci Technol*, vol. 52, pp. 6914–6926. https://doi.org/10.1007/s13197-015-1794-3.
[9] Y. Kumar, A.K. Dubey, R.R. Arora et al. (2020) Multiclass classification of nutrients deficiency of apple using deep neural network. *Neural Comput & Applic.* https://doi.org/10.1007/s00521-020-05310-x.
[10] S.K. Vidyarthi, R. Tiwari, S.K. Singh, and H.W. Xiao (2020) Prediction of size and mass of pistachio kernels using random Forest machine learning. *J. Food Process Eng,* vol. 43, p. e13473. https://doi.org/10.1111/jfpe.13473.
[11] Alaa Tharwat, Tarek Gaber, Abdelhameed Ibrahim, and Abdelhameed Ibrahim (2017) Linear discriminant analysis: A detailed tutorial. *AI Communications*, vol. 30, no. 2, pp. 169–190.
[12] S. Mika, G. Ratsch, J. Weston, B. Scholkopf, and K. R. Mullers (1999) Fisher discriminant analysis with kernels. In: *Neural Networks for Signal Processing IX: Proceedings of the 1999 IEEE Signal Processing Society Workshop (Cat. No. 98TH8468)*, pp. 41–48. doi: 10.1109/NNSP.1999.788121.
[13] S. Alawadi, M. Fernández-Delgado, D. Mera et al. (2019) Polynomial kernel discriminant analysis for 2D visualization of classification problems. *Neural Comput & Applic,* vol. 31, pp. 3515–3531. https://doi.org/10.1007/s00521-017-3290-3.
[14] Jian Yang, D. Zhang, A.F. Frangi, and Jing-yu Yang (2004) Two-dimensional PCA: A new approach to appearance-based face representation and recognition. *IEEE Transactions on Pattern Analysis and Machine Intelligence*, vol. 26, no. 1, pp. 131–137, Jan. doi: 10.1109/TPAMI.2004.1261097.
[15] D. Wang and S. Wang (2014) Improved 2DLDA algorithm and its application in face recognition. In: *IEEE 13th International Conference on Trust, Security and Privacy in Computing and Communications*, pp. 707–713. doi: 10.1109/TrustCom.2014.92
[16] Daoqiang Zhang and Zhi-Hua Zhou (2005) (2D) 2PCA: Two-directional two-dimensional PCA for efficient face representation. *Neurocomputing*, vol. 69, pp. 224–231.
[17] S. Noushath, G. Hemantha Kumar, and P. Shivakumara (2006) (2D) 2LDA: An efficient approach for face recognition, *Pattern Recognit*, vol. 39, pp. 1396–1400.
[18] J. Yu, H. Liu, and X. Zheng (2020) Two-dimensional joint local and nonlocal discriminant analysis-based 2D image feature extraction for deep learning. *Neural Comput & Applic,* vol. 32, pp. 6009–6024. https://doi.org/10.1007/s00521-019-04085-0.
[19] Wankou Yang, XiaoyongYan, Lei Zhang, and Changyin Sun (2010) Feature extraction based on fuzzy 2DLDA. *Neurocomputing,* vol. 73, no. 10–12, pp. 1556–1561.
[20] M. Wan and W. Zheng (2013) Fuzzy two-dimensional local graph embedding discriminant analysis (F2DLGEDA) with its application to face and palm biometrics. *Neural Comput & Applic*, vol. 23, pp. 201–207. https://doi.org/10.1007/s00521-012-1317-3.
[21] Bindu Garg, Shubham Aggarwal, and Jatin Sokhal (2018) Crop yield forecasting using fuzzy logic and regression. *Computers and Electrical Engineering*, vol. 67, pp. 383–403.

[22] Yogeswararao Gurubelli, Malmathanraj Ramanathan, and Palanisamy Ponnusamy (2019) Fractional fuzzy 2DLDA approach for pomegranate fruit grade classification. *Computers and Electronics in Agriculture*, vol. 162, pp. 95–105.
[23] F. Zhong and J. Zhang (2013) Linear discriminant analysis based on L1-norm maximization. *IEEE Trans. Image Process*, vol. 22, no. 8, pp. 3018–3027, Aug.
[24] H. Wang, X. Lu, Z. Hu, and W. Zheng (2014) Fisher discriminant analysis with L1-norm. *IEEE Trans. Cybern*, vol. 44, no. 6, pp. 828–842, Jun.
[25] Xi Li, Weiming Hu, Hanzi Wang, and Zhongfei Zhang (2010) Linear discriminant analysis using rotational invariant L1 norm. *Neurocomputing,* vol. 73, no. 13–15, pp. 2571–2579.
[26] F. Zhang, J. Yang, J. Qian, and Y. Xu (2015) Nuclear norm-based 2-DPCA for extracting features from images. *IEEE Trans. Neural Netw. Learn. Syst*, vol. 26, no. 10, pp. 2247–2260, Oct.
[27] Y. Lu, C. Yuan, Z. Lai, X. Li, D. Zhang, and W.K. Wong (2018) Horizontal and vertical nuclear norm-based 2DLDA for image representation. *IEEE Transactions on Circuits and Systems for Video Technology*, vol. 29, no. 4, pp. 941–955.
[28] Y. Gurubelli, M. Ramanathan, and P. Ponnusamy (2021) A combination of 2DLDA and LDA conference on recent trends in machine learning, IoT, smart cities and applications. Advances in approach for fruit-grade classification with KSVM. In: V.K. Gunjan and J.M. Zurada (eds.), *Proceedings of International Intelligent Systems and Computing*, vol 1245. Springer, Singapore. https://doi.org/10.1007/978-981-15-7234-0_14.
[29] A. Khoshroo, A. Keyhani, R. Zoroofi, S. Rafiee, Z. Zamani, and M.R. Alsharif (2009). Classification of pomegranate fruit using texture analysis of MR images. *Agricultural Engineering International: The CIGR Journal*, vol. 11.
[30] Y. Gurubelli, R. Malmathanraj, and P. Palanisamy (2020) Texture and colour gradient features for grade analysis of pomegranate and mango fruits using kernel-SVM classifiers. 2020 6th International Conference on Advanced Computing and Communication Systems (ICACCS), pp. 122–126. doi: 10.1109/ICACCS48705.2020.9074221.
[31] Ankur M Vyas, Bijal Talati, and Sapan Naik (2014) Quality inspection and classification of mangoes using color and size features. *International Journal of Computer Applications (0975–8887)*, vol. 98, no. 1, July.
[32] Varsha Bhole and Arun Kumar (2020) Mango quality grading using deep learning technique: Perspectives from agriculture and food industry. Proceedings of the 21st Annual Conference on Information Technology Education, pp. 180–186. https://doi.org/10.1145/3368308.3415370.
[33] Dameshwari Sahu and Ravindra Manohar Potdar (2017) Defect identification and maturity detection of mango fruits using image analysis. *American Journal of Artificial Intelligence,* vol. 1, no. 1, pp. 5–14. doi: 10.11648/j.ajai.20170101.12.
[34] S. Naik and Bankim Patel (2017) Thermal imaging with fuzzy classifier for maturity and size based non-destructive mango (Mangifera Indica L.) grading. International Conference on Emerging Trends & Innovation in ICT (ICEI), pp. 15–20.
[35] Sapan Naik (2019) Non-destructive mango (Mangifera Indica L., CV. Kesar) grading using convolutional neural network and support vector machine (February 23, 2019). Proceedings of International Conference on Sustainable Computing in Science, Technology and Management (SUSCOM), Amity University Rajasthan, Jaipur – India, Feb. 26–28.
[36] S. Gu, L. Zhang, W. Zuo, and X. Feng (2014) Weighted nuclear norm minimization with application to image denoising. IEEE Conference on Computer Vision and Pattern Recognition, Columbus, OH, pp. 2862–2869.

IPv6 Adaptation in Ultrasonic NFC IoT Communications

8

Rolando Herrero
College of Engineering, Northeastern University, Boston, MA, USA

Contents

8.1 INTRODUCTION

It is widely agreed that there are two main requirements in any IoT solution: (1) end-to-end IPv6 support and (2) interaction with multiple assets including databases and other sources. The first requirement, the support of end-to-end IPv6 connectivity, is highly dependent on the technologies under consideration. Specifically, Low Power Wide Area Networks (LPWANs) and Wireless Personal Area Networks (WPANs) constitute the two main categories of low power IoT technologies. Their key differences are related to the trade-off between signal coverage and transmission rate. Essentially, LPWANs redirect

DOI: 10.1201/9781003244714-8

power consumption to maximize the signal coverage while WPANs do the same to maximize the transmission rate. LPWAN mechanisms include LoRa, NB-IoT, SigFox, Weightless and D7AP among many others. Because LPWAN technologies are associated with low transmission rates, and in order to minimize latency, Maximum Transmission Unit (MTU) sizes are quite small. This imposes restrictions that prevent the use of IPv6. On the other hand, WPANs involve short range IoT solutions that exhibit higher transmission rates and natively support, by means of adaptation mechanisms, end-to-end IPv6 connectivity. WPANs are representative of indoor IoT scenarios that range from physical security and home automation to connected health and smart cities. WPAN mechanisms include Bluetooth Low Energy (BLE), IEEE 802.15.4, ITU-T G.9959 and Digital Enhanced Cordless Telecommunications Ultra Low Energy (DECT ULE). Most IoT NFC technologies fall under the umbrella of WPAN solutions. Their main goal is to provide ways for devices to communicate with each other by means of contactless transactions. The signal coverage is particularly short, typically in the order of centimeters. The main IoT NFC implementation relies on the ISO/IEC 18000–3 standard that enables passive Radio Frequency Identification (RFID) by means of electromagnetic induction between two antennas located within their near fields [1].

Almost all these technologies, LPWANs, WPANs and NFC WPANs, rely on physical layers that support the transmission of frames over non-licensed ISM bands. Moreover, these bands are also shared by other communication devices ranging from wireless phones to remote control toys. These bands are becoming increasingly more polluted, reducing the channel Signal-To-Noise Ratio (SNR) and incrementing the media access contention delay. Low channel SNR results in network packet loss that causes, when supported by the link layer, retransmissions. These retransmissions along with the longer media access produce abnormally high application layer latency. And, because in the context of real-time IoT solutions, packets that take too long to arrive are considered lost, the overall effect of overcrowded ISM bands is application layer packet loss.

In this chapter, we introduce an alternative scheme that provides IoT WPAN NFC communication by means of a physical layer that is ultrasonic and therefore it is not affected by the limitations of the ISM bands. Subultrasonic and ultrasound communications provide a wireless scenario that relies on the acoustic modulation of frames over frequencies of 14 KHz and above. On the other hand, mechanisms of acoustic modulation have been around for decades with one of such schemes standardized by the ITU-T V.23 recommendation to support half duplex digital communication over Plain Old Telephone Service (POTS) channels [2]. In this context, we introduce the Ultrasonic V.23 (UV23) physical layer. This physical layer combines multiple ITU-T V.23 modulation/demodulation (modem) schemes that are shifted to different subultrasonic and ultrasonic frequencies to support the aggregated transmission of data. The UV23 link layer includes two mechanisms: (1) Media Access Control (MAC) that follows standard Carrier Sense Multiple Access Collision Avoidance (CSMA/CA) and (2) a novel data link layer that provides a very simple frame format the relies on 16-bit link layer addresses. Because the frame sizes are small, when compared to the minimum MTU size requirements of IPv6, adaptation is needed. One of the most popular IPv6 adaptation mechanisms, 6LoWPAN, enables the compression of IPv6 header fields as well as the lowering of MTU size requirements to make it compatible with well-known IoT link layers like IEEE 802.15.4 and BLE [3]. In this chapter we use 6LoWPAN to adapt the UV23 frames in order to support end-to-end IPv6 connectivity. Besides compressing IPv6 headers to improve the throughput, 6LoWPAN also compresses UDP headers. Because TCP is not compatible with IoT Low Rate, Low Power Networks (LLNs) because of its overhead and its retransmissions that increase application layer latency, UDP is the preferred transport protocol in this context. This leaves CoAP as the session layer protocol of choice in this scenario [4]. The application layer results from considering the asset data that is being sensed or actuate upon. Putting all these layers together, the stack corresponding to this scenario is shown in Figure 8.1.

From a topology perspective, thc scenario addressed in this chapter consists of a central gateway that forwards traffic back-and-forth to and from two devices by means of IoT based NFC supported by the aforementioned stack. Figure 8.2 shows this topology. The gateway includes two interfaces: (1) an ultrasonic interface that connects to the devices and (2) a wireline one that connects through the IP cloud to applications that perform analytics.

The use of subultrasonic and ultrasonic channels to provide NFC communications have evolved throughout the years to support applications ranging from plain point-to-point scenarios to full network connectivity. All these approaches, however, are based on proprietary mechanisms and none of them complies with the IoT paradigm. For the most part these solutions involve legacy mechanisms that accomplish very low transmission rates on dedicated hardware [5] [3–4] [6–7] [8–9]. Some other scenarios rely on smartphones that exploit their audio interfaces to modulate and demodulate ultrasonic waves. In this context, a study of the ultrasonic modulation in smartphones is presented in [10] and a novel Room Area Network (RAN) that relies on ultrasonic communications is introduced in [11]. A similar concept supports embedding data streams modulated as ultrasonic waves transmitted along with regular television audio in [12] [13]. Hybrid scenarios that combine both subultrasonic and ultrasonic communications with radio technologies are presented in [14–15]. An acoustic modem is adapted to support ultrasonic modulation

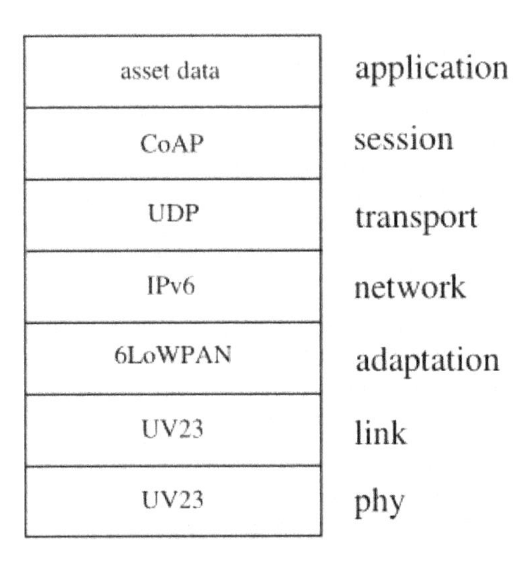

FIGURE 8.1 NFC IoT protocol stack.

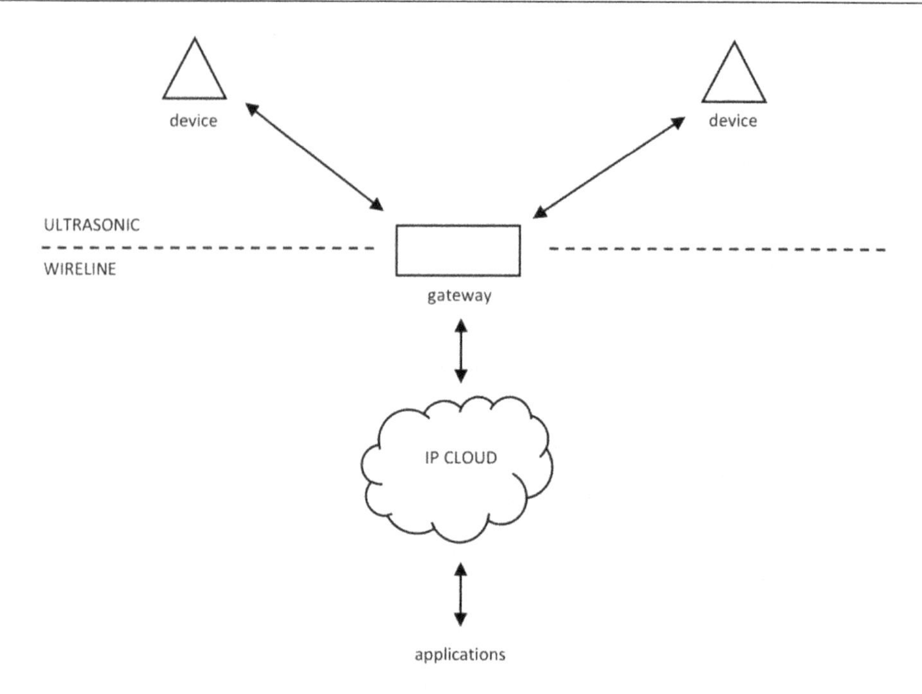

FIGURE 8.2 Topology.

in [16]. Similarly, multiple channels are combined in an Orthogonal Frequency Division Multiplexing (OFDM) scheme in [17]. IoT is addressed in [18], as the authors introduce physical and link layers that support subultrasonic and ultrasonic channels. IoT stacks with CoAP support have been investigated in a multiple number of papers. All of them are based on traditional radio based physical link layers including IEEE 802.15.4, BLE, Ethernet and IEEE 802.11 but ignoring subultrasonic and ultrasonic technologies. Application packet loss and latency are measured in a CoAP based LLN in [19]. The effect of different transport layers on CoAP applications is investigated in [20]. CoAP and other IoT session layer protocols are compared with respect to latency, loss and network bandwidth in [21]. CoAP as a mechanism for media screaming is presented in [22]. CoAP congestion control technologies are analyzed from a perspective of packet loss, latency and goodput in [23]. Similarly, CoAP is used as a service discovery mechanism in [24]. Note that neither of these papers combine an IoT full stack scenario with both subultrasonic and ultrasonic support and application layer protocols like CoAP. Specifically, those papers that address the subultrasonic and ultrasonic communications typically ignore upper layers and overall end-to-end IPv6 support. Moreover, none of them introduce models to predict behavior and better estimate real-time communication parameters. To address these issues, this chapter introduces a full IoT stack that is based on physical and link layers. The stack is modeled from a perspective of performance to predict the relationship between packet loss, latency and physical distance between devices. One of the main advantages of this scheme is that a full IoT stack enables the integration with legacy radio based IoT infrastructure.

The structure of this chapter is as follows: the physical layer of UV23 is described in Section 8.2. Details of the link layer are presented in Section 8.3. In Section 8.4, the IPv6 adaptation layer is presented. Session and application layers are indicated in Section 8.5. The stack is modeled in Section 8.6. An experimental framework to validate the stack is introduced in Section 8.7. Finally, conclusions are provided in Section 8.8.

8.2 PHYSICAL LAYER

ITU-T V.23 is a now obsolete legacy scheme that enables the modulation and demodulation of acoustic signals over POTS to accomplish very low nominal transmission rates of around 1.2 Kbps. The modulation scheme relies on basic Binary Frequency Shift Keying (BFSK) with carriers centered at 1.3 KHz and 1.7 KHz for the transmission of zeros and ones respectively. ITU-T V.23 was not defined to work in a networking environment so there are no specific MAC mechanisms associated with it.

In this chapter the ITU-T V.23 scheme serves as a building block of a more complex physical layer that, by means of OFDM, enables the multiplexing of link layer frames. As shown in Figure 8.3, the scenario involves four ITU-T V.23 modems, each of which is tuned to a pair of subultrasonic and ultrasonic frequencies. These new frequencies are the shifted

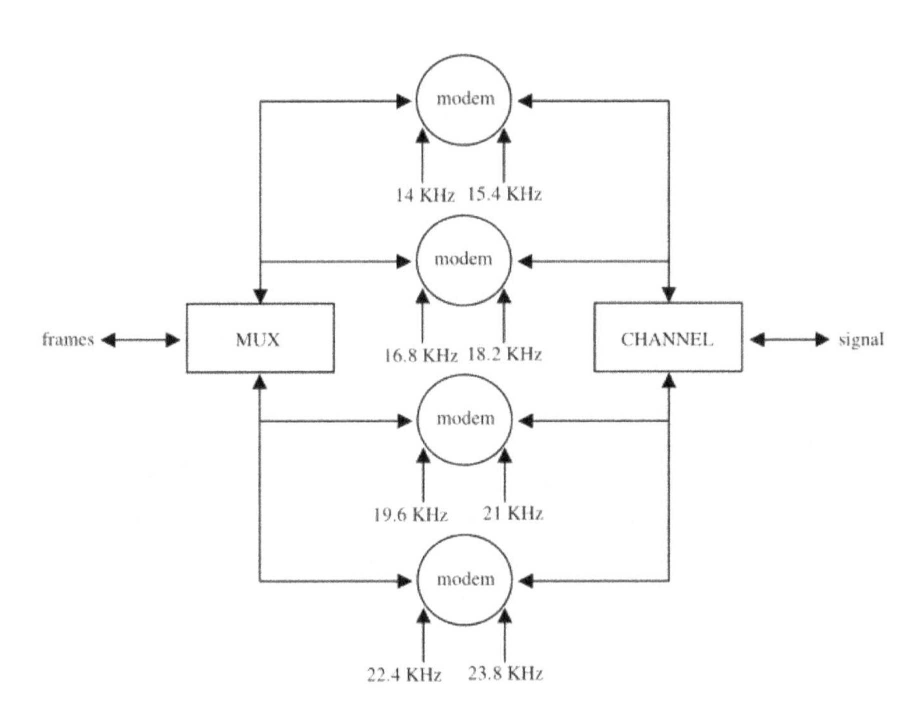

FIGURE 8.3 Physical layer block diagram.

versions of the frequencies of the original ITU-T V.23 modem. The carrier frequency separation is such that guaranties orthogonality for the symbol duration of the scheme of 714.29 microseconds. Note that these frequency changes require modifications of the sampling rate of the modulation mechanism from 8000 to 48000 samples per second.

When aggregating the transmission rate of all four modems, the maximum nominal rate of the physical layer is around 5.6 Kbps. One advantage of recycling a well-known modulation scheme like ITU-T V.23 is that implementations of the modem are widely available and optimized to run on constrained devices like those associated with IoT scenarios [25].

8.3 LINK LAYER

The functional blocks corresponding to the physical layer shown in Figure 8.3, require for each modem to access the channel by means of CSMA/CA [26].

The CSMA/CA algorithm, shown in Figure 8.4, gives transmitters the opportunity of sending a frame after a random back-off time that minimizes the collisions between

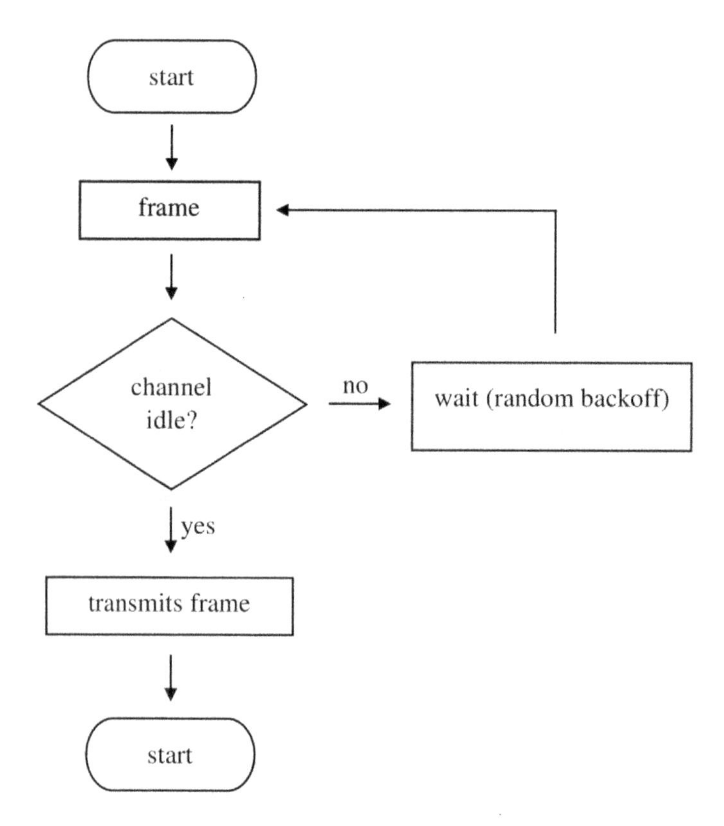

FIGURE 8.4 CSMA/CA algorithm.

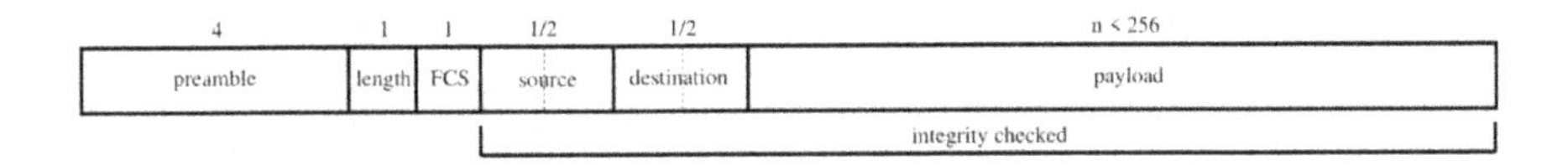

FIGURE 8.5 Link layer frame.

TABLE 8.1 MAC Address Encoding

INITIAL	*FINAL*	*LENGTH (BYTES)*	*ADDRESSES*
00	7F	1	0–127
80 00	FF FF	2	128–32767

contending devices. When a transmitter wants to send frames, it transmits it right away otherwise it waits for a random period of time, after which if the channel is found idle, it starts transmitting. Alternatively, if the channel is found busy, the device waits for another random period of time before attempting to retransmit.

Besides media access, the link layer is responsible for framing. Figure 8.5 shows a link layer frame. The frame starts with a 4-byte preamble, it continues with a 1-byte length and a 1-byte Frame Checksum (FCS). The preamble is a field that contains the 0xAA55AA55 sequence. The length field indicates the size of the payload and cannot be larger than 256. The FCS is calculated over the payload and the MAC addresses by means of an eXclusive OR (XOR) 1-byte operation. The FCS field is generated on transmission and verified on reception.

The MAC addresses can be one or two bytes long and support values between 0 and 32767. As shown in Table 8.1, 1-byte MAC addresses encode values between 0 and 127, while 2-byte MAC addresses encode values between 128 and 32767. MAC addresses with zero value are used for broadcast transmissions. At nominal transmission rates, an empty frame takes 11.43 milliseconds to be transmitted. On the other hand, a maximum size of 2120-bit frame, takes 378.6 milliseconds to be transmitted.

8.4 6LOWPAN

As IoT is intended to connect billions of devices, end-to-end IPv6 support is clearly one of the main requirements. In the end, the use of IPv6 provides a very large address space that guarantees support for this connectivity. Long addresses associated with a large address space imply long headers that are not typically compatible with the low transmission rates of most IoT scenarios. In this context, IPv6 is responsible of frame size requirements that are characterized by a minimum MTU size of around 1280 bytes. This prevents and limits the direct use of IPv6 over the UV23 physical and link layers.

6LoWPAN, that was initially designed for IEEE 802.15.4 physical and link layers, has been extended with minor changes to other technologies like BLE and ITU-T G.9903 in order to overcome the limitations of direct IPv6 support. To summarize, 6LoWPAN introduces IPv6 and UDP header compression, efficient fragmentation, and multicast support. With 6LoWPAN, IPv6 addresses are derived from the link layer addresses by means of the Stateless Address Autoconfiguration (SAA) mechanism [27]. This enables the most efficient way to compress addresses as they are not transmitted. The IEEE introduces the use of 64-bit Extended Unique Identifier (EUI-64) addresses to identify link layers, that in turn, can be used to support SAA [28]. This provides a uniform mechanism to convert link layer MAC addresses into 64-bit data structures that can be combined with network prefixes obtained from Router Advertisements (RAs) in order to generate IPv6 addresses as shown in Figure 8.6. Note that there are two 32-bit blocks that identify the network prefix and other two 32-bit blocks that identify the EUI-64.

In the context of the UV23 link layer, the conversion from 1-byte and 2-byte MAC addresses to EUI-64 fields is shown in Figure 8.7. Each EUI-64 is all zeros with the last bytes used to carry the UV23 address.

For the UV23 layers, we rely on traditional stateful 6LoWPAN IP Header Compression (IPHC) and Next Header Compression (NHC) for UDP transport layer compression. In Figure 8.8, a UV23 frame is shown being carried by a 6LoWPAN datagram. A 6LoWPAN packet starts with a 011 dispatch value that identifies IPHC. 6LoWPAN assumes that the link layer, like that of UV23, does not include a payload type identifier. The dispatch value, at the beginning of each 6LoWPAN datagram, identifies the type of 6LoWPAN datagram, in this case IPHC. After the dispatch value, the 6LoWPAN packet includes the encoding of the IPv6 traffic flow (TF) field. A value of TF = 11 indicates that neither the traffic class nor the flow label is encoded. A single 1-bit field N, that indicates that the NHC header is present, follows. The 2-bit Hop Limit (HLM) field of value 01 signals that a packet can be forwarded up to 64 times. Because IPHC supports stateful compression, the 1-bit field C of value 0 indicates that the datagram compression is stateless. Stateless compression implies that a single 6LoWPAN datagram contains all the information it needs to be decoded. The following three bits specify how the source address is encoded. This corresponds to the S and Source Address Mode (SAM) fields that together with value 011 indicate that the

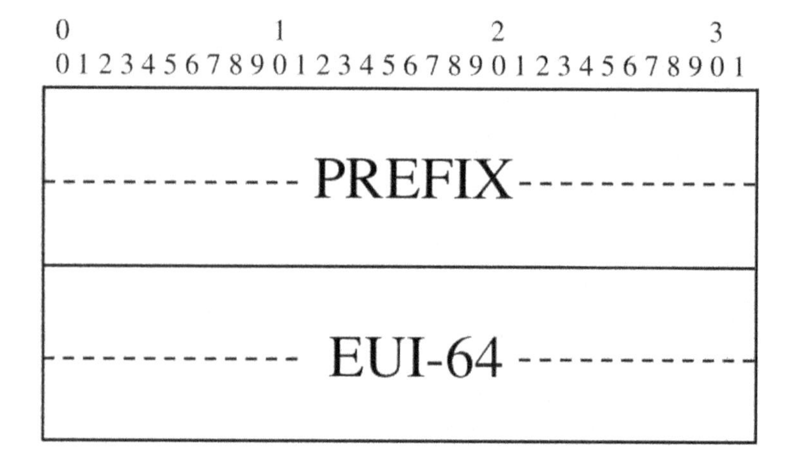

FIGURE 8.6 IPv6 address generation.

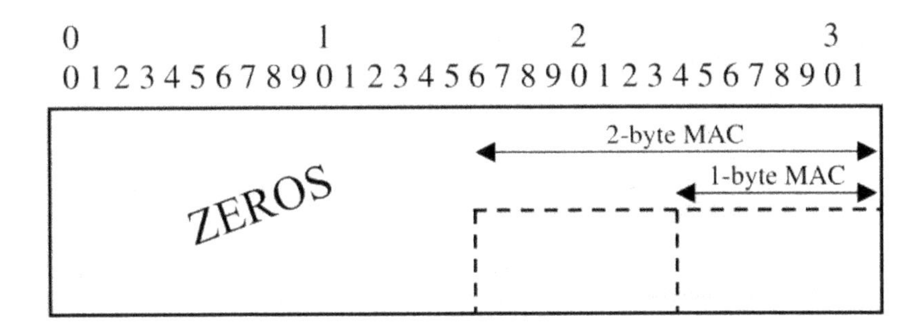

FIGURE 8.7 EUI-64 UV23 encoding.

0	1	2	3	5	6		8	9	10	12	13	14	16	21	22 23	24	28	32	48
0	1	1	11	1	0	1	0	0	11	0	0	11	11110	0	11	0000	0000	1110110000111100	UV23

dispatch | IPHC | NHC | in-line

TF	11	neither traffic class not flow label encoding
N = 1	1	NHC follows
HLM	01	hop limit is 64
C	0	no context information
S	0	no context based source address encoding
SAM	11	link layer source address
M	0	unicast destination address
D	0	no context based destination address encoding
DAM	11	link layer destination address
UDP	11110	UDP transport
K	0	transmit checksum in-line
P	11	4-bit source and destination port encoding
source	0000	UDP source port (61440)
destination	0000	UDP destination port (61440)
checksum	11 .. 00	checksum

FIGURE 8.8 UV23 carried by 6LoWPAN.

source IPv6 address of the datagram is derived from the link layer source address. The following 1-bit field M with value 0 specifies that the destination address is not multicast. The 3-bit D and Destination Address Mode (DAM) fields specify how the destination address is encoded. A value of 011, as with the source address, specifies that the destination IPv6 address of the datagram is derived from the link layer destination address. The compressed UDP header, that is signed by the 11110 sequence, follows. The following 1-bit K field of value 0, indicates that the UDP checksum is sent in-line after the UDP header. Two 4-bit fields encode the UDP source and destination ports right after. Note that this limits the range of ports to 16 possible values that are typically statically assigned to support the session and application mechanism (i.e. CoAP). The 16-bit checksum field is followed by the UV23 frame described in Figure 8.5.

IPHC and NHC headers do not contain all the IPv6 and UDP header fields. They include the minimum number of fields that are required for successful connectivity and, in this particular scenario, they are six bytes long. A subset of the available UDP ports are dynamically allocated by out-of-band means to support 16 addresses. Essentially, UDP ports are encoded as 4-bit fields. All things considered, and because IPv6 and UDP headers are 40 and 8 bytes long respectively, 6LoWPAN is responsible for a compression rate of one to eight.

8.5 SESSION AND APPLICATION LAYERS

Interaction with IoT devices is done by means of Representational State Transfer (REST) message exchanges. Specifically, REST requests and responses support connectivity between clients and servers. There are many session layer protocols that support REST Application Program Interfaces (APIs) including the well-known HyperText Transfer Protocol (HTTP) and the aforementioned CoAP. From all of them, it is this latter protocol the one that best works with 6LoWPAN. CoAP is an extra lightweight session layer protocol that was designed for the transmission of sensor and actuation traffic over LLNs. CoAP was standardized by the Internet Engineering Task Force (IETF) as RFC 7252 in 2014 [4]. CoAP is expected to become in the coming years the default mechanism for transmission of device data in the context of access IoT network. CoAP applications range from smart grid and building automation to asset tracking and Industry 4.0. CoAP also includes built-in mechanisms for resource discovery and management.

Figure 8.9 shows the two-sublayer structure that CoAP relies on. The lower sublayer is called the message sublayer that provides an interface to the UDP transport layer. This happens because CoAP is specifically designed to deal with constrained devices that cannot support standard TCP transport. The upper sublayer is the request/response sublayer that takes care of building and parsing the application messages.

The message sublayer, and in order to overcome some of the shortcomings of UDP, additionally introduces two modes of operation: (1) confirmable where requests and responses must be acknowledged by the far end to signal a good reception and (2) non-confirmable, that is a fire-and-forget mechanism, where requests and responses are

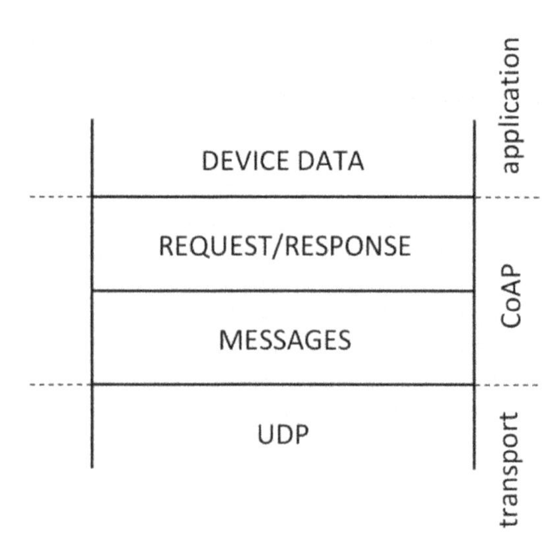

FIGURE 8.9 CoAP two-layer structure.

transmitted but never acknowledged by the far end. Confirmable CoAP brings some of the retransmission capabilities of transport protocols like TCP to the session layer.

Figure 8.10 shows the interaction between an application and a sensor. The application requests temperature readouts by transmitting CoAP GET messages with a /temperature Uniform Resource Locator (URL). These messages are forwarded to the sensor by a gateway located between the application and the sensor itself. The gateway serves as a proxy that forwards CoAP application-to-sensor session messages but regenerates all fields in the 6LoWPAN and UV23 layers. A more complex scenario involves two sessions, one between the application and the gateway and another one between the gateway and the sensor.

Once the CoAP GET request arrives at destination, the sensor transmits a single message that carries the temperature readout as part of a CoAP 2.05 Content response piggybacked with the acknowledgment. Built-in as part of confirmable CoAP, whenever a message is lost retransmissions are generated at exponentially increasing intervals as indicated in RFC 7252 'The Constrained Application Protocol (CoAP)' [4].

Figure 8.11 shows the encoding of CoAP requests (above) and responses (below). The messages contain a 2-byte Message Identifier (MID) with requests including the aforementioned URLs (/temperature) and responses carrying the readouts as plain-text. Note

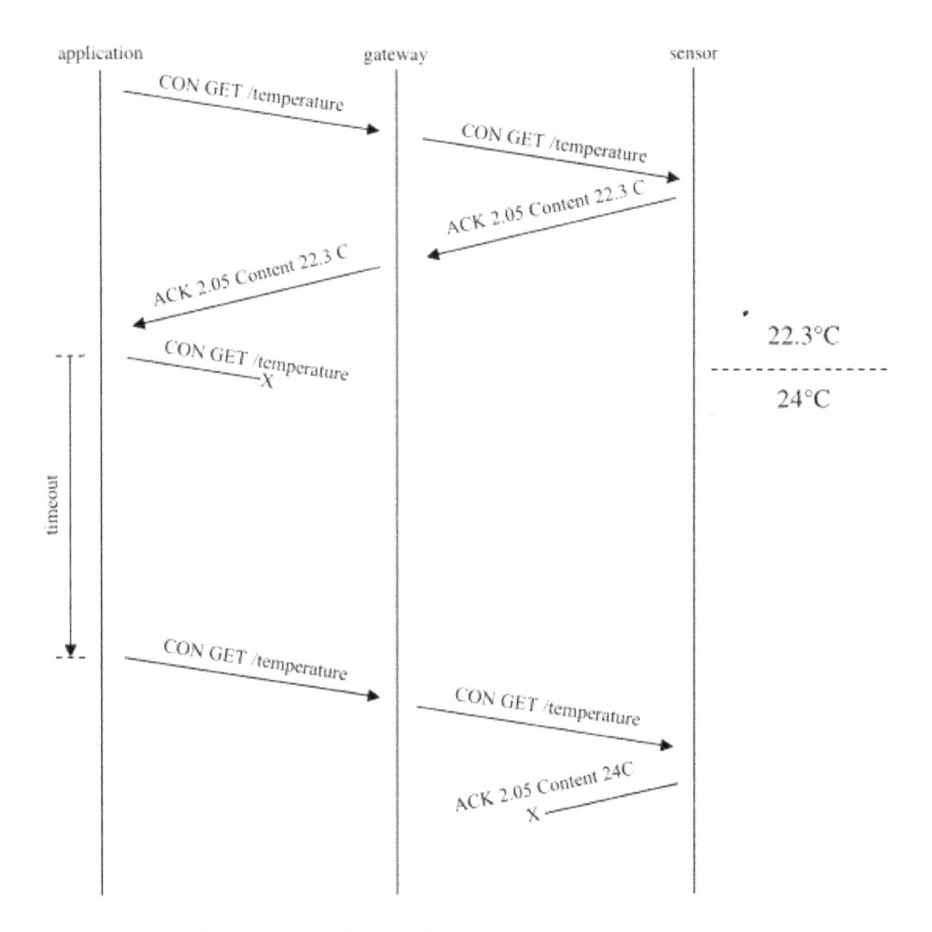

FIGURE 8.10 Application-sensor interaction.

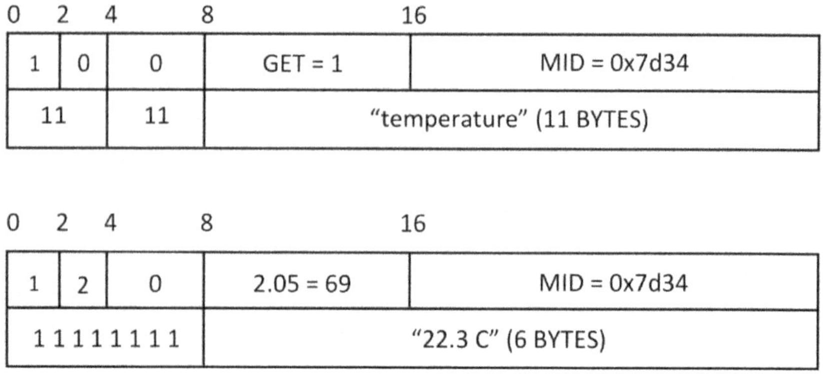

0	2	4	8	16
1	0	0	GET = 1	MID = 0x7d34
11		11	"temperature" (11 BYTES)	

0	2	4	8	16
1	2	0	2.05 = 69	MID = 0x7d34
1 1 1 1 1 1 1 1			"22.3 C" (6 BYTES)	

FIGURE 8.11 CoAP encoding.

that the CoAP requests and responses have nominal lengths of 17 and 12 bytes respectively. This leads to full stacks that are 31 and 26 bytes long respectively for requests and responses. At a maximum transmission rate, the delay it takes to transmit a request and response if 44.28 and 37.14 milliseconds respectively.

8.6 ANALYTICAL MODEL

An analytical model of the flow in Figure 8.10 can be used to determine the relationship between latency, packet loss and throughput. For each session, the total amount of traffic sent between the application and the sensor is given by:

$$\bar{L} = 2L_{req} + 2L_{resp}$$

where L_{req} and L_{resp} are the request and response message sizes respectively. Additionally, the network packet loss, P_L, results from:

$$P_L = 1 - P_S = \bar{L}P_e$$

where P_e is the bit error rate probability. On the other hand, the latency Δ is computed from considering up to $M = 3$ retransmissions with exponentially increasing timeouts. This latency, that is associated with a single session is given by:

Equation 8.1: Latency

$$\begin{aligned}\Delta &= \frac{\bar{L}}{R}\sum_{n=0}^{M} 2^n P_S P_L^n \\ &= \frac{\bar{L}}{R} P_S \sum_{n=0}^{M} (2P_S)^n \\ &= \frac{\bar{L}}{R} P_S \frac{(2P_L)^{M+1} - 1}{2P_L - 1}\end{aligned}$$

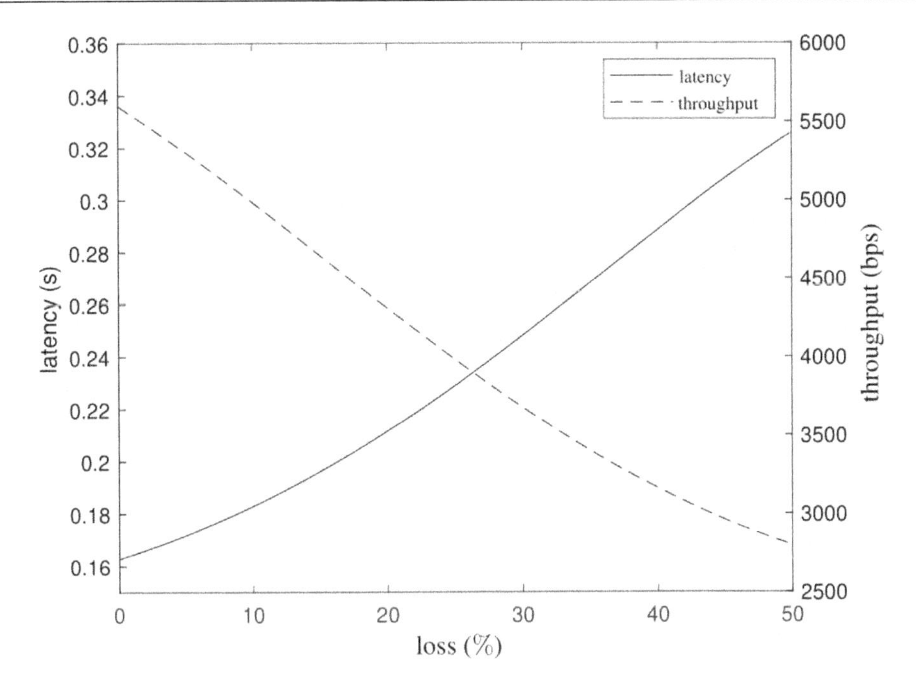

FIGURE 8.12 Theoretical: Latency and throughput versus network packet loss.

where R is the transmission rate of the system and the packet loss P_L must be $P_L < 0.5$ for the series above to converge. The throughput, T, is therefore given by

Equation 8.2: Throughput

$$T = \frac{\bar{L}}{\Delta}$$

$$= \frac{R(2P_L - 1)}{P_S\left[(2P_L)^{M+1} - 1\right]}$$

Note that, as expected, when $P_L \approx 0$, then $T = R$ and similarly, when $P_L \approx 0.5$, then $T = 2\frac{R}{M+1}$. For M =3, Figure 8.12 shows the latency and throughput as a function of the network packet loss based on the results obtained from Equation 8.1 and Equation 8.2.

Similarly, Figure 8.13 shows the throughput as a function of latency obtained from the results of Figure 8.12.

8.7 EXPERIMENTAL FRAMEWORK

The performance of the NFC IoT communication stack presented in this Chapter can be evaluated by carrying out the interaction between the devices and the gateway shown in

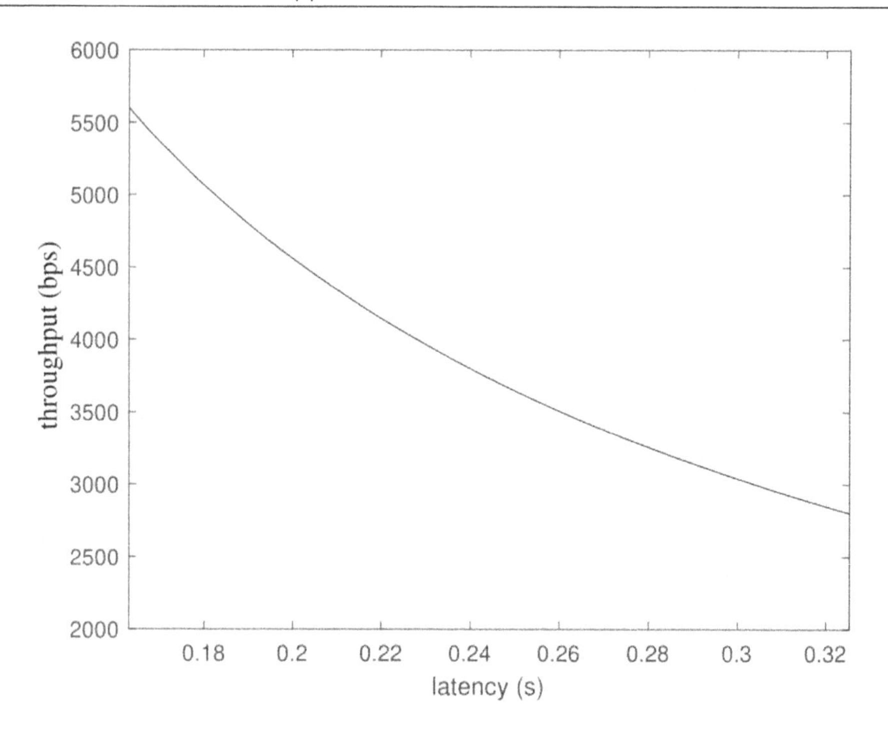

FIGURE 8.13 Throughput versus latency.

the message flow in Figure 8.10. In this scenario, the application sends confirmable CoAP requests that when they arrived at the gateway, they are forwarded to the sensors. Similarly, responses generated by the sensor that arrive at the gateway, are forwarded to the application.

This scenario is implemented by means of the well-known Netualizer framework [27]. Netualizer provides IoT protocol virtualization as well as customization and can be used for deployment in physical and virtual devices. All upper layers of the stack presented in this chapter, namely 6LoWPAN and CoAP, are mainstream so no special action is needed.

The UV23 physical and link layers, however, are implemented by means of a specially built plugins for Netualizer [29]. One plugin implements modem capabilities that are based on the linmodem open-source implementation of the ITU-T V.23 modulator [26]. Another plugin implements the basic CSMA/CA state machine shown in Figure 8.4 [30]. The stack is installed and deployed on two Android phones that play the roles of application and sensor. Additionally, a laptop plays the role of gateway. The interaction between these components is by means of the standard audio interfaces (that natively support ultrasound playback and recording) of these devices. A photograph of the scenario is shown in Figure 8.14.

The experimental framework relies on a Netualizer script that controls not only the application but also the emulated temperature sensor and the gateway itself. The script is responsible of driving, by means of Wi-Fi, agents in the different components. The application agent generates CoAP requests while the sensor agent generates CoAP responses. The gateway automatically forwards these messages back-and-forth between devices. This setup enables the measurement of the performance parameters.

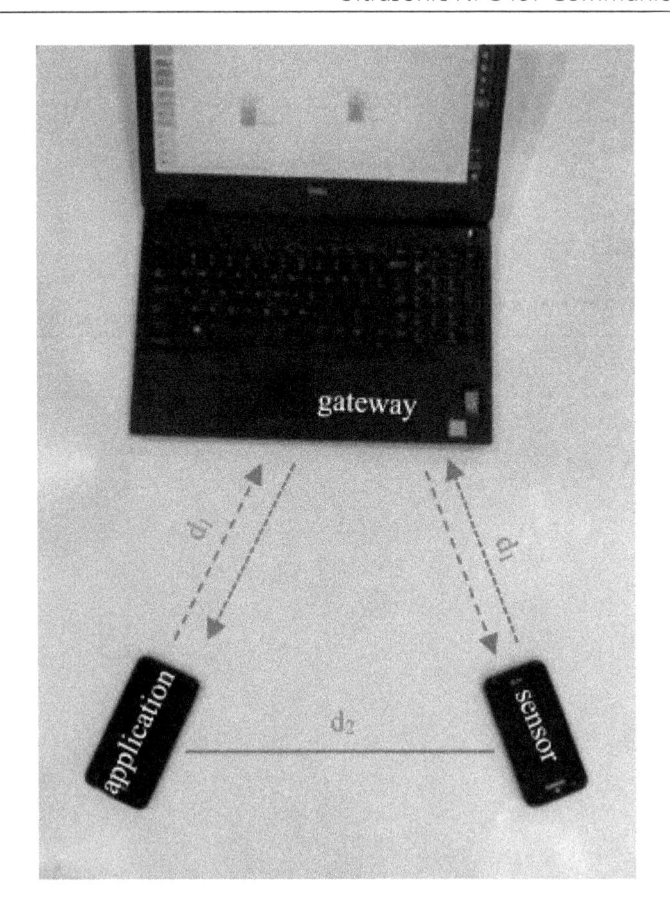

FIGURE 8.14 Experimental setup.

The experimental framework supports the transmission and reception of 300 temperature readouts. The separation between the application and the sensor is set fixed at d_2 = 30 centimeters. The control variable is the distance d_1 between the gateway and the devices. In this scenario several parameters are measured including the transmission rate, throughput and application layer packet loss and latency. Note that the transmission rate and throughput values are averaged over all 300 sessions.

Table 8.2 shows the different performance parameters as a function of the distance d_1 between the gateway and the devices.

Note that the transmission rates are considerably lower than the nominal transmission rates partially due to channel access contention. As expected, the throughput is lower than the transmission rate because of the packet loss. This network packet loss leads to retransmissions, latency and, potentially, to application layer packet loss. Formally, the application packet loss is calculated as the percentage of transmitted packets that are never received at the application layer. On the other hand, the latency is measured by how long it takes for all 300 readouts to be received by the application. It is clear that the overall performance degrades quite fast as the distance between the devices and the gateway

TABLE 8.2 Performance versus Distance d_1

D_1 (CM)	RATE (BPS)	THROUGHPUT (BPS)	MESSAGE LOSS (%)	LATENCY (S)
5	4974.41	4860.99	2.33	56.16
10	4977.96	4752.16	4.91	57.56
15	4921.43	4499.93	10.69	60.80
20	4942.31	3760.32	32.63	72.54
25	4998.24	3266.82	53.54	101.32
30	4819.01	2612.68	67.12	144.20

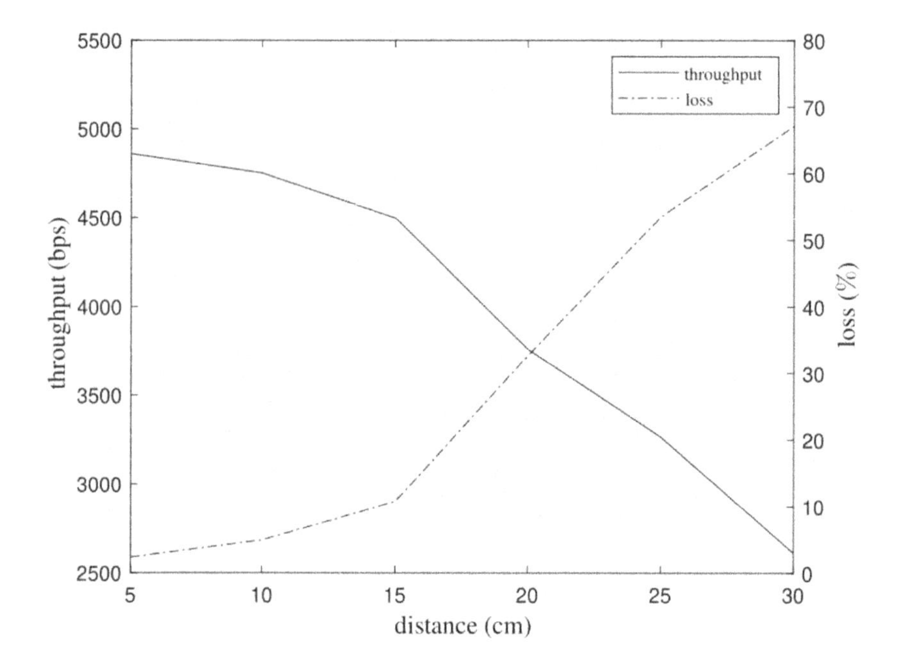

FIGURE 8.15 Experimental throughput versus packet loss versus distance

increases. When such distance is longer than 30 centimeters the degradation is quite dramatic and renders the scheme unreliable for practical scenarios. This is clearly consistent with NFC scenarios.

Based on the results indicated in Table 8.2, Figure 8.15 shows throughput and packet loss as a function of the distance d_1 between the gateway and the devices. Similarly, Figure 8.16 shows latency and packet loss as a function of the distance. Note that the throughput is measured in bps units, loss as a percentage, distance in centimeters and latency in seconds. Both Figure 8.15 and Figure 8.16 can be combined into a single figure that eliminates the distance. This leads to Figure 8.17 where throughput and latency are shown as a function of the packet loss. Compare these experimental curves against their theoretical counterparts in Figure 8.12.

Figures 8.18 and Figure 8.19 show the theoretical and the experimental latency and throughput respectively as a function of the network packet loss. Note that the theoretical

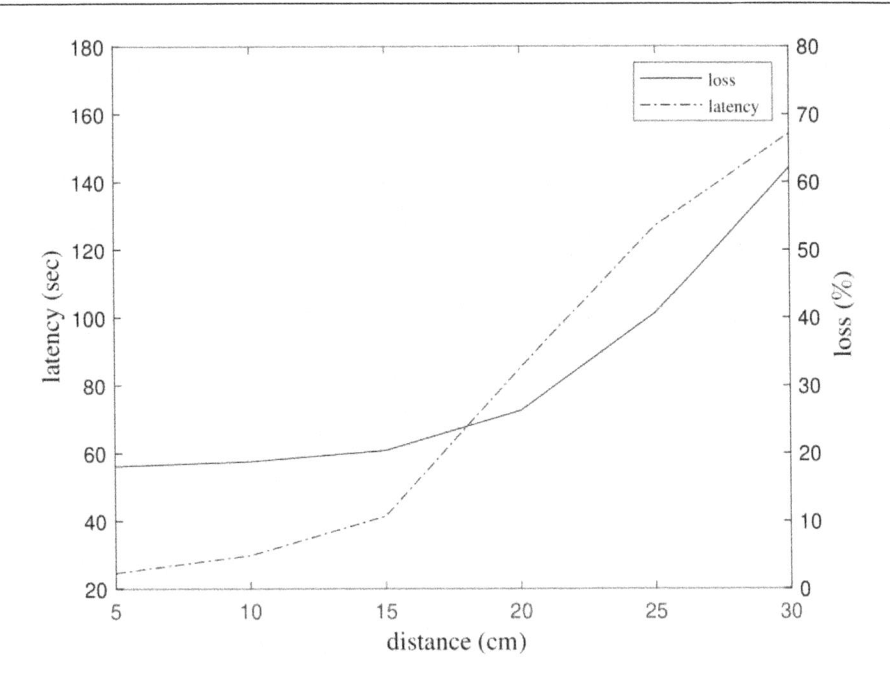

FIGURE 8.16 Experimental latency versus packet loss versus distance.

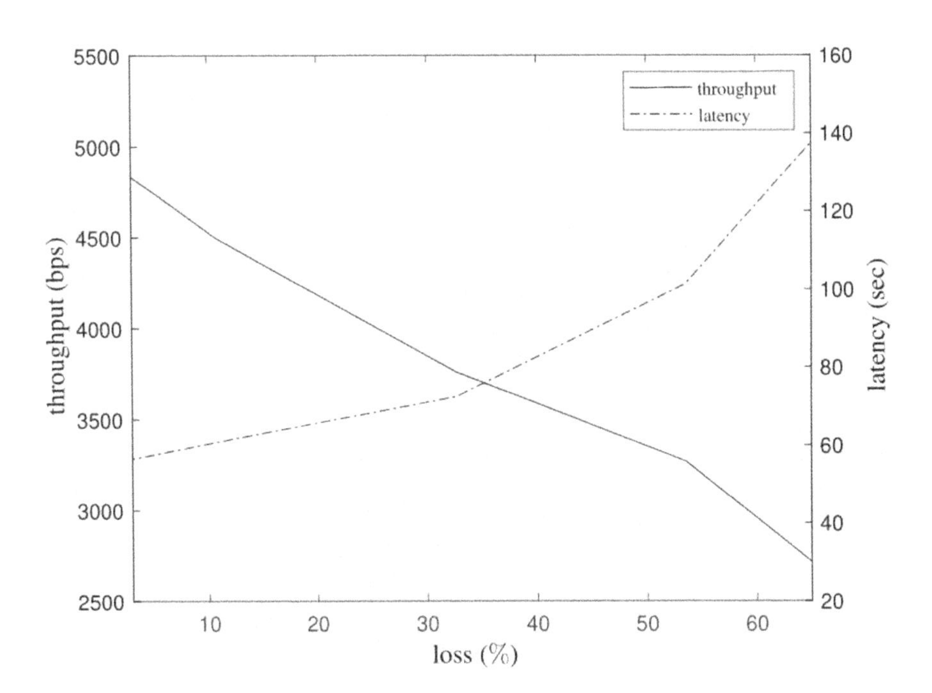

FIGURE 8.17 Experimental: Latency and throughput versus network packet loss.

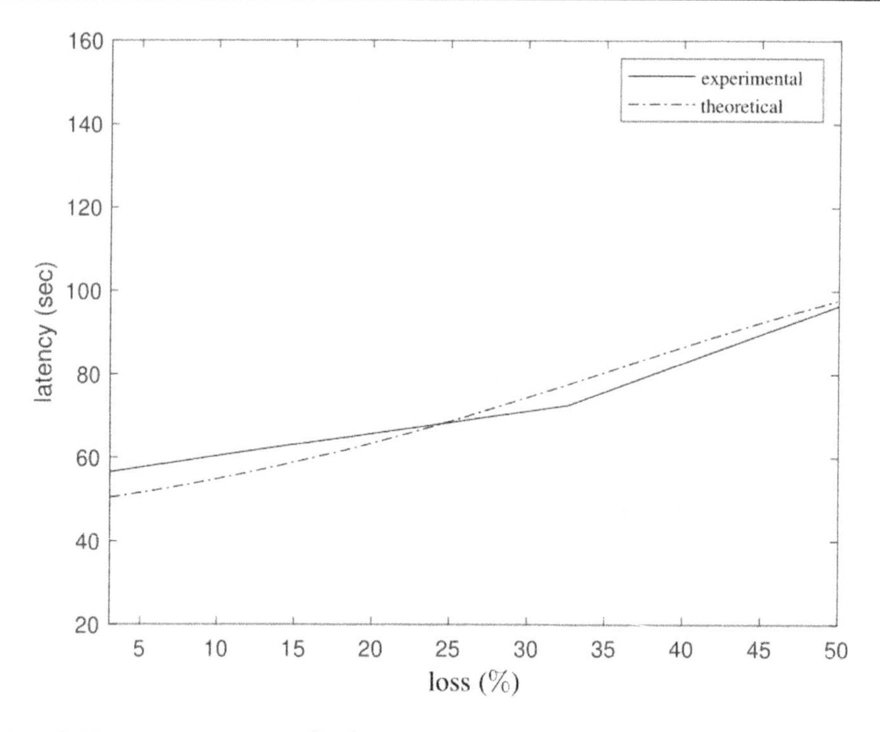

FIGURE 8.18 Latency versus packet loss.

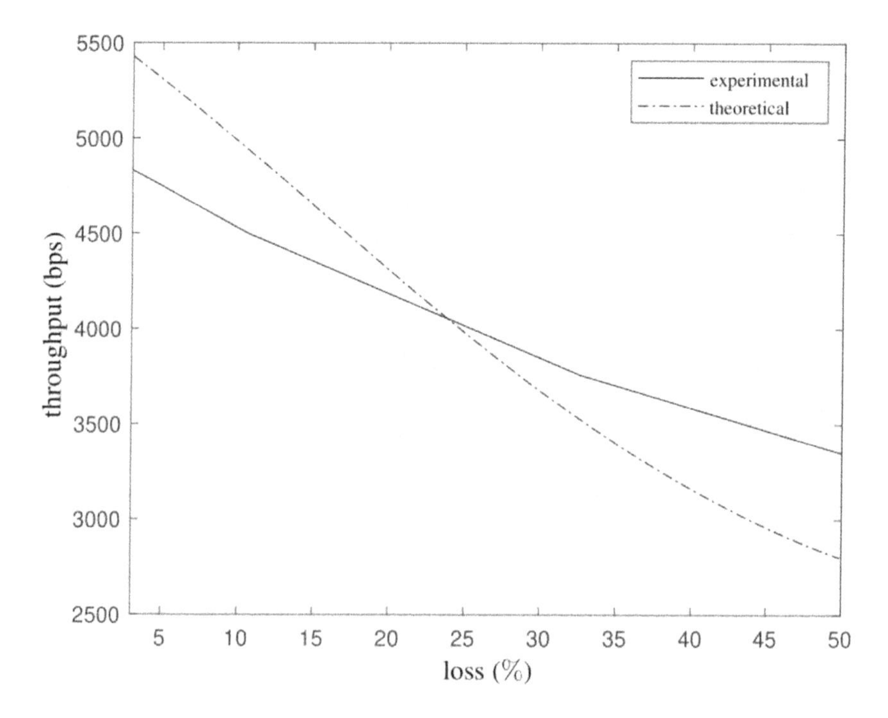

FIGURE 8.19 Throughput versus packet loss.

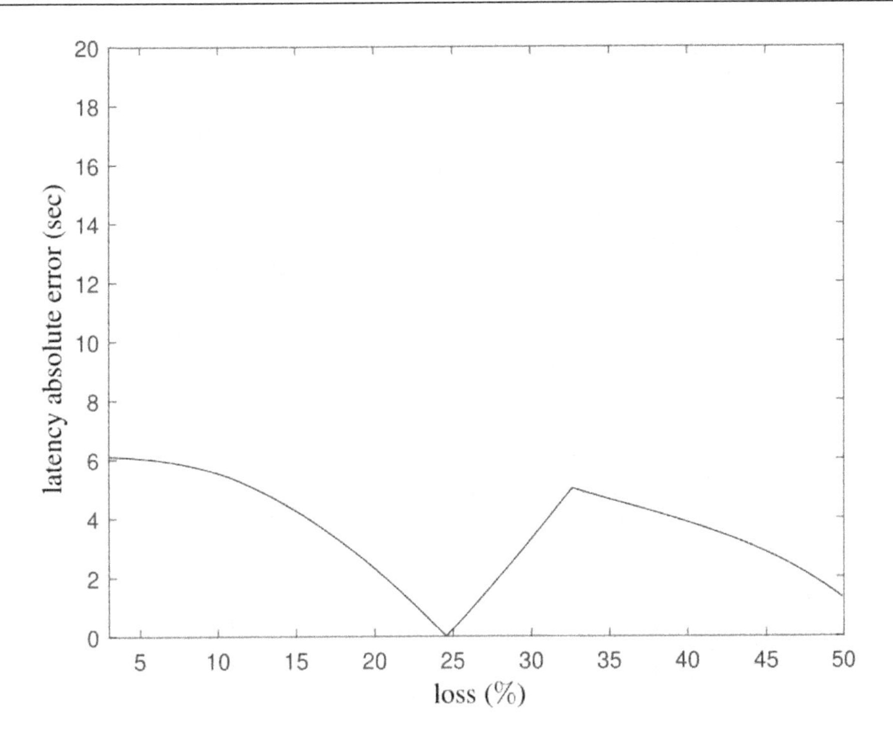

FIGURE 8.20 Latency absolute error.

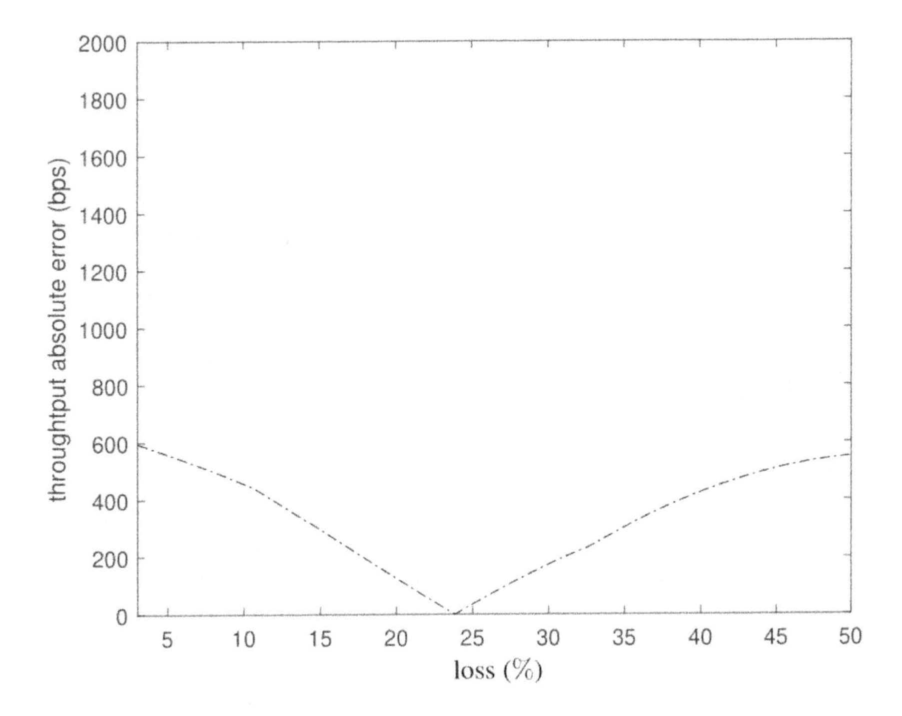

FIGURE 8.21 Throughput absolute error.

latency and throughput values are calculated based on results from Equation 8.1 and Equation 8.2 respectively. Because the mathematical model presented in Section 8.6 does not take into account the effects of CSMA/CA contention, the experimental and theoretical results differ.

To understand the accuracy of the model, Figure 8.20 and Figure 8.21 show the latency and throughput absolute errors as a function of the network packet loss. When taking averages, the latency model relative error is 5.08% while the throughput model relative model is 7.82%. It can be seen than for $P_L \approx 0.25$, the model estimation is very accurate. The model becomes less accurate for extreme values of the loss when $P_L \to 0$ and $P_L \to 0.5$. This upper limit has to do with the fact that, as indicated in Section 8.6, the model converges only when $P_L < 0.5$.

8.8 CONCLUSIONS

In this chapter, we introduced a full NFC IoT protocol stack can be used in scenarios where the use of traditional radio-based communications is neither possible nor reliable. UV23, a novel physical and link layer scheme, that combines multiple ITU-T V.23 modems tuned to transmit over ultrasonic bands was detailed. In order to comply with IoT requirements, 6LoWPAN was selected as the preferred adaptation mechanism in order to enable end-to-end IPv6 connectivity. A CoAP session layer was introduced to support REST interaction between the devices. All these components were put together in the context of an experimental framework that includes an application, a sensor, and a gateway. This setup is used, in turn, to obtain performance parameters. Clearly the distance d_1 between the devices and the gateway is critical to determine the level of degradation of the scheme. A coverage hard limit of 30 centimeters makes the solution an excellent candidate for NFC in the context of IoT. On the other hand, this chapter introduced a mathematical model that can be used to estimate the performance of the stack as a function of the network packet loss. The network packet loss can be measured and the throughput as well as the latency can be calculated and compared against specific goals. This can be used to trigger real-time protocol stack parameter adjustments like configuring Forward Error Correction (FEC) levels [29].

To summarize, this chapter presented a mechanism to support NFC IoT communication that (1) introduces a full stack that is based on an ultrasonic physical layer that repurposes the well-known ITU-T V.23 acoustic modulation standard, (2) relies on a new link layer that introducing framing to support the transmission of ultrasonic data, (3) uses 6LoWPAN to provide adaptation in order to efficiently support end-to-end IPv6 connectivity, (4) presents a mathematical model that can be used to estimate latency and throughput as a function of measured network loss and, thus, supports dynamic stack configuration and (5) supports an experimental framework that is used to compare the performance of the scheme against the mathematical model.

8.9 APPENDIX

TABLE 8.3 Acronyms

ACRONYM	*MEANING*
6LoWPAN	IPv6 over Low Power Wireless Personal Area Networks
API	Application Program Interface
BLE	Bluetooth Low Energy
BFSK	Binary Frequency Shift Keying
CoAP	Constrained Application Protocol
CSMA/CA	Carrier Sense Media Access/Collision Avoidance
DAM	Destination Address Mode
DECT ULE	Digital Enhanced Cordless Telecommunications Ultra Low Energy
EUI	Extended Unique Identifier
FCS	Frame Checksum
FEC	Forward Error Correction
HLM	Hop Limit
HTTP	Hypertext Transfer Protocol
IEEE	Institute of Electrical and Electronics Engineers
IETF	Internet Engineering Task Force
IoT	Internet of Things
IPHC	IP Header Compression
ISM	Industrial Scientific and Medical
ITU-T V.23	International Telecommunications Union Telecommunication Sector V.23
LLN	Low Power Low Rate Network
LoRa	Long Range
LPWAN	Low Power Wide Area Network
MAC	Media Access Control
MID	Message Identifier
modem	Modulator/Demodulator
MTU	Maximum Transmission Unit
NB-IoT	Narrowband IoT
NFC	Near Field Communication
NHC	Next Header Compression
OFDM	Orthogonal Frequency Division Multiplexing
POTS	Plain Old Telephone Service
RA	Router Advertisement
RAN	Room Area Network
REST	Representational State Transfer

(Continued)

TABLE 8.3 (Continued)

ACRONYM	*MEANING*
RFID	Radio Frequency Identification
SAA	Stateless Address Autoconfiguration
SAM	Source Address Mode
SNR	Signal-to-Noise Ratio
TCP	Transport Control Protocol
TF	Traffic Flow
UDP	User Datagram Protocol
URL	Uniform Resource Locator
UV23	Ultrasonic V.23
Netualizer	Virtual Protocol Stack+
WPAN	Wireless Personal Area Network
XOR	eXclusive OR

BIBLIOGRAPHY

[1] R. Herrero, *Fundamentals of IoT Communication Technologies*, New York, NY, Springer-Nature, 2021.
[2] ITU-T, *Recommendation V.23: 600/1200-Baud Modem Standardized for Use in the General Switched Telephone Network*, International Telecommunication Union, Geneva, Switzerland, 1988.
[3] C. Bormann, Z. Shelby, S. Chakrabarti and E. Nordmark, *RFC 6775: Neighbor Discovery Optimization for IPv6 over Low-Power Wireless Personal Area Networks (6LoWPANs)*, Freemont, CA, IETF, 2012.
[4] C. Bormann, Z. Shelby and K. Hartke, *RFC 7252: The Constrained Application Protocol (CoAP)*, Freemont, CA, IETF, 2015.
[5] A. Madhavapeddy, D. Scott and R. Sharp, "Context-Aware Computing with Sound," in *UbiComp 2003: Ubiquitous Computing*, Berlin, Heidelberg, Springer, 2003, pp. 315–332.
[6] M. Uddin and T. Nadeem, "A2PSM: Audio Assisted Wi-Fi Power Saving Mechanism for Smart Devices," in *Proceedings of the 14th Workshop on Mobile Computing Systems and Applications*, Association for Computing Machinery, New York, 2013.
[7] W. Jiang and W. Wright, "Ultrasonic Wireless Communication in Air Using OFDM-OOK Modulation," in *Proceedings of the 2014 IEEE International Ultrasonics Symposium*, IEEE International Ultrasonics Symposium, Chicago, 2014.
[8] V. Gerasimov and W. Bender, "Things That Talk: Using Sound for Device-to-Device and Device-to-Human Communication," *IBM Systems Journal*, vol. 39, no. 3, pp. 530–546, 2000.
[9] A. Madhavapeddy, R. Sharp, D. Scott and A. Tse, "Audio Networking: The Forgotten Wireless Technology," *IEEE Pervasive Computing*, vol. 4, no. 3, pp. 55–60, 2005.
[10] Q. Wang, K. Ren, M. Zhou, D. Koutsonikolas and L. Su, "Messages behind the Sound: Real-Time Hidden Acoustic Signal Capture with Smartphones," in *Proceedings of the 22nd Annual International Conference on Mobile Computing and Networking*, MobiCom, New York, 2016.

[11] P. Iannucci, R. Netravali, A. Goyal and H. Balakrishnan, "Room-Area Networks," in *Proceedings of the 14th ACM Workshop on Hot Topics in Networks*, Association for Computing Machinery, New York, NY, 2015.
[12] A. S. Nittala, X.-D. Yang, S. Bateman, E. Sharlin and S. Greenberg, "PhoneEar: Interactions for Mobile Devices That Hear High-Frequency Sound-Encoded Data," in *Proceedings of the 7th ACM SIGCHI Symposium on Engineering Interactive Computing Systems*, EICS, New York, 2015.
[13] H. Lee, H. T. Kim, W. J. Choi and S. Choi, "Chirp Signal-Based Aerial Acoustic Communication for Smart Devices," IEEE International Conference on Computer Communications, 2015.
[14] L. Zhang, X. Zhu and X. Wu, "No More Free Ridcrs: Sharing WiFi Secrets with Acoustic Signals," International Conference on Computer Communication and Networks, 2019.
[15] B. Zhang, Q. Zhan, S. Chen, M. Li, K. Ren, C. Wang and D. Ma, "PriWhisper: Enabling Keyless Secure Acoustic Communication for Smartphones," *IEEE Internet of Things Journal*, vol. 1, pp. 33–45, 2014.
[16] M. Hanspach and M. Goetz, "On Covert Acoustical Mesh Networks in Air," *CoRR*, vol. 1406, p. 1213, 2014.
[17] H. Matsuoka, Y. Nakashima and T. Yoshimura, *Acoustic Communication System Using Mobile Terminal Microphones*, NTT Docomo, Tokyo, Japan, 2007.
[18] E. Novak, Z. Tang and Q. Li, "Ultrasound Proximity Networking on Smart Mobile Devices for IoT Applications," *IEEE Internet of Things Journal*, vol. 6, no. 1, pp. 399–409, 2019.
[19] S. Thombre, R. Islam, K. Andersson and M. Hossain, "Performance analysis of an IP based protocol stack for WSNs," IEEE Conference on Computer Communications Workshops, 2016.
[20] R. Herrero, "Analysis of the Constrained Application Protocol Over Quick UDP Internet Connection Transport," *Internet of Things*, vol. 12, 2020.
[21] Y. Chen and T. Kunz, "Performance Evaluation of IoT Protocols under a Constrained Wireless Access Network," International Conference on Selected Topics in Mobile Wireless Networking, 2016.
[22] R. Herrero, "Analysis of IoT Mechanisms for Media Streaming," *Internet of Things*, vol. 9, 2020.
[23] M. Collina, M. Bartolucci, A. Bartolucci and G. Corazza, "Internet of Things Application Layer Protocol Analysis Over Error and Delay Prone Links," 7th Advanced Satellite Multimedia Systems Conference and the 13th Signal Processing for Space Communications Workshop, 2014.
[24] R. Herrero, "Mobile Shared Resources in the Context of IoT Low Power Lossy Networks," *Internet of Things*, vol. 12, pp. 172–191, 2020.
[25] T. Narten, T. Jinmei and S. Thomson, *RFC 4862: IPv6 Stateless Address Autoconfiguration*, Freemont, CA, IETF, 2007.
[26] J. Arkko, C. Jennings and Z. Shelby, *RFC 9039: Uniform Resource Names for Device Identifiers*, Freemont, CA, IETF, 2021.
[27] Netualizer, Visual Protocol Stack Emulator, www.l7tr.com, 2022.
[28] F. Bellard, Linux Modem, https://bellard.org/linmodem/, 2000.
[29] Z. Dahham, A. Sali, B. Ali and M. Jahan, "An efficient CSMA-CA algorithm for IEEE 802.15.4 Wireless Sensor Networks," International Symposium on Telecommunication Technologies, Kuala Lumpur, MA, 2012.
[30] R. Herrero, "Modeling and Comparative Analysis of Forward Error Correction in the Context of Multipath Redundancy," *Telecommunication Systems: Modelling, Analysis, Design and Management*, vol. 65, pp. 783–794, 2017.

An AI-Assisted IoT-Based Framework for Time Efficient Health Monitoring of COVID-19 Patients

9

Punitkumar Bhavsar[1] and Vinal Patel[2]

[1]Department of Electronics and Communication Engineering, Visvesvaraya National Institute of Technology, Nagpur, India

[2]Department of Information and Communication Technology, ABV-Indian Institute of Information Technology and Management, Gwalior, India

Contents

DOI: 10.1201/9781003244714-9

9.1 INTRODUCTION

COVID-19 is a pandemic that started in China's Wuhan City and spread around the world in a matter of months. By August 23rd, 2021, this highly contagious disease had affected over 212 million people around the world. It is caused by severe acute respiratory syndrome coronavirus 2 (SARS-CoV-2), which attacks the respiratory tracts. The micro droplets from infected people transfer the disease to healthy individuals through coughing, sneezing, or exhalation in the vicinity. Healthy individuals are also at risk of exposure to the infected surfaces where the virus stays alive for a certain duration. Incubation period of 14 days is considered for this disease in which the exposed one may start developing the symptoms. The exact symptoms are still ambiguous and quite analogous to the normal influenza, hence it requires a special test for identification. The World Health Organization (WHO) has categorized a broad range of symptoms into three different categories – mild symptoms, moderate symptoms, and serious symptoms (WHO, 2021). The mild symptoms are fever, dry cough, and tiredness. The moderate symptoms include aches and body pain, sore throat, diarrhea, conjunctivitis, loss of taste and smell, rash on skin, etc. The serious symptoms involve difficulty in breathing or shortness of breath due to hypoxia, chest pain, and loss of speech or movement. Though this disease is identified into a highly infectious category with early reproduction number between 2.2–3.6, the mortality is low and ranges between 0.5%–5% depending on the various factors of prior health condition, age and level of treatment. Majority of the patients with common symptoms recover without the need for extra care. However, the patients with serious symptoms are at higher risk and require continuous monitoring from a medical practitioner.

Several countries implemented lockdown by restricting people's movement to contain the spread of the virus. It is observed that an implementation of lockdown certainly reduces the number of cases by reducing the rate of infections. However, such implementations for longer duration may adversely affect several aspects including economy. Though relaxations in lockdown through restricted movement and social distancing reduce the rate of increase of the number of patients, it has been found to be unfavorable to financial activities (Baldwin & Mauro, 2020). Several industries suffer economic backlash due to unavailability or inadequate timely supply of resources. It has led to a challenging decision-making process for the governing authorities to control the spread of the disease and yet maintain the economic activity running in order to avoid a foreseeable financial crisis. Also the increasing number of infected people thread to flood the medical infrastructure which may potentially degrade the quality of medical services. In such scenarios, adoption of work-from-home culture has become an alternative in many companies and organizations.

The health sector is at the greatest crisis due to this global pandemic. Ever increase in the demand of medical resources such as ventilators, medical masks, sanitizers and medicines has impacted the supply chain. Apart from pertinent commodities, the surging demand of additional medical staff is another challenge to cope with. Developing countries are already deprived of the availability of doctors, which is less than the standard ratio 1000:1 (population to doctor proportion) as suggested by WHO (Campbell et al., 2013). The statistics reveal over 45% of the member countries of WHO reports less than 1 doctor per 1000 population (Campbell et al., 2013). Only 11 out of 28 states in India meet the WHO recommendations. This shortage exacerbates when the doctors in small hospitals and clinics with limited nursing facilities are not able to treat the COVID-19 patient due to increased likelihood of getting the infection or not having enough resource and infrastructure to avoid the spread. It has created further shortage of medical staff and personnel available for the treatment. The available doctors find it difficult to properly allocate their time to treat an increasing number of patients. In such a time, smart technology with mobile devices and artificial intelligence can help doctors to allocate their time and attention efficiently. The government of India has released 'Aarogya Setu', a mobile application that tracks the real-time location of COVID-19 patients and provides information about the risk of exposure to other users in order to prevent the spreading of infection. This mobile application also provides self-assessment tests to check the status of their risk level by answering a few MCQ based questionnaires. Similarly, Singapore's health ministry advised the citizens to use 'TraceTogether' mobile application that traces the people who were exposed to infected patients within the last 21 days (Wei et al., 2020). If a person is found infected, the application provides information about people who had close contact with the patient. Germany, in order to control the spread of the disease after relieving the restrictions, launched 'Corona-warn-app' that warns the users if they have been in the vicinity of infected individuals (Reelfs et al., 2020). These mobile applications support in containing the spread of the disease by raising an alert at the right time. Though such mobile applications potentially reduce the time compared to manual tracing of a contact, their usage provoked global debate on jeopardization of privacy of users (Darbyshire, 2020). Though the recovery rate for COVID-19 is approximated to around 98% as declared by the WHO, the exponentially increased number of patients increases the number of deaths on a daily basis. Initially, patients feel mild symptoms which are overlapping with other flu conditions hence in the absence of proper awareness they may turn into moderate ones. The moderate symptoms, if not treated in time, may turn into severe ones which have to be managed and monitored under the supervision from a medical practitioner. However, the shortage of medical staffing and resources, it becomes difficult for the doctor to properly allocate their time to the serious patients.

This chapter shows the potential of IoMT based framework which incorporates smart sensors with mobile devices and artificial intelligence that can help patient-doctors to get along in a time efficient manner to reduce the foreseeable serious health complications. This framework utilize timely data as measured from the medical devices to classify the health status in different categories. In addition, it can help the doctor to judge the health status of several patients clustered in a specific category. Section 9.2 presents the general structure and components of IoMT for health monitoring. Section 9.3 provides case study of IoMT framework for COVID-19. Section 9.4 briefs about the futuristic trends and corresponding challenges for IoMT. Conclusion is provided in Section 9.5.

9.2 GENERAL STRUCTURE OF IOMT-BASED FRAMEWORK FOR HEALTH MONITORING

IoMT, when integrated with other technological solutions, has brought about a paradigm shift in the medicine field for saving and improving people's lives. It developed new application fields and altered established ones, such as clinical decision making, data collecting,

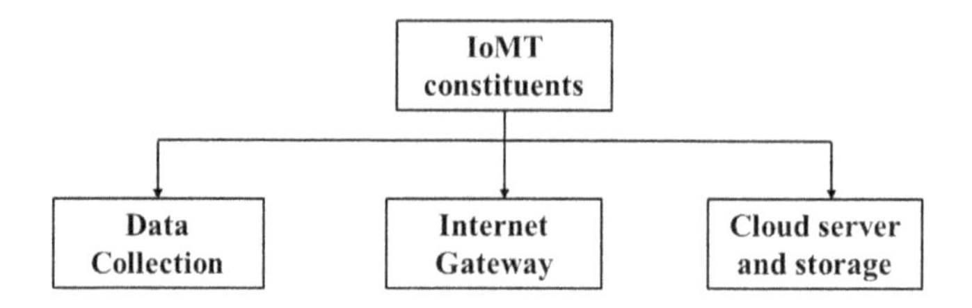

FIGURE 9.1 Components of IoMT.

and patient record management. IoMT consist of three different layers for its functionality starting from data collection layer at abstract level to cloud storage and server at highest level with internet gateways in the intermediate level as shown in Figure 9.1.

9.2.1 Data Collection

Unlike other IoT applications, the IoMT requires to collect biological data for various applications. These biosensors are broadly classified as either wearable sensors or non-wearable sensors. Wearable sensors is a subset of smart biosensors that can be implanted or used as extensions on the human body. These devices have a good prospect for producing data and communicating with other devices without the need for human interaction. In non-wearable technology, the devices that cannot be fixed on the human body are classified as non-wearable. Without human intervention, these devices can also generate large amounts of data and communicate with other devices. Some of the wearable and non-wearable devices (described next) are also utilized and under practice for continuous health monitoring under the framework of IoMT.

1. Wearable devices
   a. **Wearable as hearing aid:** This is a small electrical device worn behind the ears. A microphone, speaker, and amplifier are all part of a standard hearing aid. The count is in millions who need aid or support due to loss of hearing. Hearing aids allow users to hear clearly in both quiet and noisy environments. Microphone converts the sound wave into electrical signals. Electrical signals are then amplified by the amplifier and sent to the speaker. Hearing aids are the most basic form of wearable technology, but they are becoming increasingly complex and intelligent. Smart hearing aids are rechargeable devices that may be connected to users' smartphones, televisions, laptops, tablets, and other electronic devices.
   b. **Wearable for fitness tracking**: Fitness trackers, often known as fitness bands, are wrist-worn gadgets that may monitor a person's physical activity as well as other data such as body surface temperature, heart rate, and footfall counts. These devices can be easily connected to the smart phones via Bluetooth. Wearable technology users have been observed to be more physically active (Pourzanjani et al., 2016). According to research, fitness trackers encourage people to

create health goals and stick to them by monitoring and recording daily activities (Neupane et al., 2020).

c. **Wearables for chronic pain:** Chronic pain is a condition of central nervous system in which the pain is persistent without any apparent reason. Back pain and headache are two common examples. More than 100 million alone in the US are suffering from such chronic pain (Dellwo, 2020). Wearable technologies were developed by a number of inventive companies to cure chronic pain without the need for hospitalizations or medicine. Although this is a relatively new sector in IoMT, certain products have already been permitted for use while resting (neurometrix, 2002). A thin slim band device is positioned on the upper calf to block persistent discomfort. The device sends neurological impulses to the brain.
d. **Wearables for skincare:** These devices include built-in sensors that capture skin data (usually ph levels) and communicate it through Bluetooth to a smartphone app (JUNG, 2017). Based on the skin type, a mobile app analyzes the data and selects an appropriate solution.
e. **Wearables for detection of cancer:** Cancer is usually identified by obtaining a sample of cells. A rigorous examination is then followed to detect and identify the presence of any cancer or tumor cells. Such procedures, often known as biopsy, are intrusive. Sometimes the circulating tumor cells (CTC) are present in the blood which are identified from the sample of blood. This process is also intrusive and not suitable for continuous monitoring. However, the researchers could develop the non-invasive wearable CTC detector to collect large amount of data from CTCs. Research from the University of Michigan suggests the potential of IoMT-based wearable CTC detector has efficiency as high as 3.5 times more compared to traditional methods.

2. Non-wearable devices
   a. **Smart pillbox:** Due to a hectic schedule, many people mix up their pills. Untimely medication has a negative impact on one's health. Patients with chronic disorders, in particular, must take treatment over time. As a result, it's critical to take the proper medication at the right time. Elders are more likely to forget to take prescriptions or to overdose, which might have negative consequences. A methodical approach to resolving such a problem is required. Minaam et al. proposed a smart pillbox design in (Abdul Minaam & Abd-ELfattah, 2018). It's a small computerized gadget that can automatically categorize various medicines.
   b. **Human activity recognition:** The number of people with diabetes, cardiovascular diseases, and stroke have increased due to sedentary position for longer duration or lack of required level of physical activity. The wearable solution provide information to tackle such issues. These devices tracks person's activity throughout the day or week or month (Nweke et al., 2019). They also alert the users for physical activity suggestion to meet the target in order to reduce the potential health risks.
   c. **Smart beds:** The medical practitioners or caregivers can use smart beds to monitor their patients from distant. The biosensors are part of this equipment which help to monitor patient's vital parameters like temperature, heartbeat blood

oxygen levels, blood pressure, and respiratory rates (Timothy M. Sauer, 2014). These gadgets can create an alarm or send an alert to the responsible persons in an emergency, allowing them to take rapid action.

Bio-signals are quite often weak and noisy, making them challenging to interpret. As a result, they are first amplified with the proper amplifiers. A typical amplifier is made up of an electrical circuit that generates a high-amplitude output signal from a weak signal fed into it. It enhances the signal without changing other aspects like frequency or shape. Resistors, wires, and transistors make up a conventional amplifier circuit. A single-stage amplifier has only one transistor, while a multi-stage amplifier has many transistors. Multi-stage amplifiers are frequently used in practical applications. Signals are cleaned after they have been amplified to eliminate undesirable portions. The typical filtering process involves low pass, high pass or band pass filtering. Low pass filter attenuates high frequency components, low-pass filter attenuates high frequency components, whereas band-pass filter attenuates both low-pass and high-pass components of the signal (Matzik & Anwar, 2016). Adaptive filtering procedures are also adopted in case of in-band noise corrupting important information (Allwood et al., 2019).

Analog signals are converted to digital ones using A/D converters. A series of numeric numbers makes up a digital signal. These values can be readily stored and processed by modern computers. Sampling and quantization are the two basic stages in A/D conversion. To convert continuous time to discrete time, the sampling method is utilized. It entails determining the value of a signal at regular intervals. Every measurement is now referred to as a sample. If x(t) is an analog signal, it is sampled every T seconds, and x(kT) represents the amplitude value, with k (k = 0,1,2,3, . . . n) being the data sequence's sample number and T being the sampling interval. The reciprocal of the sampling time (1/T) is the sampling frequency fs. To avoid distortions, the sampling frequency must be set higher than the maximum frequency of the source signal. The sample rate should be equal to or more than twice the highest frequency of the original signal, according to Nyquist's theorem. If the maximum frequency in the signal is F_{max} then minimum sampling rate is given by $F_s = 2F_{max}$.

The main objective of the data collection layer (device layer) is to develop a reliable and precise sensing technology for collecting various forms of health-related data. Wearable sensor technologies are shown in Table 9.1.

TABLE 9.1 Wearable Sensors and Corresponding Technology

SENSOR TYPE	*EXAMPLE*
Movement measurement sensor	Accelerometer and Gyroscopes
Position sensor	Global Positioning System
Electro-physiological sensor	Electrocardiogram (ECG), Galvanic Skin Response (GSR), Electroencephalogram (EEG) Electrooculography (EOG)
Image sensor	Camera
IR sensor	Body surface temperature
Optical sensor	Photoplethysmography

9.2.2 Internet of Things Gateway

The biosensors, intelligent devices, and cloud servers are all connected together through the IoT gateway. This gateway acts as the bridge between the sensor devices and cloud server to transfer data between them. The primary role of the gateway is to perform data normalization, data preprocessing and provide network connectivity. Data normalization is the process of data conversion to a predefined unique format in order to perform the further processing as the raw data from all the devices may not be in the same format initially. Data preprocessing mainly involves filtering operations and adjustment of data suitable for processing/storage in the cloud server. Biosensors are usually equipped with poor network connectivity. They cannot directly connect to the extensive network like WAN. This gateway provides a platform for data connectivity with such an extensive network by Bluetooth or ZeeBee like short range communication technology. Table 9.2 shows application of such communication protocol in an IoMT framework.

Zigbee: This protocol is developed to provide low-power local wireless area network to accommodate large number of devices in an industrial applications. It consumes relatively less power from the battery. The IEEE specification 802.15 forms the foundation for Zigbee. 802.15 include monitoring applications that requires relatively

TABLE 9.2 Different Wireless Protocols Corresponding Applications in IoMT

WIRELESS STANDARD	*RANGE*	*DATA RATE*	*OPERATING FREQUENCY RANGE*	*APPLICATION IN IOMT*
ZigBee	10m-100m	250 kb/s	868/915 MHz, 2.4 GHz	Heart rate monitoring device (Filsoof et al., 2014) e-stethoscope and remote healthcare applications (Lee et al., 2009) (Buthelezi et al., 2019)
Bluetooth	10m-100m	1 Mb/s	2.4GHz-2.5 GHz	Wearable activity monito (Fitbit, 2015); e-stethoscope (Tang et al., 2010)
RFID	0–3m	Depends on frequency	LF/HF/UHF microwave	Wearable RFID tag antenna (López-Soriano & Parrón, 2015); Medical supply tracking system (Tsai et al., 2019), (Kang et al., 2021)
Wi-Fi	45m-90m	Up to 54 MBPS	900 MHz-60 GHz	Monitoring vital health parameters (Shao et al., 2014), (Bocicor et al., 2017) (Lou et al., 2013)

modest data throughput. With this unique feature, it has been widely accepted in machine to machine communication and IoT. Typical communication range is 10 m to 100 m. Currently, the health industries have been increasingly adopting IoT technology in which the wireless communication has essential role to play by improving accuracy, efficiency and cost. Zigbee has been used in potential IoMT applications described in (Guntur et al., 2018).

Bluetooth: Ericsson invented Bluetooth technology in 1994, which relies on physical proximity to maintain connections between devices even without a password. Bluetooth employs UHF radio waves between 2.400 and 2.5 GHz to communicate between two devices across a distance of up to 164 feet. Many electronic gadgets, such as computers, cell phones, speakers, and printers, use this technology. IoMT is another area where Bluetooth is widely used. The advancement of Bluetooth in healthcare systems is still ongoing which can be observed by the development of monitoring devices in (Zhang et al., 2014) (Frank & Meng, 2017)(Suvarna et al., 2016). Bluetooth is relatively safer than Wi-Fi as it connects the devices in the physical proximity only and reduces the thread from the remotely located device.

Wi-Fi: This wireless protocol has been widely used for wireless communication between two devices. It requires the router to be positioned at a fixed location as it requires relatively large power. Wi-Fi operated though radio wave at 2.4 Ghz to 2.5 GHz for data transport between the devices. The common example of Wi-Fi connectivity include indoor networks. In IoMT applications also, Wi-Fi plays an important role of wireless connectivity to biosensors which can be observed in (Manzoor et al., 2020) (Tan et al., 2018). When it comes to decision making for the choice of the wireless protocol between Wi-Fi and Bluetooth, the Bluetooth is relatively cheaper and easier to use whereas Wi-Fi provide more data rates, security at more cost. Hence, the choice depends on the requirement of the applications and cost constraints.

RFID: RFID (radio frequency identification system) is an automated technology that uses radio waves to help machines or computers recognize objects and control targets. Usually, RFID is connected to external network in order to detect, track, and monitor tags-attached objects automatically in real time. RFID has become a new trend in the healthcare industry. Tracking supplies in hospitals can save a lot of money and time. Hospitals continue to buy products they already have due to a lack of efficient inventory record keeping (Kang et al., 2021). Tracking systems are one of the medical applications of RFID technology in IoMT framework (Karthi & Jayakumar, 2020)(Huang et al., 2017).

9.2.3 Distributed Cloud Server and Storage

Servers are the central component of IoMT systems that detect and analyze abnormal activity. The gateway transfers the digital data to the servers or cloud for data processing from biosensors. The process of identifying crucial information regarding bio-signals is known as data mining in IoMT. Various algorithms from signal processing and artificial intelligence are the part of cloud server. The server processes data received from various sources in order to provide meta-analysis. Also, the distributed cloud services sometimes used as data storage for records which may be provided to the users on request.

One of the main objectives from data processing algorithm in the cloud server is either enhance the quality of the signal or feature extraction for classification. A description of all the data processing techniques is beyond the scope of this chapter. However, some of the frequently used techniques and well-known classification algorithms are briefly explained here.

The classification algorithms are broadly categorized into supervised and unsupervised classification algorithms. In case of supervised classification algorithms, the training input and output data is labeled. For each input, the labeled output is mentioned. The algorithm attempts to learn the hidden pattern in the input data to best describe the output. In case of unsupervised classification, the labeled data is not available for the algorithm to learn the pattern. Instead, the algorithms here attempts to cluster the data based on the observed input pattern. Support Vector Machine (SVM), Nave Bayes (NB), K-Nearest Neighbor (K-NN), Artificial Neural Network (ANN), and Linear Discriminant Analysis (LDA) are some of the well-known classification algorithms.

SVM is known to provide better accuracy in case of noisy and complex data, however, its performance is dependent on the quality of input (Burges). Over the time, the use of SVM is extended from classification to regression and ranking of the elements. One of the major drawback of SVM is heavy computational requirement. The NB and ANN algorithms take a completely different approach for classification. The NB algorithm is a probability-based classifier that has been around for a long time. It employs Bayes' theorem in conjunction with the independence supposition among the features. NB is a robust and simple classifier. Despite its simplicity, it frequently yields surprisingly accurate results (Wu et al., 2008)(Jiang et al., 2009). ANN algorithms, on the other hand, are based on simulating the neural system of the human brain. ANNs are extremely effective at solving data processing tasks with greater precision. However, ANN-based algorithms are too complex, requiring a large amount of computation to solve a problem with greater accuracy (Eglen et al., 1992) (Bigus, 2009) (Kamruzzaman & Jehad Sarkar, 2011). The revolutionary learning algorithms from deep learning concept are an extension of ANNs. Deep ANNs have extreme learning ability, can process massive amounts of data, and produce highly accurate results that other traditional machine learning and data mining algorithms cannot (Schmidhuber, 2015).

As mentioned earlier, it is beyond the scope of this Chapter to provide details of all the data processing and classification algorithm used in IoMT framework. However, two widely used classification algorithms (SVM and KNN) and feature selection methodology are described in detail to understand the case study presented in this Chapter.

9.2.3.1 Recursive Feature Elimination-Support Vector Machine (RFE-SVM)

SVM-RFE method uses SVM classifier and ranks the features according to their contribution in improving the classification accuracy. The output of this method is a ranked feature list. Features with highest importance are selected first. The ranking of the feature is evaluated based on the weight vector associated with the input. SVM manipulates the optimal hyperplane which can maximize the distance between the classes. Input data is mapped to the higher dimensional space using suitable kernel where an optimal hyper plane can be constructed. The hyperplane is given by

$$S(x) = wF(x) + b$$

where $x \in R^m$ is the input features, w and b are the parameters and F(.) is the kernel function which maps the vector x to higher dimensional space R^d, $d > m$. The distance between the S and x is given by $S(x)/\|w\|$. The margin M is given by $2/\|w\|$, (here, b=1). Figure 9.2 shows the linear hyperplane and margin M with two-dimensional feature vector x. Here, maximizing the margin M is equivalent to minimizing $\|w\|$. So, SVM solves following optimization problem,

$$\min \frac{1}{2}\|w\|^2 + C\sum_{i=1}^{N} \mu_i$$

$$\text{s.t } y_i(w.x_i + b) \geq 1 - \mu_i, \; i = 1, 2, \ldots, N$$

$$\varepsilon_i \geq 0, \; i = 1, 2, \ldots, N$$

where $y_i \in \{-1, 1\}$, ε_i is a slack variable and C is a hyper parameter which is tuned to balance between the misclassification of labels and the maximization of margin M. By incorporating the Lagrangian multiplier and solving the dual of the optimization problem gives,

$$w = \sum_{i=1}^{n} \alpha_i y_i x_i$$

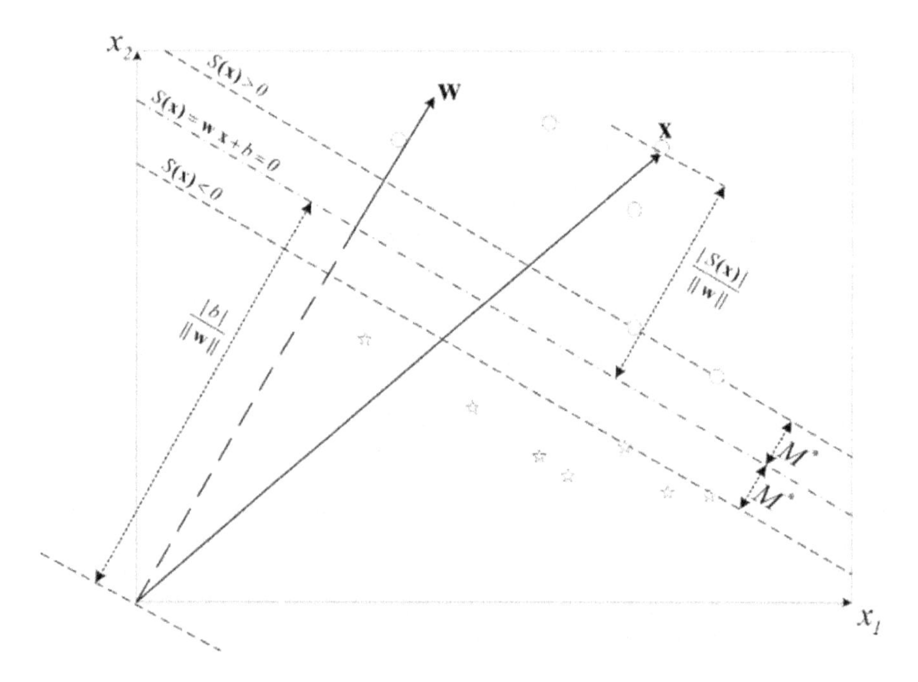

FIGURE 9.2 SVM with linear hyper plane for decision boundary S(x) and margin M.

where α_i is Lagrange multipliers.

The ranking of feature k is evaluated for kth element of w by,

$$J(k) = w_k^2$$

In RFE-SVM method, an SVM model is trained with all the features in the first iteration. Then, the feature with least ranking criterion from J is removed due to least effect in the classification accuracy. A new model with removed features is trained for classification. Again, the least contributing feature is eliminated from further iteration. This process is repeated till all the features are iteratively removed. The features are then sorted in descending order from the order of removal. The features which are removed at a later stage are important.

9.2.3.2 K-Nearest Neighbor (KNN)

K-Nearest Neighbor (KNN) is one of the simplest forms of supervised learning algorithms. This algorithm is of particular use when a little or no knowledge about the underlying distribution of data points is available. It utilizes all the training samples without generalization for the classification of testing data. The new or test data point is classified based on the majority of training classes from the K nearest data points.

Following steps are carried out to perform KNN classification.

1. Initialize the value of k for the data set
2. Select a test data sample and find its distance from all the training data samples.
3. Arrange the distance in ascending order.
4. Select top K samples from the sorted array.
5. Get the most frequent class from the selected K samples.
6. Assign the most frequent class to the test data sample.

The method to select the value of K lacks a predefined statistical method. The selection is either based on the satisfactory level of performance from arbitrary selected K or from the curve of rate of error with K for the testing data samples. Empirically, the value of K is chosen as the square root of the number of training data samples (Duda et al., 2016).

9.3 CASE STUDY

9.3.1 IoMT-Based Framework for COVID-19

The proposed method presents a framework that facilitates the medical staff and patients to get along in a time efficient way and reduces the risk of infection. This framework consists of three layers with distinct functionality as shown in Figure 9.3. The first layer interacts with the patient to collect raw data and submits it to the second layer. Second layer acts as a mediator between the first layer and third layer. It collects raw data from the first

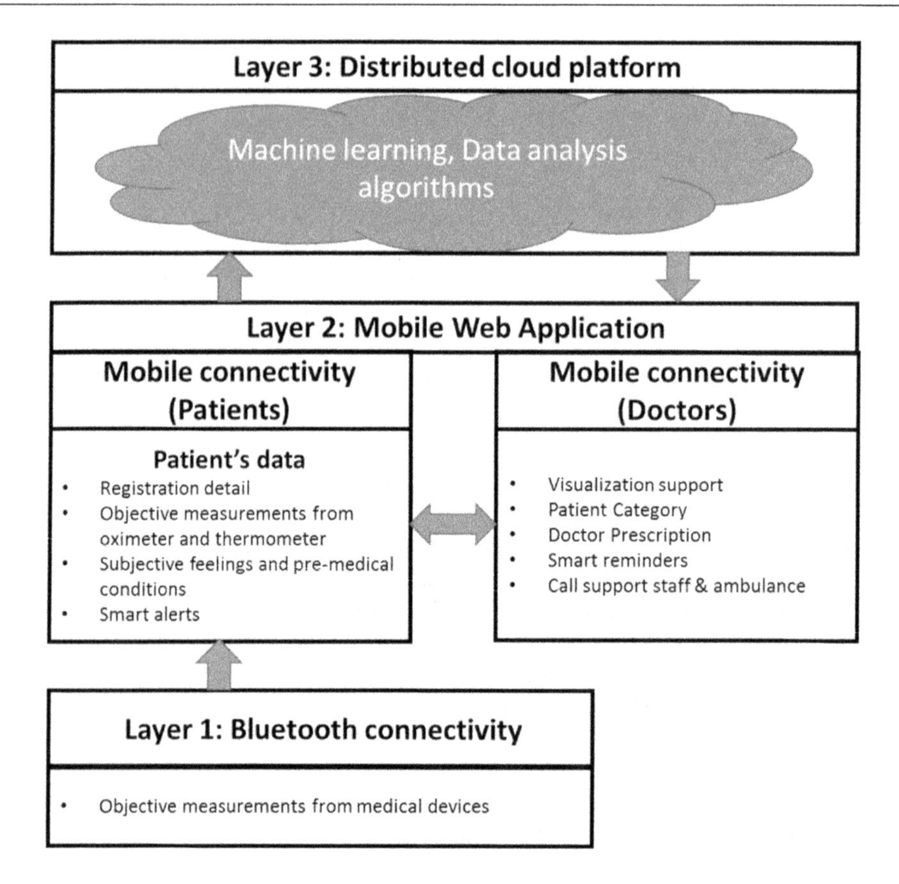

FIGURE 9.3 Framework of the proposed architecture.

layer, validates it through preprocessing and transmits them to the third layer for further processing. The third layer is the cloud platform which inherits advanced data processing and machine learning algorithms to provide better insights for decision making.

Layer 1 collects subjective and objective forms of data through means of questionnaires and physiological sensors. Subjective data consists of information about patient's symptoms and pre-medical conditions which are collected through means of questionnaires. Information about the symptoms include the presence of fatigue, cough, difficulty in breathing, loss of test, loss of smell and other COVID-19 symptoms. Information about pre-medical conditions include the presence of diabetes, heart disease, and other chronic diseases. The objective information about oxygen saturation level and presence of fever is collected through means of biosensors like pulse oximeter and thermometer. These devices are capable enough to provide accurate, high end clinical information due to advanced electronics and local computing power. They have short range communication to the next layer with the use of Bluetooth. They can provide the data continuously with high time precisions, however, the symptoms and diagnosis can also be effectively inferred from the samples collected in the interval of a few minutes. Also, the longer intervals between consecutive measurements reduce the energy consumption resulting in relatively less battery recharge. These devices are generally the low-cost stand-alone systems that require less computing power and communication bandwidth. They require a middle layer known as inter connected gateways to transmit data to the high-end computing resources.

The smartphone or tablet equipped with a dedicated application which provides a platform for the second layer. These devices are connected to the biosensors at the lower end and receive the raw measurement through short range communication technology like Bluetooth and exchange the data to third layer through heterogeneous networks such as Wi-Fi or GSM. Second layer also contains algorithms which evaluate the clinical relevance of the measurement and filter the unwanted noise before transmitting the data to the next layer. The subjective data collected through application-interface are also processed in this layer and submitted to the next layer. This layer consists of a cloud computing infrastructure where the time series data from all the registered patients are collected, stored and managed for further processing. This layer has advanced functionality of machine learning algorithms, data mining and visualization techniques. It converts data into meaningful information to help doctors make important decisions. For example, the observed data can be used to form the clusters of patients who represent a particular category of health conditions. The category specific health consultation or treatment is then provided to the patients. If the patient is found to be closer to the category representing critical health conditions, the doctor can alleviate the effort focusing on the patient's previous health data records. If the patient belongs to moderate condition, the doctor can prescribe the medicine without needing to directly come in physical contact. The patients who belong to the cluster representing mild conditions can get health consultation from the doctor with the use of mobile application as well.

In order to categorize the health condition, machine learning algorithms use the selected features as input and translate them into specific classes. This categorization relies on the classification algorithms that automatically estimate the class of the data as represented by a feature vector. The property of the classifier should be known to decide the most appropriate classifier for a given set of features (Lotte et al., 2007).

9.3.2 Data Set and Statistics of Symptoms in Mortality Cases

In this preliminary analysis, the data set of 472 patients who tested positive for COVID-19 is considered in this study. It consists of information about symptoms and premedical conditions present at the time of hospitalization (Karimy et al., 2020). Mean age of the patients is observed as 57.4 ± 17.2 years. Information about the patient's symptoms were available in 'yes' or 'no' format. If the symptom or preexisting medical condition is present '1' available in corresponding data entry otherwise '0' is available in corresponding data entry. Only the patients with positive RT-PCR are considered in this analysis. Out of 472 such cases 398 patients recovered, while 74 resulted in mortality. Nineteen features are available in the data representing either symptoms or the preexisting medical conditions. It is assumed that the data set is available in the cloud server after receiving it from biosensor (first layer) and preprocessing from mobile application (second layer).

Table 9.3 lists the distribution of symptoms present in the population of mortality cases. Though the classification of health status is based on the presence or absence of certain symptoms, other symptoms, preexisting medical conditions and age may be responsible for the higher likelihood of outcome as mortality. Low oxygen level is present in 66.2% (49 out of 74) of mortality cases. The remaining 33.8% (N=25) do not produce this symptom, however, they are observed to have higher age, diabetes, shortness of breath, and fatigue. Sixteen out of those 25 such mortality cases belonged to the age group above 60 years with a mean age of 74.6 years. These 16 elderly patients produced the symptoms of shortness of breath and fatigue without any dip in the oxygen levels. The remaining nine patients were less than 60 years of age. Among them, five patients produced the symptoms of shortness of breath without producing any symptom of low oxygen level. The remaining four morality cases produced only fever as a dominant symptom. They did not produce any shortness of breath or low oxygen levels and did not have any other chronic conditions. It is surprising to observe that 90 out of total 176 patients who produced low oxygen level did not report shortness of breath. Silent hypoxia is one of the confusions among medical practitioners, who report the patients do not feel low oxygen level at a conscious level (Couzin-Frankel, 2020).

Apart from the symptom of low blood oxygen level, which is predominantly present in the majority of the cases, other symptoms like fever, coughing, myalgia heart disease, chronic diseases, diabetes, and heart disease are also present in a significant number of cases. Shortness of breath, coughing, fever, and low blood oxygen level are present either alone or in combination in more than 90% of the cases. Apart from such severe symptoms, age is also considered a significant factor contributing to the likelihood for recovery. Figure 9.4 shows the histograms of the patient's population with age for (1) those who recovered and (2) those who had the final outcome of mortality. It is evident from the histogram that age is also one of the significant factors in a patient's likelihood of recovery. More than 64% of the recovered cases were below the age of 60. On the other side, more than 70% of the mortality cases are found to be more than 60 years of age. It is likely that the old age people also suffered from some kind of chronic or other medical condition which worsened the infection due low immunity level. 20% of the mortality cases had preexisting heart disease, whereas 17% were diabetic patients. None of the cases below the age of 20 years was available for analysis in this data set. However, studies reveal the mortality rate for children below this age is low enough (Bhopal et al., 2020).

Figure 9.5 shows the Venn diagram representing the relative proportion of symptoms of low oxygen level, fever and the number of mortalities caused by COVID-19. More than 80% of the mortality cases had these two symptoms present in them either alone or in combination. According to WHO, almost 81% of the patients have mild infections or asymptomatic whereas 14% of the cases have severe infection in which patients require oxygen while 5% of the cases may turn into critical and the patients require ventilation support. Monitoring of oxygen level at regular intervals of time even for the patients with mild symptoms is crucial to help them to avoid reaching severe or critical conditions (WHO, 2021). Hence, in this framework, we suggest using oximeter reading and body temperature as objective measurements along with subjective measurements and preexisting medical conditions from the patients.

As explained in Section 9.1, WHO categorizes the health condition of COVID-19 patients into three categories based on severity. The mild condition with symptoms of only

TABLE 9.3 Distribution of Symptoms Present in the Mortality Cases

	BLOOD OXYGEN LEVEL	*SHORTNESS OF BREATH*	*COUGH*	*FEVER*	*HEART DISEASE*	*OTHER CHRONIC DISEASE*	*FATIGUE*	*DIABETES*	*LOSS OF CONSCIOUSNESS*
Number of patients	49 66.2%	36 48.6%	23 31.0%	29 39.2%	11 14.9%	12 16.2%	10 13.5%	13 17.6 %	4 5.4%
	cancer	Liver Disease	Hematologic disease	HIV	Heart Disease	Renal Disease	Asthma	COPD	Other Chronic Diseases
Number of patients	0	0	2 2.7%	0	15 20.3%	2 2.7%	1 1.3%	2 2.7%	17 23.0%

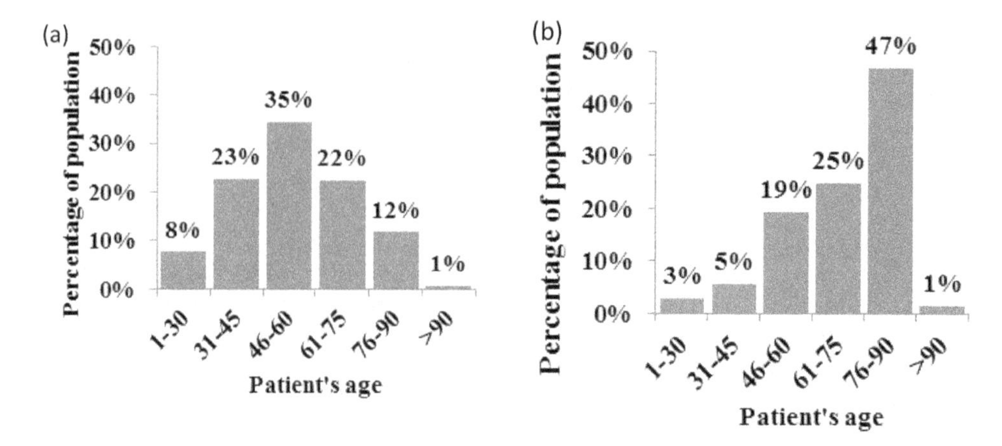

FIGURE 9.4 Histogram of patients' ages for those who had the final outcome as (a) recovered (b) mortality.

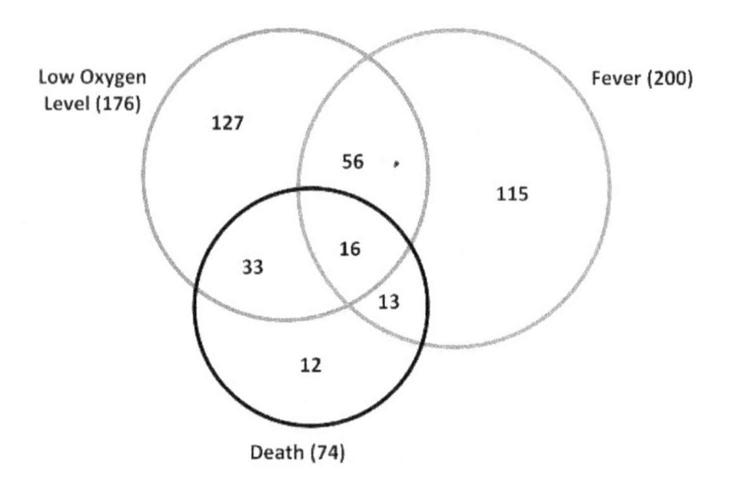

FIGURE 9.5 Venn diagram representation of symptoms of fever, low oxygen level, and mortality cases.

fever, dry cough, and fatigue is at lower risk of severity, whereas shortness of breath and low PO_2 levels are considered as serious symptoms.

9.3.3 Result

As discussed in Section 9.3.2, the data set consist of 18 different sign and symptoms of 472 COVID-19 patients. In this section, we have identified predominant symptoms which can categorize the final outcome into either recovery or mortality. In order to determine the relative contribution and weight of different features, RFE-SVM is used. The parameters used for tuning the algorithm is described in Yan & Zhang (2015). Based on the ranking, eight features are identified that resulted in the highest accuracy of the SVM classifier. Incorporation of the features beyond the selected ones do not significantly contribute to the accuracy hence discarded from the further analysis. The list of features according to the ranking is (1) blood oxygen level, (2) shortness of breath, (3) coughing, (4) fever, (5) heart disease, (6) other chronic diseases, (7) myalgia, and (8) diabetes.

This list is selected based on the maximum times the features is ranked on a particular position with tenfold cross validation. Literature suggests high correlations between blood oxygen level (PO_2) and oxygenation of tissues (SpO_2) which is monitored noninvasively with the use of oximeter (Baudelet & Gallez, 2002).

In this study, we aim to classify the health condition into three different categories based on the information available from the selected features. Fever and oxygen levels can be objectively measured whereas other features can be collected through subjective forms available through mobile application. Next, the patient's health status is labeled into three categories. Category I represents the patients with mild symptoms which include the presence of either fever, or cough or fatigue. This category of patients do not have any other preexisting medical condition and symptoms. Category III represents the patients with severe symptoms which include the lower oxygen level as well as other symptoms and medical conditions. Classification of Category I and Category III is in line with categorization declared by WHO (WHO, 2020). The rest of the patients who did not belong to either Category I or Category III are classified into Category II. The health condition in category II patients is neither mild nor severe, however, they may move into Category I or Category III depending on the evolution of health condition with time. Table 9.4 shows the number of patients in each category. It is observed that the Category I has the lowest mortality rate whereas Category III has produced highest mortality rate. Similarly, the highest recovery rate is observed for Category I, whereas lowest recovery rate is observed for Category III. Category II produced intermediate numbers of mortality and the recovery.

The data set is divided into training (80%) and testing (20%) for the classification using SVM, KNN, and Naïve bayes. Tenfold cross verification used to reduce the effect of bias in the data.

Support vector machine with linear kernel shows better performance compared to KNN and Naïve Bayes in terms of all the performance measures. The performance of KNN is comparable with linear SVM as majority of the assessment metrics are within ±0.05%. Poor performance Naïve Bayes classification method may have resulted from the unknown distribution and limited number of samples. Repeated assessment of above measures does not change significantly. Hence, it is concluded that the SVM with linear

TABLE 9.4 Distribution of Patients in Each Category

CATEGORY	*NUMBER OF PATIENTS*	*NUMBER OF MORTALITY*	*NUMBER OF RECOVERED PATIENTS*
I	161	8 (5%)	153 (95%)
II	135	17 (13%)	118 (87%)
III	176	49 (28%)	127 (72%)

TABLE 9.5 Performance Verification with Various Classification Models

	KNN (K=22)			*NAÏVE BAYES (GAUSSIAN KERNEL)*			*SVM (LINEAR KERNEL)*
	Cat I	Cat II	Cat III	Cat I	Cat II	Cat III	Cat I/ Cat II/ Cat III
Accuracy	0.99	0.97	0.98	0.65	0.28	0.37	1.00
Sensitivity	1.00	0.85	0.97	0	0.94	0.45	1.00
Specificity	0.98	0.99	0.98	0.66	0.13	0.05	1.00
Precision	0.98	0.92	0.97	0	0.20	0.65	1.00
F1 score	0.99	0.95	0.97	0	0.33	0.53	1.00

kernel combined with one versus all classification methods better suits for the automatic categorization of health status of patients based on the selected features.

9.4 CHALLENGES FOR FUTURISTIC TRENDS IN IOMT

This section points out the current challenges involved for scalability of the IoMT framework for achieving personalized high end health services to individuals.

1. **Low-cost and non-invasive biosensors:** It is difficult to design low-cost biosensors to work non-invasively and measure vital signals from the patients (Barata et al., 2019).
2. **Big data analysis:** The current framework is bombarded with huge amount data from several wearable and/or non-wearable sensors. Analysis and interpretation of such remarkably high amount of data requires different level of storage, processing algorithms, and analytical techniques (Rghioui et al., 2020) (Al-khassaweneh & AlShorman, 2020).
3. **Privacy and security issues:** The patient data has to travel through wireless protocols. Any security issue may result in potential threat for leakage of private information (Mohanty et al., 2020) (Garg et al., 2020).
4. **Interference from surrounding environment:** Majority of the wearable data measures the body vitals from the skin surface. Such biosensors are vulnerable to various noise from the measurement environment and artefacts. Controlling the interference from surrounding environment is difficult task due to high level of random factors that may appear at any time (Atlam & Wills, 2020).
5. **Energy demand:** Data acquisition from biosensors and its wireless transmission requires the measuring devices to be highly energy efficient for their long duration usage in a single charge of battery (Mittal et al., 2019).
6. **Accurate and reliable algorithms:** The algorithm in the cloud server should be reliable and accurate enough to establish trust for the medical practitioners and patients (Schueler et al., 2019).

7. **Feasibility of wearable device:** The wearable device should allow the patients to perform other activities without need to adjust the device. This feature allows long duration high quality recording of the data (Antolín et al., 2017).
8. **Integrity of wireless protocols:** The integrity of wireless protocol allows to discard or introduce other devices without significant effort (Panarello et al., 2018).

9.5 CONCLUSION

This chapter provided the comprehensive overview of health monitoring through IoMT framework. Remote monitoring of health parameters is the foundation of IoMT era. In addition, current era is in high demand for continuous monitoring of vital health parameters for the slowly progressing many physical and mental diseases. It is necessary to combine the advantage of the state of the art communication technology, sensing technology and computational techniques to improve the effectiveness of IoMT. This chapter explored state of the art sensing technologies, communication protocols and computational algorithms. The case study presented in this chapter provided the power of IoMT to tackle large number of COVID-19 patients with limited resources. In the end, even after exploring the strength and utility of IoMT, it has long way to travel and fulfil promises by addressing the challenges described in Section 9.4.

REFERENCES

Abdul Minaam, D. S., & Abd-ELfattah, M. (2018). Smart drugs: Improving healthcare using smart pill box for medicine reminder and monitoring system. *Future Computing and Informatics Journal*, *3*(2), 443–456. https://doi.org/10.1016/j.fcij.2018.11.008.

Al-khassaweneh, M., & AlShorman, O. (2020). Frei-Chen bases based lossy digital image compression technique. *Applied Computing and Informatics*. https://doi.org/10.1016/j.aci.2019.12.004.

Allwood, G., Du, X., Webberley, K. M., Osseiran, A., & Marshall, B. J. (2019). Advances in acoustic signal processing techniques for enhanced bowel sound analysis. *IEEE Reviews in Biomedical Engineering*, *12*, 240–253. https://doi.org/10.1109/RBME.2018.2874037.

Antolín, D., Medrano, N., Calvo, B., & Pérez, F. (2017). A wearable wireless sensor network for indoor smart environment monitoring in safety applications. *Sensors (Switzerland)*, *17*(2). https://doi.org/10.3390/s17020365.

Atlam, H. F., & Wills, G. B. (2020). IoT security, privacy, safety and ethics. In *Internet of Things* (Issue March). Springer International Publishing. https://doi.org/10.1007/978-3-030-18732-3_8.

Baldwin, R., & Mauro, B. W. di. (2020, March). Economics in the time of COVID-19: A new eBook. *Human Vaccines and Immunotherapeutics*, *123*.

Barata, F., Kipfer, K., Weber, M., Tinschert, P., Fleisch, E., & Kowatsch, T. (2019). Towards device-agnostic mobile cough detection with convolutional neural networks. 2019 IEEE International Conference on Healthcare Informatics, ICHI 2019. https://doi.org/10.1109/ICHI.2019.8904554.

Baudelet, C., & Gallez, B. (2002). How does blood oxygen level-dependent (BOLD) contrast correlate with oxygen partial pressure (pO2) inside tumors? *Magnetic Resonance in Medicine: An Official Journal of the International Society for Magnetic Resonance in Medicine*, *48*(6), 980–986.

Bhopal, S., Bagaria, J., & Bhopal, R. (2020). Children's mortality from COVID-19 compared with all-deaths and other relevant causes of death: Epidemiological information for decision-making by parents, teachers, clinicians and policymakers. *Public Health*, *185*, 19.

Bigus, J. P. (2009). Data mining with neural networks. *New York*, *5*(1), 1–154. http://dl.acm.org/citation.cfm?id=231007.

Bocicor, M. I., Frau, D. C., Draghici, I. C., Goga, N., Molnar, A. J., Pérez, R. V., & Vasilateanu, A. (2017). Cyber-physical system for assisted living and home monitoring. Proceedings – 2017 IEEE 13th International Conference on Intelligent Computer Communication and Processing, ICCP 2017, 487–493. https://doi.org/10.1109/ICCP.2017.8117052.

Buthelezi, B. E., Mphahlele, M., DuPlessis, D., Maswikaneng, S., & Mathonsi, T. (2019). ZigBee healthcare monitoring system for ambient assisted living environments. *International Journal of Communication Networks and Information Security*, *11*(1), 85–92.

Campbell, J., Dussault, G., Buchan, J., Pozo-Martin, F., Guerra Arias, M., Leone, C., Siyam, A., & Cornetto, A. (2013). *A Universal Truth: No Health without a Workforce*. World Health Organization. www.who.int/workforcealliance/knowledge/resources/hrhreport2013/en/.

Couzin-Frankel, J. (2020). The mystery of the pandemic's 'happy hypoxia'. *Science*, *368*, 455–456.

Darbyshire, T. (2020). Do we need a coronavirus (Safeguards) act 2020? Proposed legal safeguards for digital contact tracing and other apps in the COVID-19 crisis. *Patterns*, *1*(4), 100072. https://doi.org/10.1016/j.patter.2020.100072.

Dellwo, A. (2020). *What Is Chronic Pain?* Verywell Health. www.verywellhealth.com/what-is-chronic-pain-4134684.

Duda, R. O., Hart, P. E., & Stork, D. G. (2016). *Pattern Classiflcation* (2nd ed., p. 41). Computational Complexity.

Eglen, S. J., Hill, A. G., Lazare, F. J., & Walker, N. P. (1992). Using neural networks. *GEC Review*, *7*(3), 146–155.

Filsoof, R., Bodine, A., Gill, B., Makonin, S., & Nicholson, R. (2014). Transmitting patient vitals over a reliable ZigBee mesh network. 2014 IEEE Canada International Humanitarian Technology Conference, IHTC 2014, 2–6. https://doi.org/10.1109/IHTC.2014.7147531.

Fitbit. (2015). *Fitbit Official Site for Activity Trackers & More*. Fitbit. www.fitbit.com/#i.1mxltfgou4fqdt.

Frank, L. P. W., & Meng, M. Q. H. (2017, August). A low cost bluetooth powered wearable digital stethoscope for cardiac murmur. 2016 IEEE International Conference on Information and Automation, IEEE ICIA 2016, 1179–1182. https://doi.org/10.1109/ICInfA.2016.7831998.

Garg, S., Kaur, K., Batra, S., Kaddoum, G., Kumar, N., & Boukerche, A. (2020). A multi-stage anomaly detection scheme for augmenting the security in IoT-enabled applications. *Future Generation Computer Systems*, *104*, 105–118. https://doi.org/10.1016/j.future.2019.09.038.

Guntur, S. R., Gorrepati, R. R., & Dirisala, V. R. (2018, February). Internet of medical things. *Medical Big Data and Internet of Medical Things*, 271–297. https://doi.org/10.1201/9781351030380-11.

Huang, S., Guo, Y., Zha, S., Wang, F., & Fang, W. (2017). A real-time location system based on RFID and UWB for digital manufacturing workshop. *Procedia CIRP*, *63*, 132–137. https://doi.org/10.1016/j.procir.2017.03.085.

Jiang, L., Zhang, H., & Cai, Z. (2009). A novel bayes model: Hidden naive bayes. *IEEE Transactions on Knowledge and Data Engineering*, *21*(10), 1361–1371. https://doi.org/10.1109/TKDE.2008.234.

JUNG, S. (2017). *Samsung S Skin Analyzes and Improves Your Skin*. Medgadget. www.medgadget.com/2017/01/samsung-s-skin-analyzes-improves-skin.html.

Kamruzzaman, S. M., & Jehad Sarkar, A. M. (2011). A new data mining scheme using artificial neural networks. *Sensors*, *11*(5), 4622–4647. https://doi.org/10.3390/s110504622.

Kang, J., Fan, K., Zhang, K., Cheng, X., Li, H., & Yang, Y. (2021). An ultra light weight and secure RFID batch authentication scheme for IoMT. *Computer Communications*, *167*(2019), 48–54. https://doi.org/10.1016/j.comcom.2020.12.004.

Karimy, Mahmood, Araban, Marzieh, Mesri, Mehdi, Armoon, Bahram, Koohestani, Hamid Reza, & Azani, Hadi. (2020). Data on epidemiological and clinical characteristics of coronavirus-infected (COVID-19) individuals in Iran. Figshare. Dataset. https://doi.org/10.6084/m9.figshare.12446120.v1

Karthi, P., & Jayakumar, M. (2020). Smart integrating digital contact tracing with IoMT for COVID-19 using RFID and GPS. *Journal of Xi'an Shiyou University, Natural Science Edition*, *16*(12).

Lee, H. J., Lee, S. H., Ha, K. S., Jang, H. C., Chung, W. Y., Kim, J. Y., Chang, Y. S., & Yoo, D. H. (2009). Ubiquitous healthcare service using Zigbee and mobile phone for elderly patients. *International Journal of Medical Informatics*, *78*(3), 193–198. https://doi.org/10.1016/j.ijmedinf.2008.07.005.

López-Soriano, S., & Parrón, J. (2015). Wearable RFID tag antenna for healthcare applications. Proceedings of the 2015 IEEE-APS Topical Conference on Antennas and Propagation in Wireless Communications, IEEE APWC, 287–290. https://doi.org/10.1109/APWC.2015.7300156.

Lotte, F., Congedo, M., Lécuyer, A., Lamarche, F., & Arnaldi, B. (2007). A review of classification algorithms for EEG-based brain–computer interfaces. *Journal of Neural Engineering*, *4*(2), R1.

Lou, D., Chen, X., Zhao, Z., Xuan, Y., Xu, Z., Jin, H., Guo, X., & Fang, Z. (2013). A wireless health monitoring system based on android operating system. *IERI Procedia*, *4*, 208–215. https://doi.org/10.1016/j.ieri.2013.11.030.

Manzoor, S., Karmon, P., Hei, X., & Cheng, W. (2020). Traffic aware load balancing in software defined WiFi networks for healthcare. 2020 Information Communication Technologies Conference, ICTC, 81–85. https://doi.org/10.1109/ICTC49638.2020.9123278.

Matzik, A., & Anwar, S. (2016). Review of electrical filters. *IJISET-International Journal of Innovative Science, Engineering & Technology*, *3*(4), 543–556. www.ijiset.com.

Mittal, M., Tanwar, S., Agarwal, B., & Goyal, L. M. (2019). Energy conservation for IoT devices. In *Concepts, Paradigms and Solutions, Studies in Systems, Decision and Control, in Preparation*. Springer Nature Singapore Pte Ltd. (Issue May). https://doi.org/10.1007/978-981-13-7399-2.

Mohanty, S. N., Ramya, K. C., Rani, S. S., Gupta, D., Shankar, K., Lakshmanaprabu, S. K., & Khanna, A. (2020). An efficient lightweight integrated blockchain (ELIB) model for IoT security and privacy. *Future Generation Computer Systems*, *102*, 1027–1037. https://doi.org/10.1016/j.future.2019.09.050.

Neupane, A., Hansen, D., Sharma, A., Fails, J. A., Neupane, B., & Beutler, J. (2020). A review of gamified fitness tracker apps and future directions. CHI PLAY 2020 – Proceedings of the Annual Symposium on Computer-Human Interaction in Play, 522–533. https://doi.org/10.1145/3410404.3414258.

neurometrix. (2002). *Product & Services*. Machinery and Production Engineering. https://doi.org/10.1089/gen.32.15.07.

Nweke, H. F., Teh, Y. W., Mujtaba, G., & Al-garadi, M. A. (2019, June). Data fusion and multiple classifier systems for human activity detection and health monitoring: Review and open research directions. *Information Fusion*, *46*(2018), 147–170. https://doi.org/10.1016/j.inffus.2018.06.002.

Panarello, A., Tapas, N., Merlino, G., Longo, F., & Puliafito, A. (2018). Blockchain and iot integration: A systematic survey. In *Sensors (Switzerland)* (Vol. 18, Issue 8). https://doi.org/10.3390/s18082575.

Pourzanjani, A., Quisel, T., & Foschini, L. (2016). Adherent use of digital health trackers is associated with weight loss. *PLoS One*, *11*(4), 1–14. https://doi.org/10.1371/journal.pone.0152504.

Reelfs, J. H., Hohlfeld, O., & Poese, I. (2020). *Corona-Warn-App: Tracing the Start of the Official COVID-19 Exposure Notification App for Germany* (pp. 19–21). https://doi.org/10.1145/3405837.3411378.

Rghioui, A., Lloret, J., & Oumnad, A. (2020). Big data classification and internet of things in healthcare. *International Journal of E-Health and Medical Communications*, *11*(2), 20–37. https://doi.org/10.4018/IJEHMC.2020040102.

Schmidhuber, J. (2015). Deep learning in neural networks: An overview. *Neural Networks*, *61*, 85–117. https://doi.org/10.1016/j.neunet.2014.09.003.

Schueler, M., Cox, A., Yuan, S., Crouch, S., Graham, J., & Hall, W. (2019). From observatory to laboratory: A pathway to data evolution. IET Conference Publications (CP756), 1–8. https://doi.org/10.1049/cp.2019.0137.

Shao, D., Yang, Y., Liu, C., Tsow, F., Yu, H., & Tao, N. (2014). Noncontact monitoring breathing pattern, exhalation flow rate and pulse transit time. *IEEE Transactions on Biomedical Engineering*, *61*(11), 2760–2767. https://doi.org/10.1109/TBME.2014.2327024.

Suvarna, R., Kawatkar, S., & Jagli, D. (2016). Internet of medical things [IoMT]. *International Journal of Advance Research in Computer Science and Management Studies*, *4*(6), 173–178.

Tan, B., Chen, Q., Chetty, K., Woodbridge, K., Li, W., & Piechocki, R. (2018). Exploiting WiFi channel state information for residential healthcare informatics. *IEEE Communications Magazine*, *56*(5), 130–137. https://doi.org/10.1109/MCOM.2018.1700064.

Tang, Y., Cao, G., Li, H., & Zhu, K. (2010). The design of electronic heart sound stethoscope based on Bluetooth. 2010 4th International Conference on Bioinformatics and Biomedical Engineering, ICBBE, 8–11. https://doi.org/10.1109/ICBBE.2010.5516342.

Timothy, M., & Sauer, J. A. B. (2014). The efficacy of goodmark medical's solution using the BAM labs® smart bed technologyTM in the prevention of pressure ulcers. *Journal of Aging Science*, *2*(2). https://doi.org/10.4172/2329-8847.1000123.

Tsai, M. H., Pan, C. S., Wang, C. W., Chen, J. M., & Kuo, C. B. (2019). RFID medical equipment tracking system based on a location-based service technique. *Journal of Medical and Biological Engineering*, *39*(1), 163–169. https://doi.org/10.1007/s40846-018-0446-2.

Wei, Y., Zhou, J., Wang, Y., Liu, Y., Liu, Q., Luo, J., Wang, C., Ren, F., & Huang, L. (2020). A review of algorithm & hardware design for AI-based biomedical applications. *IEEE Transactions on Biomedical Circuits and Systems*, *14*(2), 145–163. https://doi.org/10.1109/TBCAS.2020.2974154.

WHO. (2020). Health Topics-Coronavirus: Symptoms. Available at: https://www. who.int/healthtopics/coronavirus# tab= tab_3 (accessed 19 October 2020).

WHO. (2021). *Coronavirus Disease (COVID-19) Pandemic*. WHO.

Wu, X., Kumar, V., Ross, Q. J., Ghosh, J., Yang, Q., Motoda, H., McLachlan, G. J., Ng, A., Liu, B., Yu, P. S., Zhou, Z. H., Steinbach, M., Hand, D. J., & Steinberg, D. (2008). Top 10 algorithms in data mining. In *Knowledge and Information Systems* (Vol. 14, Issue 1). https://doi.org/10.1007/s10115-007-0114-2.

Yan, K., & Zhang, D. (2015). Feature selection and analysis on correlated gas sensor data with recursive feature elimination. *Sensors and Actuators B: Chemical*, *212*, 353–363.

Zhang, T., Lu, J., Hu, F., & Hao, Q. (2014). Bluetooth low energy for wearable sensor-based healthcare systems. 2014 IEEE Healthcare Innovation Conference, HIC, 251–254. https://doi.org/10.1109/HIC.2014.7038922.

Cognitive Analytics of Social Media for Industrial Manufacturing under Internet of Things Technology

10

Zhihan Lv
Uppsala University, Sweden

Contents

DOI: 10.1201/9781003244714-10

10.1 INTRODUCTION

As the information technology develops, the deep integration of industrial manufacturing and information technology has accelerated the transformation of its production mode and industrial chain, and information technology has gradually penetrated into all key links of industrial manufacturing. In this context, 'Industry 4.0' is proposed [1]. The single industrial manufacturing has also been transformed into a new production service-oriented manufacturing mode based on the industrial Internet, and the collaborative development mode and the intelligent manufacturing mode have developed together. Whereas, this transformation is not achieved overnight, and various setbacks encountered in this process have gradually emerged [2–3]. Thus, to realize the rapid development and transformation of industrialization, the understanding and application of the new connotation of intelligent industrial manufacturing has become the most critical part, which has successfully attracted the attention of researchers in related fields.

In the Internet era, the combination of traditional industrial manufacturing and the internet has become a new normal. The industrial Internet is network-based, which combines traditional factories with information technology [4]. In traditional factories, most of the production equipment and office equipment are simply connected with gateways through the intranet, while in today's industrial Internet factories, industrial manufacturing fully realizes the data-based business process, and introduces new equipment and new business process to create digital and intelligent inside and outside the factory [5–6]. Of course, how to balance the production cost and customer satisfaction is also an urgent problem to be solved. With the current service transformation of manufacturing enterprises, the improvement of intelligent internet technology and the implementation of enterprise ecological strategy, all fields are developing towards building the ecosystem of intelligent product service [7]. Meanwhile, with the upgrading of industrial systems worldwide, higher performance computing, intelligent analysis, low-cost data acquisition, and the interconnection and integration of all things, innovation and change have become the driving force and wind vane of industrial development in the new era. The Internet of Things (IoT) for industrial manufacturing, namely industrial Internet of Things (IIoT), is the product of the combination of traditional factories and information technology based on the network. The traditional factory intranet only uses gateways to connect production equipment and office equipment. In the IIoT factory, it fully realizes the digitization of business processes, and introduces new equipment (robots, automated guided vehicle (AGV), and mobile portable devices) and new business processes (asset performance information management, personnel and material positioning, predictive maintenance, etc.). It connects the network with new equipment, making the production and sales inside and outside the factory more digital and intelligent [8–9]. In addition, ant colony algorithm (ACO), as an intelligent optimization algorithm, is characterized by global optimization performance, strong universality, and parallel computing. When solving the problem of path planning in IoT, set virtual ants to explore different routes. The final preferred path is shorter and more pheromone. According to the principle of 'more pheromone, closer route', the best route can be selected [10], which is of great significance to improve the network performance of social media in industrial manufacturing.

To sum up, as the service-oriented process accelerates and the intelligent technology progresses in industrial manufacturing, a situational intelligent network environment is built based on network interconnection to accelerate the intelligent transformation of industrial manufacturing. Moreover, a cognitive service analysis model of industrial manufacturing is constructed, and the performance of the improved algorithm is compared with other algorithms.

10.2 RELATED WORK

10.2.1 Development Status of Intelligent Industrial Manufacturing

At present, the information science and technology develop fast, and is gradually integrated and penetrated in industrial production and manufacturing, which greatly impact the reform of the industrial field. The intelligence of industrial production is the most obvious, which has been studied by many researchers. Terziyan et al. (2018) suggested that, under the trend of Industry 4.0, intelligent factory became a trend. In this trend, using Pi-Mind to carry out the intelligent reform of industrial manufacturing, it was finally found that this technology was a set of models, technologies, and tools, which was built based on the principle of value-based biased decision making and creative cognitive computing, so that the rationality of decision making in industry was significantly enhanced [11]; Han et al. (2019), according to the fact that the converter mouth flame is the comprehensive external manifestation of physical and chemical reactions in the process of steelmaking, used USB4000 spectrometer to obtain the continuous spectral information of converter mouth flame. In the framework of Internet of things (IoT), a dynamic prediction model of carbon content and temperature value of molten steel in the later stage of steelmaking was established. The results showed that the steelmaking model realized a key control of steelmaking process [12]. He et al. (2020) established an intelligent manufacturing framework suitable for microcircuit module manufacturing by studying the characteristics and requirements of microcircuit module manufacturing. After upgrading the software and hardware conditions, the intelligent factory had many intelligent functions, and the order information and processing program could be automatically sent to the production equipment. Key data of production process could be uploaded automatically in real time. The scientific, efficient, and agile production organization and management was effectively improved [13]. Peng et al. (2020) believed that the intelligent health management of industrial equipment was an extension of traditional fault diagnosis. They analyzed the definition and application fields of intelligent management of industrial equipment, knew the difficulties and bottlenecks of various technologies through comparison, and finally discussed the feasible scheme of future health management and integration [14].

10.2.2 Development Trend of IIoT

As the integration of intelligent devices, intelligent systems, and intelligent decision-making integrates with the latest information technology, Li et al. (2017) applied industrial internet technology to improve the productivity of the entire industrial economy, and reduce cost and waste. In addition, they discussed the intelligent technologies of industrial network, industrial intelligent perception, cloud computing, big data, intelligent control, and security management to fully exploit the potential of industrial internet system [15]. Sisinni et al. (2018) clarified the concepts of IoT, Industrial IoT, and Industry 4.0 in the rapid development of IoT, and analyzed the opportunities brought by this paradigm change, and finally systematically summarized the latest research results and potential research directions of Industrial IoT [16]. Zhang et al. (2019) found that, with the emergence of many industrial Internet platforms, the sharing of manufacturing services among multiple stakeholders was more frequent than ever before. Then, according to this situation, an improved non-dominated sorting genetic algorithm (GA) II was proposed, which combined tabu search and improved k-means mechanism to find the optimal solution set. Finally, it was found that the diversity, convergence, and stability of the solution all verified the validity, and it was further found that the change of consumer preference had little effect on the long-term utility of suppliers [17]. Narayanan et al. (2020) analyzed the emerging technologies related to IIoT pervasive edge computing (PEC) supported by 5G and later communication networks, and emphasized the view that PEC paradigm was a very suitable and important deployment mode of industrial communication networks [18].

10.2.3 Application Status of Social Media in Industrial Manufacturing

Patel et al. (2018) proposed a semantic web of things for Industry 4.0 (SWeTI). Through the real use case scenario, they also showed how the technology of SWeTI could meet the challenges of Industry 4.0, promote cross-departmental and cross-domain integration of the system, and develop intelligent services for intelligent manufacturing [19]. Gulati et al. (2019) discussed the significant opportunities brought by the introduction of new concept of Social Internet of Things (SIoT) into manufacturing industry, and proposed the reference architecture of SIoT. They discussed the architecture from the semantic point of view, proposed a new method of manufacturing resource (asset) relationship management, and designed a use-case scenario and simulation model for the performance management of automatic filling industry equipment. Finally, the model proposed provided a reliable basis for the future development of IIoT applications for researchers and developers [20]. Xu et al. (2020) proposed a collaborative method for the quantification and placement of edge servers (ESs) of industrial cognitive Internet of vehicles (CioV) social media services. Meanwhile, they estimated the approximate number of ESs by the population initialization strategy of canopy and K-medoids clustering. Then, they used non-dominated sorting genetic algorithm to obtain higher quality of service (QoS). Finally, CQP was evaluated based on the real ITS social media data set in China [21]. Bashir et al. (2021) focused on the cognitive analytics of social media in industrial manufacturing, and combined smart factories, smart machines, networked processes, and big data to promote industrial growth and change models. Eventually, they found that social media as an important part of the industrial value chain, collaboration perception or crowd perception could help manufacturers, suppliers, and customers understand and use the insights obtained from tremendous perception data, thus obtaining competitive advantage [22].

In conclusion, the preceding analysis reveals that there are many studies related to IIoT, but there are few researches on the combination of social media, industrial manufacturing, and IoT, such as building a characteristic IIoT platform by using structural restructuring, increasing optimization strategies, and improving service quality. Thus, the cognitive analytics of industrial Internet services and social media under situational intelligence proposed is of great significance.

10.3 METHOD

10.3.1 Cognitive Service Demand Analysis of IoT in Intelligent Manufacturing

As the intelligent manufacturing is continuously popularized, digital sensors, social media, mobile devices, enterprise portals and other data sources generate great data. By 2019, the data volume has exceeded TB (terabyte). But due to the heterogeneity of data sources, the collected data present many forms like structured, semi-structured, and unstructured, among which unstructured data account for the majority. For example, data are generated by customers' purchase behavior, lifestyle, time and space state, and operation state of the enterprise, which are affected by noise, attack, failure, and other factors [23]. Thus, it is necessary to understand the cognitive services of IoT in industrial manufacturing. In general, the role of IoT in intelligent manufacturing is shown in Figure 10.1.

As Figure 10.1 suggests, the impact of IoT in intelligent manufacturing is mainly analyzed from the scale, context perception, reconfigurability, and complexity. From the perspective of scale, traditional intelligent manufacturing information systems include network communication equipment, terminals, servers, software, data, processes, and people, most of which only exist in cyberspace. In the intelligent manufacturing mode, IoT blurs the boundary between cyberspace and physical space, and adds many entities from physical space such as sensors, machines, smart phones, and radio frequency identification (RFID) to the intelligent manufacturing field through wired and wireless technologies such as Ethernet, Zigbee and Wi-Fi, so that the data scale in its system increases explosively.

Regarding the situational perception, it is mainly divided into four stages: situational acquisition, situational simulation, situational reasoning, and situational communication. In the situational acquisition stage, IoT-enabled physical entities mainly enable themselves to collect tremendous original data by embedding sensors, brakes and RFID, that is, original context data. In the situational simulation stage, based on sensor ontology, sensor devices and their behavior can be simulated, and the obtained context is transformed into

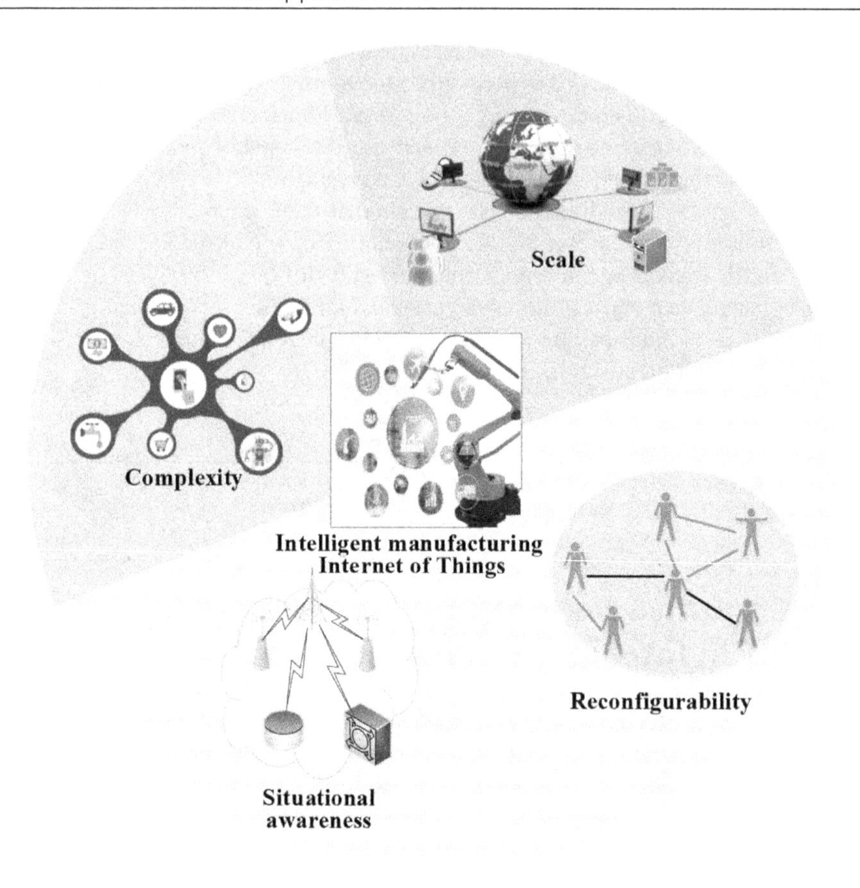

FIGURE 10.1 The role of IoT in intelligent manufacturing.

ontology combined with semantic technology. In the situational reasoning stage, since IoT-enabled equipment has the ability of self-monitoring, self-comparison, and self-configuration, higher-level context information can be inferred from the original context data of industrial manufacturing, such as the remaining service life of the equipment [24]. Of course, physical entities have social ability, and other higher-level situational information can be inferred through social activities. In the situational communication stage, according to the advanced context information obtained by reasoning, cyberspace transmits relevant information to users in the form of query, push, subscription, etc.

For reconfigurability, IoT entity is a typical agent, which is characterized by autonomy, reactivity, initiative, sociality, and so on. When the entity fails or needs change, the entity will conduct flexible and timely self-organizing negotiation to reconstruct its own behavior or position. The reconfigurability of IoT mainly appears in the data acquisition layer of intelligent manufacturing in the industrial field.

The complexity of intelligent manufacturing information system is reflected in the number and types of its constituent elements, the diversity and variability of the interaction between constituent elements, and the difficulty of cognitive system. Driven by IoT, the large amount of access of traditional workshop level equipment of intelligent manufacturing system and the big data generated by the equipment have led to a sharp increase in the number and types of constituent elements of intelligent manufacturing system. The self-organization and cooperation between equipment and other organizations have made their interaction more complex and changeable, which also led to an increase in the uncertainty and unpredictability of the structure and behavior of intelligent manufacturing system. It makes it much more difficult for people to understand, use, operate, and improve the intelligent manufacturing system.

Hence, to meet the needs of intelligent development of industrial manufacturing and the application of IoT technology, it is of great significance to construct cognitive service model. Figure 10.2 presents the framework of intelligent manufacturing cognitive service model based on IoT technology constructed.

Figure 10.2 suggests that the data acquisition layer is at the bottom, which can collect data generated by management system, sensors, social media, and mobile devices. In addition, aided by physical sensors (such as edge nodes and contact sensors) and virtual sensors

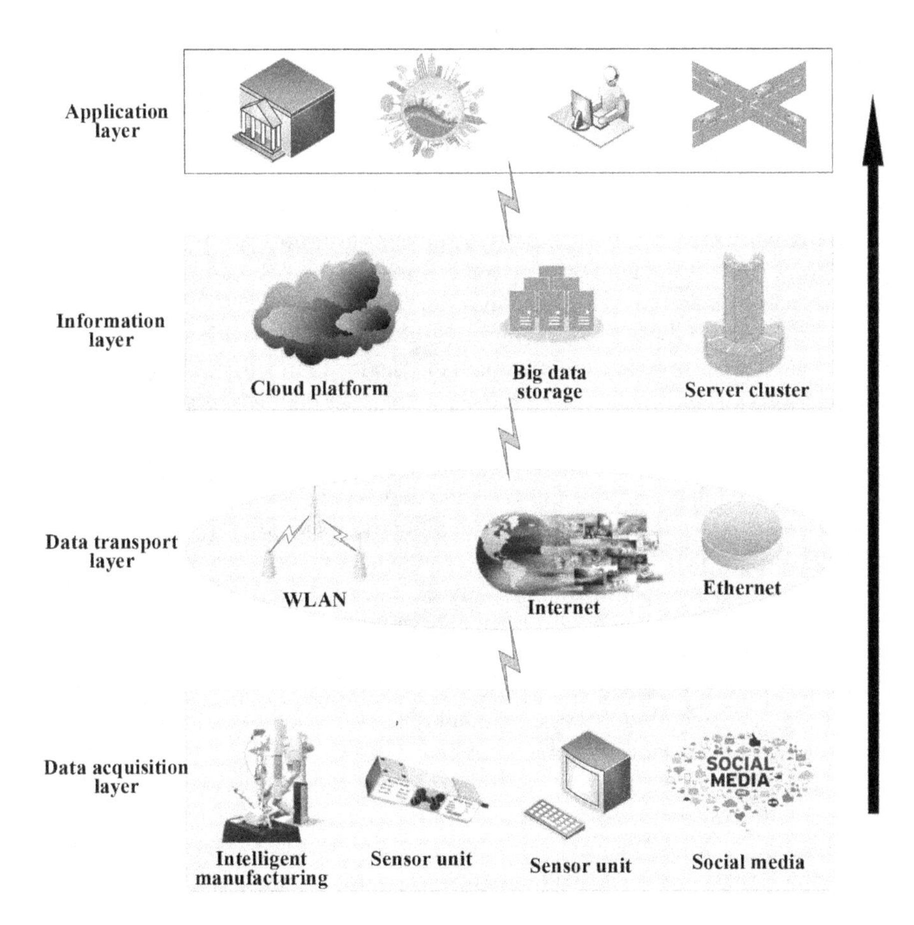

FIGURE 10.2 Schematic diagram of intelligent manufacturing technology based on IoT.

(such as social media platform), the data acquisition layer can perceive the state, market preference, customer behavior, and demand of intelligent manufacturing enterprises. Due to the heterogeneity between different data sources, their requirements for data transmission are also different. For example, notebook computers have high bandwidth requirements, so wireless local area networks (WLAN) and other powerful networks are selected as the access network. Thus, many communication modes appear in the data transmission layer, such as Ethernet and 4G network. In the information layer, as the massive data generated in intelligent manufacturing has far exceeded the storage and computing capacity of traditional information systems, cloud platform has to be used to store and analyze these data. Finally, in the application layer, based on the acquired knowledge, data scientists assist other industrial manufacturing IoTs at all levels to make real-time decisions and take actions. These decisions and actions have feedback on the data acquisition layer, resulting in new data generation in the data acquisition layer.

10.3.2 Analysis of Industrial Manufacturing and Its Intelligent Development

As science and technology develops, the intelligence of industrial manufacturing is rising rapidly. Regarding the manufacturing scope, the scope of intelligent industrial manufacturing is constantly expanding. Vertically, it gradually develops from intelligent equipment to intelligent workshop and rapidly expands from intelligent workshop to the whole intelligent enterprise. Horizontally, it moves from no leap to intelligent supply chain and from intelligent supply chain to the whole intelligent manufacturing industry chain. In the end-to-end, it develops from no step to the whole intelligent manufacturing ecosystem around the whole product life cycle [25–26]. The traditional study of information system is usually based on reductionism and mechanism, but it ignores the fact that information system is a complex system, so the exploration of complex system and its dynamics is very important. Figure 10.3 shows the classification of complex system theory [27].

In the actual production, in addition to the above complex system theory, there are some other algorithms suitable for describing complex system theory. For example, complex network theory, Petri net theory, cellular automata model, sand pile model, multi-Agent system are all feasible means to study complex system theory. In the industrial manufacturing, intelligence and information have become the mainstream, especially intelligence has become the development direction of industrialization at this stage and later. The intelligence of manufacturing also makes the intelligent manufacturing, an advanced manufacturing mode, develop gradually [28–29]. Broadly, manufacturing is the specific process of converting raw materials into finished products; narrowly, manufacturing is part of the production process from raw materials to final products. The combination of intelligent technology and industrial manufacturing is mainly reflected in the two aspects of product intelligence and manufacturing process intelligence [30]. But the intelligent industry is still in the exploration stage, and as various technologies develop, especially information technology, the development pace of intelligent manufacturing will also accelerate day by day.

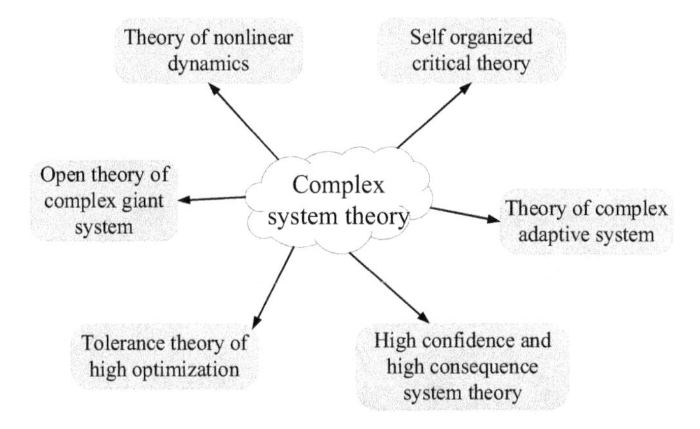

FIGURE 10.3 Classification of complex system theory.

10.3.3 Cognitive Service Analysis Model of Industrial Manufacturing Based on IoT Technology

In industrial manufacturing, cognitive service is mainly based on the matching relationship between manufacturing tasks and manufacturing services. The research on the adaptive ability of industrial internet is usually divided into the adaptive ability of environment and the adaptive ability of function. The adaptive ability of environment is usually expressed as the working ability of basic equipment in different environments. Industrial equipment uses available means to maintain its normal operation according to its working environment, thereby adapting to and optimizing the environmental changes. The adaptive ability of function refers to that all kinds of intelligent devices in industrial internet can realize the normal operation and use of the link of the controlled object by controllable means according to the sudden situation of the controlled object and the change of external environment, thus ensuring its safety [31–32]. Based on the industrial internet, a cognitive service framework of industrial Internet based on situational intelligence is designed. In this framework, based on the network interconnection, a situational intelligence network environment is built, which is divided into four layers: industrial equipment layer, situational intelligence network layer, cognitive service management layer, and application layer, which is centered on data flow and control flow. From three dimensions, namely, industrial manufacturing chain, product cycle chain, and enterprise management chain, a three-dimensional cognitive service adaptive system framework of IIoT is formed (Figure 10.4).

In this model, new complex manufacturing tasks are emerging in the process of industrial manufacturing. The equipment in the industrial equipment layer is engaged in production activities and consumes a lot of resources. The scenario intelligence network layer is responsible for data collection, cleaning, preprocessing, and storage. The cognitive services management layer obtains rich computing resources from the network layer and decomposes the instructions issued by different energy management modules, control the industrial production equipment, and finally return the execution results to the

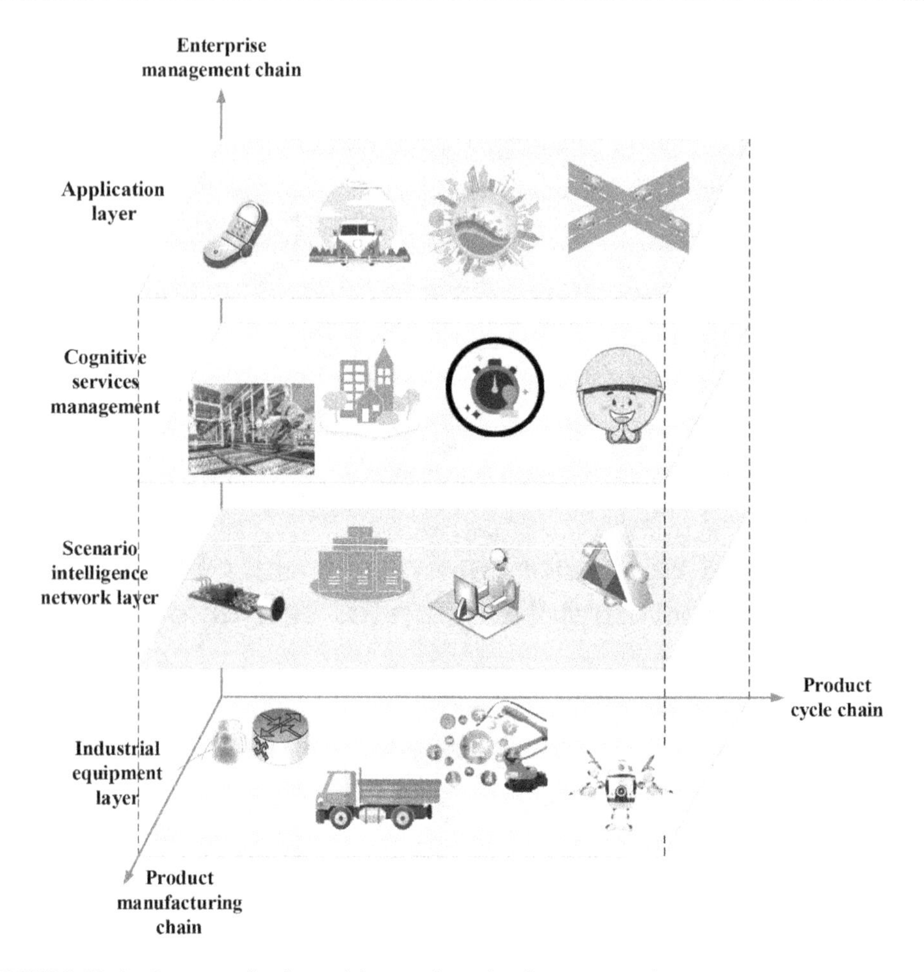

FIGURE 10.4 Framework of cognitive service adaptive system of IIoT.

corresponding energy management module. In terms of quality, service quality represents the current state of service, which is constrained by the current use environment. As the service environment may face unpredictable context changes, which may have uncertain environmental impact, the reasons affecting service quality are divided into four categories: service accession, service failure, service exit, and service evolution, and finally the most efficient service value chain is found as the best service value chain of service quality evaluation and the best goal of service quality evaluation (Figure 10.5) [33–34].

The evaluation model of service value chain is as follows:

$$Q_{SQ} = \left(SM_j, P, V\right) = \begin{pmatrix} & P_1 & V_1 \\ SM_j & P_2 & V_2 \\ & \vdots & \vdots \\ & P_n & V_{nj} \end{pmatrix} \left(j = 1, 2, \cdots, m\right) M_i \tag{10.1}$$

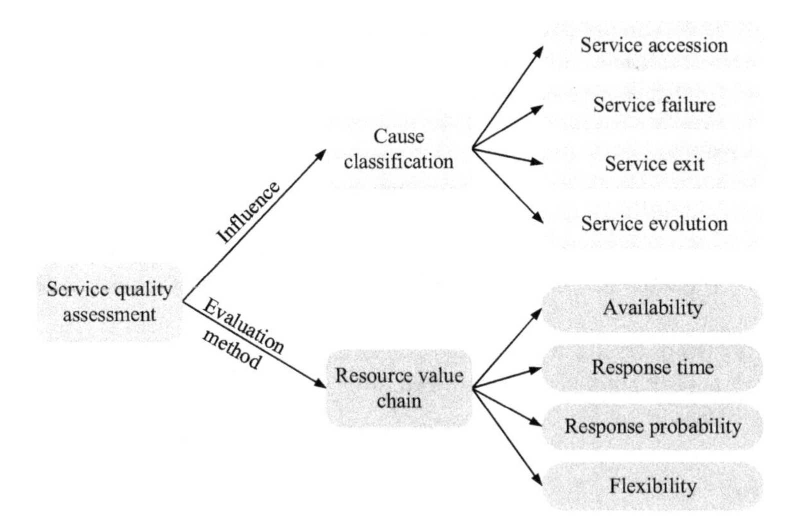

FIGURE 10.5 Framework of service quality assessment.

$$st.\ \ C = (C_1, C_2, \cdots, C_n) \tag{10.2}$$

Q_{SQ} refers to the service value chain to be evaluated, $C = (C_1, C_2, \cdots, C_n)$ is set as the measurement condition set, and $M_i = (P_i, V_i)$ is the feature element. P_i (i=*1*, *2*, . . . , *n*) refers to the characteristic factors that affect the evaluation of service value chain, i.e. evaluation index, and V_j denotes the weight of SM_j of service value chain on the quantification of index P_i [35]. However, since the goal of service adaptation is to minimize the completion time, reduce the service cost as much as possible, and decrease the total equipment load, four main parameters of the definition index P_i are defined by using DqoS dynamic quality estimation, namely availability (A), response time (RT), response probability (RP), and flexibility (F) [36]. Availability (A) suggests whether the calling service is currently available. When A=1, the calling service is available; otherwise, A=0, the calling service is unavailable. In other words, the service is not callable, and it indicates that the service is deleted or the service response times out, which is expressed as Equation 10.3.

$$A = \begin{cases} 1 & \text{Service available} \\ 0 & \text{Service not available} \end{cases} \tag{10.3}$$

RT refers to the time taken by the service call to send the request and receive the last byte response. RT is inversely proportional to service quality, and the lower RT value means better service. $T_{timeout}$ is introduced as the maximum duration allowed, which is expression as Equation 10.4.

$$T_{timeout} = \tau \cdot RT, \tau > 1 \tag{10.4}$$

RP indicates the probability that a service is ready to respond to a request within a period of not exceeding the time limit. RP is directly proportional to the overall quality of the service, whose update rule is expressed as Equation 10.5.

$$RP_{t+1} = \frac{RP_t \cdot count + AV_{t+1}}{count + 1} \tag{10.5}$$

Where, RP_{t+1} refers to RP of service call at (t+1) time, *count* denotes the number of service calls in [t, t + 1] continuous time, and AV_{t+1} suggests the value of service availability at (t+1) time.

Flexibility (F) refers to the probability that an exception can still meet the user's needs after adaptive adjustment during service execution. The update rules of F are expressed as Equation 10.6.

$$F_{t+1} = \lambda F_t + (1-\lambda)\frac{M_r}{M} \tag{10.6}$$

F_{t+1} stands for the flexibility of calling service at (t+1) time, λ is the weight, M_r indicates the number of times that the service can still meet the user's needs after adaptive adjustment in [t, t+1] continuous time, and M refers to the number of exceptions in the service execution process. Then, the DQoS attribute set of each service chain SM_j is recorded as q(SM_j)={rt, a, rp, f}. Hence, the linear weighted sum method is used to construct the total objective function of service quality assessment, expressed as Equation 10.7.

$$MaxQ = v_1\left(\sum\nolimits_{i=1}^{n} RT_i\right)^{-1} + v_2\prod\nolimits_{i=1}^{n} A_i + v_3\prod\nolimits_{i=1}^{n} RP_i + v_4\sum\nolimits_{i=1}^{n} F_i \tag{10.7}$$

The constraints are shown in Equations 10.8–10.10.

$$st.\sum_{SM_i} RT \le D_{SM_i}, \forall 1 \le j \le m, RT \le T_{timeout} \tag{10.8}$$

$$\sum_{SM_i}(A \cdot RP)_t \ge 0, \forall 1 \le j \le m, t \ge 0, 0 \le RP \le 1 \tag{10.9}$$

$$\sum_{i=1}^{n} v_i = 1, n = 4 \tag{10.10}$$

In the constraints, D_{SM_i} is the total deadline of the service chain SM_i. When evaluating the operation efficiency of service value chain SM_i, the proportion of each attribute condition is different. v_i is the weight coefficient, which refers to the relative importance of each condition.

Further, the generation process of IoT in intelligent manufacturing is optimally scheduled. After the appropriate process is arranged for each workpiece, the appropriate processing equipment is determined, and the cost, quality, and load constraints of each workpiece are met, thus optimizing the final task cost and service quality. First, in the budget cost constraint, different resources have varying costs, and so do different tasks. Next, considering the differences of resource requirements of tasks, the maximum and minimum budget cost constraint rules are formulated to reflect the costs of different tasks [37].

$$Cost_{RTast}^{\min}(t) \le Cost_{RTast}(t) \le Cost_{RTast}^{\max}(t) \tag{10.11}$$

$$Cost_{LTast}^{\min}(t) \le Cost_{LTast}(t) \le Cost_{LTast}^{\max}(t) \tag{10.12}$$

$$Cost_{STast}^{\min}(t) \le Cost_{STast}(t) \le Cost_{STast}^{\max}(t) \tag{10.13}$$

Second, deadline constraint. Deadline violation rate is an important evaluation index of task quality [38]. For users, considering the constraints of the deadline $D_i^{\max}$ on the completion time T_i, the task completion time must be less than the maximum deadline set by the user.

$$T_{RTast}(t) \le D_{RTast}^{\max}(t) \tag{10.14}$$

$$T_{LTast}(t) \le D_{LTast}^{\max}(t) \tag{10.15}$$

$$T_{STast}(t) \le D_{STast}^{\max}(t) \tag{10.16}$$

In addition, it also includes device load constraints and QoS constraints. The device load refers to the number of tasks to process the service. The maximum load $L_{j\max}$ is the maximum number of tasks that can be borne. Once the maximum load $L_{j\max}$ is exceeded, the device shall reject the task request. QoS constraints means that the service quality $Q(i)$ completed by any subtask cannot be less than the minimum service quality $Q_{i\min}$ required by the manufacturing service demander. The relationship between the two groups is displayed as Equation 10.17.

$$Q(i) \ge Q_{i\min}, i = 1, 2, \cdots, n \tag{10.17}$$

Finally, the standard ACO algorithm is improved by using multi strategies, the process service adaptive method of IIoT is formed, and the optimal intelligent scheme in IIoT is obtained adaptively. The interaction and restriction between multiple targets lead to the failure to reach the optimal value, and cannot judge the absolute merits and disadvantages of the solution by a standard. Thus, the Pareto non-inferior solution set is introduced to evaluate the multi-objective solution. The standard Pareto multi-objective optimal solution set is to compare the advantages and disadvantages of the solution in the form of minimum value. Consequently, the above two-layer adaptive model is converted into the form of minimum value, as Equation 10.18.

$$Minisize \sum_{x} F(x) = \min Cost(x), \left[2 - \max Q(x)\right] \tag{10.18}$$

Pareto multi-objective optimal solution set represents the set of solutions that are not dominated by any solution in a set of feasible solutions. For example, the Pareto dominating solution x_2 of x_1 is to satisfy that each objective of x_1 is not worse than that of x_2, and that there is at least one objective function so that the objective value of x_1 is better than that of x_2. Meantime, considering the multi QoS requirements, the study uses the normalized actual QoS attribute information calculation to optimize the heuristic function in IoT.

10.3.4 Simulation Analysis

The system is forecasted and analyzed in Matlab network simulation platform. The multi-strategy improved ant colony optimization (ACO) is used to solve the standard multi-objective optimization test function to verify the advantages and disadvantages of the algorithm, and the representative standard optimization test functions ZDT1 and ZDT3 are selected. In ZDT1 function, the mathematical expression is shown in Equations 10.19–10.21.

$$\min f_1(x) = x_1 \tag{10.19}$$

$$\min f_2(x) = g(x)\left(1 - \sqrt{x_1/g(x)}\right) \tag{10.20}$$

$$st.g(x) = 1 + 9\left(\sum_{i=1}^{n} x_i\right)\Big/ n - 1 \tag{10.21}$$

Where x_i ranges in [0,1]. In ZDT3 function, the mathematical expression is shown in Equations 10.22–10.24.

$$\min f_1(x) = x_1 \tag{10.22}$$

$$\min f_2(x) = g(x)\left(1 - \sqrt{x_1/g(x)}\right) - \frac{x_1 \sin(10\pi x_1)}{g(x)} \tag{10.23}$$

$$st.g(x) = 1 + 9\left(\sum_{i=1}^{n} x_i\right)\Big/ n - 1 \tag{10.24}$$

The value range of x_i is the same as that of ZDT1, both of which are [0,1], and only $minf_2(x)$ is different. IGD is the distance index between the Pareto front-end solution obtained by common measurement algorithms and the real Pareto front-end. The lower the IGD value is, the better the convergence and diversity of the algorithm is, and the closer it is to the real Pareto front-end [39]. IGD index is introduced to evaluate the performance of the algorithm, which is calculated as Equations 10.25–10.28.

$$IGD(R,P) = \frac{1}{|R|}\sum_{i=1}^{|R|} Dis\tan ce_i \tag{10.25}$$

$$Dis\tan ce_i = \min_{j=1}^{|p|} \sqrt{\sum_{m=1}^{M}\left(\frac{f_m(r_i) - f_m(a_j)}{f_m^{\max} - f_m^{\min}}\right)^2} \tag{10.26}$$

TABLE 10.1 Parameter Setting

PARAMETER	*VALUE*
Population size	100
Global external files	100
Maximum evaluation times	25,000
Maximum iterations	300
Number of algorithm runs	40

$$r_i \in R, i = 1,2,\cdots,|R| \tag{10.27}$$

$$a_j \in A, j = 1,2,\cdots,|A| \tag{10.28}$$

$Distance_i$ refers to the minimum normalized Euclidean distance between the true value and the approximate value, $f_m^{\max}$ and $f_m^{\min}$ represent the maximum and minimum value on the m-th target in the real Pareto optimal solution set R, and M indicates the number of targets. In addition, the improved ACO is compared with three representative algorithms, multi-objective particle swarm optimization (MOPSO), elitist non-dominated sorting genetic algorithm (NSGA-II), and the strength pareto evolutionary algorithm 2 (SPEA2) regarding completion time, service cost, and resource utilization. In the analysis of data transmission performance of the system, the algorithm is compared with three common data transmission methods of wireless network: cloud server download (CSD), edge server caching (ESC), and multiple communication paths (MCP). In the comparative test, to verify the effectiveness of the algorithm proposed, select the same parameter settings, and make the average value analysis of the calculation results for many times to reduce the impact of random in the performance analysis, as shown in Table 10.1.

10.4 RESULTS AND DISCUSSION

10.4.1 Comparative Analysis of Algorithm Performance of Cognitive Model

The completion time, service cost, and resource utilization rate of the algorithm in this chapter are compared with other algorithms (Figures 10.6 to 10.8). Figures 10.6 to 10.8 show that as the number of subtasks increases, the completion time and service cost of each algorithm are increasing, and obviously the completion time and service cost of the algorithms in this chapter are lower than those of other algorithms. Further analysis of resource utilization shows that as the scheduling times change, the resource utilization rate of the algorithms in this chapter is more than 88%, while the utilization rate of other algorithms is significantly lower. Thus, through the comparison results, it is seen that with the increase of the number of subtasks, the service cost and completion time of the

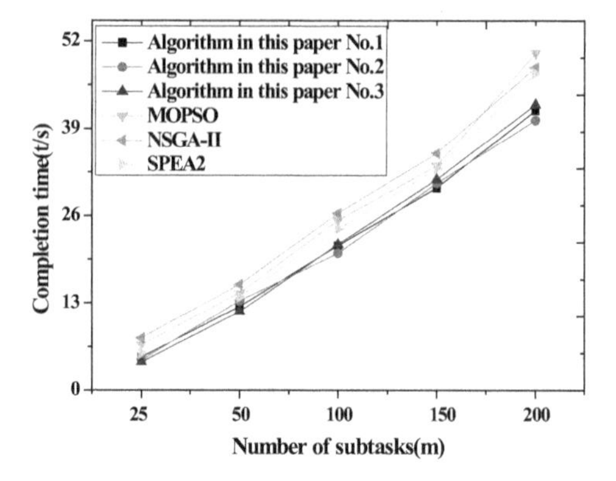

FIGURE 10.6 Comparative analysis of the completion time between the algorithms proposed and other algorithms.

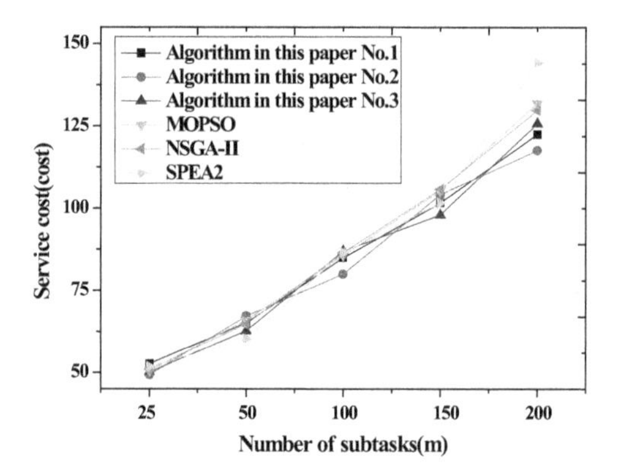

FIGURE 10.7 Comparative analysis of the service cost between the algorithms proposed and other algorithms.

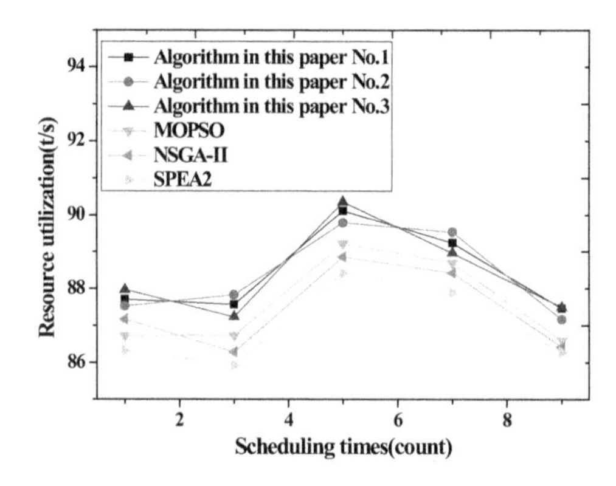

FIGURE 10.8 Comparative analysis of resource utilization between the algorithms proposed and other algorithms.

improved algorithm are lower than other algorithms, and the resource utilization rate is significantly higher.

10.4.2 Comparative Analysis of Data Transmission Performance in Cognitive Model

In the process of intelligent manufacturing, the effective transmission of information is the prerequisite for building a new intelligent industrial system, and the efficient transmission of data is vital in the information interaction between all levels and components of the system. The resulting data transmission performance is shown in Figures 10.9 to 10.16.

Figures 10.9 and 10.10 show that the data transmission delay under different data distribution and different communication rates. In different data distribution schemes, as the data volume increases, the transmission delay shows an increasing trend, but obviously the transmission delay of the algorithms in this chapter is lower than that of ESC, MCP, and CSD algorithm. It may be because the algorithm proposed can not only cache the data with cluster heads and common nodes, but also use mobile connections to quickly obtain the data strategy. In different communication rates, as the communication rate increases, the transmission delay of the four algorithms shows a downward trend, and the delay of the algorithms in this chapter is the smallest. It may be due to the use of cluster heads and common nodes for data caching, which makes the number of data transmission hops reduced.

Figures 10.11 and 10.12 show the success rate of data receiving under different amount of distributed data and different communication rates. In addition to the ESC algorithm, there is no significant change in the other three kinds of data distribution, and it is obvious that the successful data receiving ratio of the algorithm proposed is higher, close to 100%. In different communication rates, it is obvious that the successful data receiving effect of the algorithm proposed is the best. There is no obvious change between the algorithm proposed and the CSD, while ESC and MCP show a downward trend. This may be due

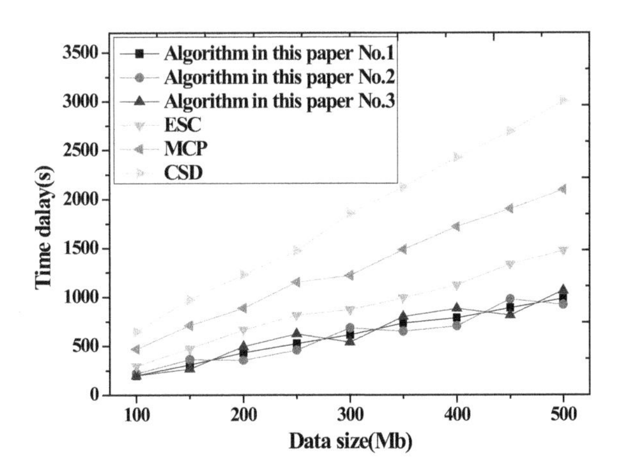

FIGURE 10.9 Transmission delay of acquired data under different amount of distributed data.

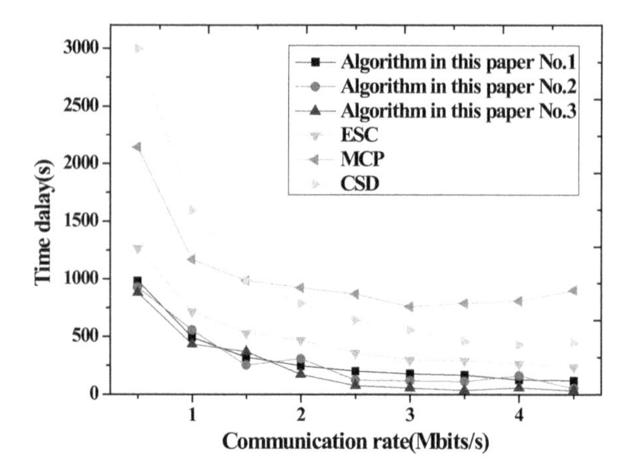

FIGURE 10.10 Transmission delay of acquired data at different communication rates.

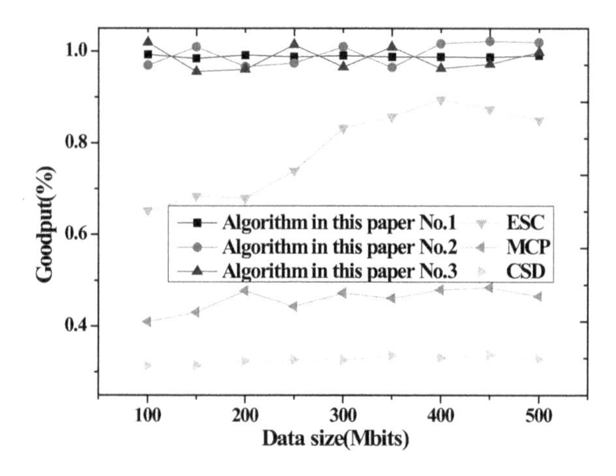

FIGURE 10.11 Comparative analysis of data successful receiving ratio under different data quantity.

to the fact that mobile nodes spend a lot of time in data acquisition, while mobile nodes in MCP and CSD spend a lot of time in communication waiting in the process of network monitoring and data acquisition.

The results of comparative analysis of network throughput are shown in Figures 10.13 and 10.14. In the analysis of the network throughput among the three data sizes of different data distribution schemes, the network throughput proposed is the largest, and its total value is greater than 0.45 Mbit/s. In the analysis of network throughput of three communication rates, it is obvious that with the increase of network communication rate, the network throughput of four strategies will increase, and the performance of the algorithm proposed is the best. When the communication rate is 4.5 Mb/s, the network throughput of this algorithm can achieve about nine, three, and one and a half times of the corresponding performance of MCP, CSD, and ESC, respectively.

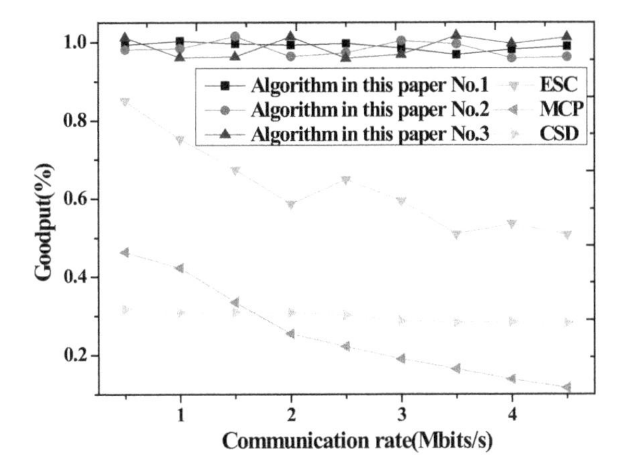

FIGURE 10.12 Comparative analysis of data successful receiving ratio under different communication rates.

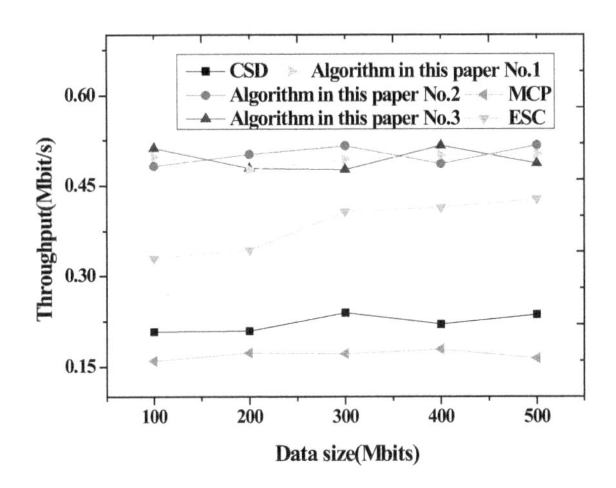

FIGURE 10.13 Comparative analysis of network throughput under different data quantity.

The results of comparative analysis of energy consumption are shown in Figures 10.15 and 10.16. In different data distribution schemes, it can be found that the energy consumption of the four algorithms increases with the increase of data volume, and the energy consumption of the algorithm proposed is the smallest. This is because many cluster heads and common nodes can cache data, reduce the waiting time of mobile nodes, listen to wireless channels, and other times, so that the utilization rate of energy consumption increases. In different communication rates, the algorithm proposed consumes the least energy. Hence, the transmission performance of the algorithm proposed is the best regarding transmission delay, data acceptance rate, network throughput, and energy consumption.

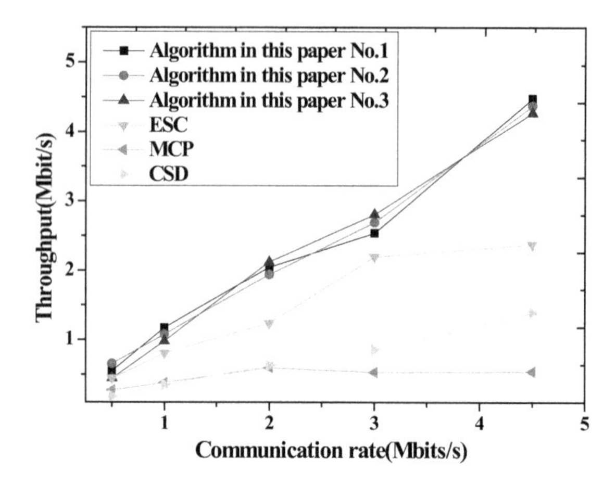

FIGURE 10.14 Comparative analysis of network throughput under different communication rates.

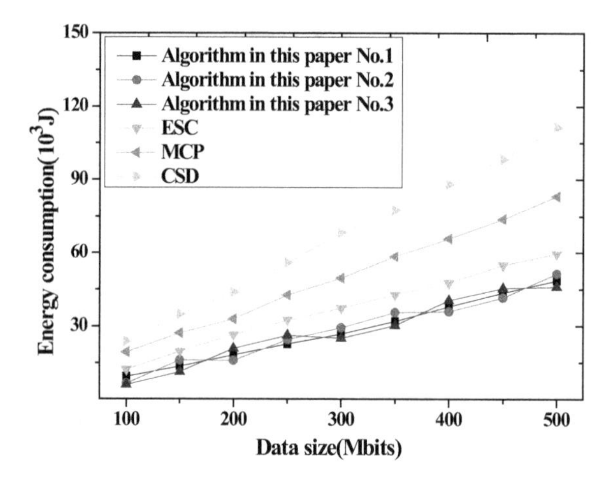

FIGURE 10.15 Comparative analysis of energy consumption under different data distributions.

10.5 CONCLUSION

Under the rapid development of artificial intelligence, to meet the different changes of industrial manufacturing needs, this study creates an intelligent production system model of situational intelligence, and simulates the model. In the comparative analysis of the algorithm proposed and other algorithms, the improved algorithm, as the number of subtasks increases, requires lower service cost and completion time than other algorithms, but higher in resource utilization. While in the comparative analysis of system transmission

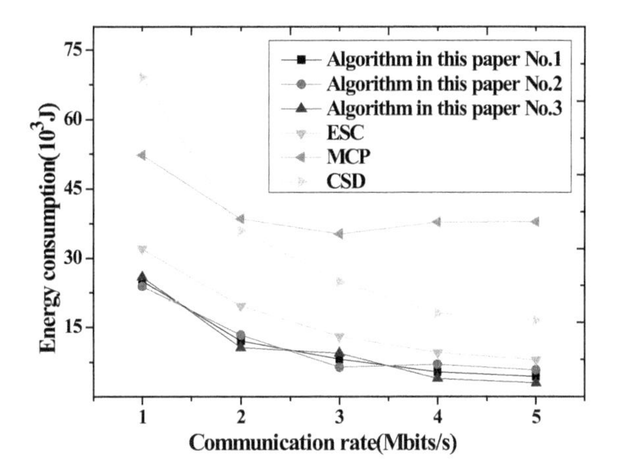

FIGURE 10.16 Comparative analysis of energy consumption at different communication rates.

performance, regarding transmission delay, data success acceptance rate, network throughput, and energy consumption, the performance of the algorithm proposed is the best.

To sum up, the system constructed can obviously have better transmission performance and higher resource utilization rate, which provides experimental reference for later industrial manufacturing and development. However, there are some shortcomings in this study. The data transmission is based on the simulation derivation and optimization under the condition of relatively stable industrial environment. In the specific actual environment, the complex environment will greatly impact the results. Thus, in the follow-up study, the design of a good scheme will be further improved so that the data transmission can adapt to more complex environmental characteristics, so that the availability and fault tolerance of the system are greatly increased.

REFERENCES

[1] Vidanapathirana, M., Meegahapola, L., & Perera, I. (2017). Cognitive analysis of 360 degree surround photos – An augmented image processing technique for photosphere analysis. IEEE Future Technologies Conference (FTC). IEEE.

[2] Psannis, K. E., Stergiou, C., & Gupta, B. B. (2018). Advanced media-based smart big data on intelligent cloud systems. *IEEE Transactions on Sustainable Computing*, 4(1), 77–87.

[3] Fletcher, K. K., Sparks, T. E., Flood, A., & Liou, F. (2017). A SOA approach to improve performance of metal additive manufacturing simulation. 2017 IEEE International Conference on Cognitive Computing (ICCC). IEEE.

[4] Khan, M., Wu, X., Xu, X., & Dou, W. (2017, May). Big data challenges and opportunities in the hype of Industry 4.0. 2017 IEEE International Conference on Communications (ICC). IEEE, pp. 1–6.

[5] Li, L., Chen, C., Wang, Y., He, T., & Guan, X. (2017). *Adaptive Beacon Transmission in Cognitive-OFDM-Based Industrial Wireless Networks*. IEEE Communications Letters.

[6] He, Y., Guo, J., & Zheng, X. (2018). From surveillance to digital twin: Challenges and recent advances of signal processing for industrial internet of things. *IEEE Signal Processing Magazine*, 35(5), 120–129.
[7] Sahoo, P.K., Mohapatra, S., & Sheu, J.P. (2017). Dynamic spectrum allocation algorithms for industrial cognitive radio networks. *IEEE Transactions on Industrial Informatics*, 14(7), 3031–3043.
[8] Cao, B., Wang, X., Zhang, W., Song, H., & Lv, Z. (2020). A many-objective optimization model of industrial internet of things based on private blockchain. *IEEE Network*, 34(5), 78–83.
[9] Hui, H., Zhou, C., Xu, S., & Lin, F. (2020). A novel secure data transmission scheme in industrial internet of things. *China Communications*, 17(1), 73–88.
[10] Vasovala, P. J., Mirchiwala, M. I., Mayank, V., & Ghanchi, V. H. (2021). Application of ant colony optimization technique in economic load dispatch of IEEE-26 bus system with valve point loading. *International Journal for Research in Applied Science and Engineering Technology*, 9(1), 51–58.
[11] Terziyan, V., Gryshko, S., & Golovianko, M. (2018). Patented intelligence: Cloning human decision models for Industry 4.0. *Journal of Manufacturing Systems*, 48, 204–217.
[12] Han, Y., Zhang, C. J., Wang, L., & Zhang, Y. C. (2019). *Industrial IoT for Intelligent Steelmaking with Converter Mouth Flame Spectrum Information Processed by Deep Learning*. IEEE Transactions on Industrial Informatics.
[13] He, Y., Hu, C., Hu, Y., Zhang, M., Zhang, J., & Shi, R. (2020). Construction and Implementation of Microcircuit Module Smart Factory. In *Proceedings of the Seventh Asia International Symposium on Mechatronics* (pp. 1–10). Springer.
[14] Peng, C., Tang, Z., Gui, W., Chen, Q., Zhang, L., Yuan, X., & Deng, X. (2020). Review of key technologies and progress in industrial equipment health management. *IEEE Access*, 8, 151764–151776.
[15] Li, J. Q., Yu, F. R., Deng, G., Luo, C., Ming, Z., & Yan, Q. (2017). Industrial internet: A survey on the enabling technologies, applications, and challenges. *IEEE Communications Surveys & Tutorials*, 19(3), 1504–1526.
[16] Sisinni, E., Saifullah, A., Han, S., Jennehag, U., & Gidlund, M. (2018). Industrial internet of things: Challenges, opportunities, and directions. *IEEE Transactions on Industrial Informatics*, 14(11), 4724–4734.
[17] Zhang, Y., Tao, F., Liu, Y., Zhang, P., Cheng, Y., & Zuo, Y. (2019). Long/short-term utility aware optimal selection of manufacturing service composition toward industrial Internet platforms. *IEEE Transactions on Industrial Informatics*, 15(6), 3712–3722.
[18] Narayanan, A., De Sena, A. S., Gutierrez-Rojas, D., Melgarejo, D. C., Hussain, H. M., Ullah, M., Bayhan, S., & Nardelli, P. H. (2020). Key advances in pervasive edge computing for industrial Internet of Things in 5G and beyond. *IEEE Access*, 8, 206734–206754.
[19] Patel, P., Ali, M. I., & Sheth, A. (2018). From raw data to smart manufacturing: AI and semantic web of things for industry 4.0. *IEEE Intelligent Systems*, 33(4), 79–86.
[20] Gulati, N., & Kaur, P. D. (2019). Towards socially enabled internet of industrial things: Architecture, semantic model and relationship management. *Ad Hoc Networks*, 91, 101869.
[21] Xu, X., Shen, B., Yin, X., Khosravi, M. R., Wu, H., Qi, L., & Wan, S. (2020). Edge server quantification and placement for offloading social media services in industrial cognitive IoV. *IEEE Transactions on Industrial Informatics*, 17(4), 2910–2918.
[22] Bashir, A. K., Mumtaz, S., Menon, V. G., & Tsang, K. F. (2021). Guest editorial: Cognitive analytics of social media for industrial manufacturing. *IEEE Transactions on Industrial Informatics*, 17(4), 2899–2901.
[23] Yan, H., Yang, J., & Wan, J. (2020). KnowIME: A system to construct a knowledge graph for intelligent manufacturing equipment. *IEEE Access*, 8, 41805–41813.
[24] Guo, D., Zhong, R. Y., Ling, S., Rong, Y., & Huang, G. Q. (2020). A roadmap for Assembly 4.0: Self-configuration of fixed-position assembly islands under graduation intelligent manufacturing system. *International Journal of Production Research*, 58(15), 4631–4646.

[25] Mcmahon, P., Zhang, T., & Dwight, R. (2020). Requirements for big data adoption for railway asset management. *IEEE Access*, 8, 15543–15564.
[26] Sengupta, A., Tiwari, A., & Routray, A. (2017). Analysis of cognitive fatigue using EEG parameters. 2017 39th Annual International Conference of the IEEE Engineering in Medicine and Biology Society (EMBC). IEEE.
[27] Allwood, G., Du, X., Webberley, K. M., Osseiran, A., & Marshall, B. J. (2018). Advances in acoustic signal processing techniques for enhanced bowel sound analysis. *IEEE Reviews in Biomedical Engineering*, 12, 240–253.
[28] Mishra, K. V., Shankar, M. B., Koivunen, V., Ottersten, B., & Vorobyov, S. A. (2019). Toward millimeter-wave joint radar communications: A signal processing perspective. *IEEE Signal Processing Magazine*, 36(5), 100–114.
[29] Ghorbanian, M., Dolatabadi, S. H., & Siano, P. (2019). Big data issues in smart grids: A survey. *IEEE Systems Journal*, 13(4), 4158–4168.
[30] Ogbodo, E. U., Dorrell, D. G., & Abu-Mahfouz, A. M. (2017). Performance analysis of correlated multi-channels in cognitive radio sensor network based smart grid. 2017 IEEE AFRICON. IEEE.
[31] Zhu, L., Yu, F. R., Wang, Y., Ning, B., & Tang, T. (2018). Big data analytics in intelligent transportation systems: A survey. *IEEE Transactions on Intelligent Transportation Systems*, 20(1), 383–398.
[32] Abeywardana, R. C., Sowerby, K. W., & Berber, S. M. (2018). Empowering infotainment applications: A multi-channel service management framework for cognitive radio enabled vehicular ad hoc networks. 2018 IEEE 87th Vehicular Technology Conference (VTC Spring). IEEE.
[33] Wang, W. J., Usman, M., Yang, H. C., & Alouini, M. S. (2017). Service time analysis for secondary packet transmission with adaptive modulation. 2017 IEEE Wireless Communications and Networking Conference (WCNC). IEEE.
[34] Fan, X., Wang, F., Wang, F., Gong, W., & Liu, J. (2019). When rfid meets deep learning: Exploring cognitive intelligence for activity identification. *IEEE Wireless Communications*, 26(3), 19–25.
[35] Liu, H., Ong, Y. S., Shen, X., & Cai, J. (2020). *When Gaussian Process Meets Big Data: A Review of Scalable GPs*. IEEE Transactions on Neural Networks and Learning Systems.
[36] Zheng, K., Liu, X. Y., Liu, X., & Zhu, Y. (2019). Hybrid overlay-underlay cognitive radio networks with energy harvesting. *IEEE Transactions on Communications*, 67(7), 4669–4682.
[37] Zhang, Y., Ma, X., Zhang, J., Hossain, M. S., Muhammad, G., & Amin, S. U. (2019). Edge intelligence in the cognitive internet of things: Improving sensitivity and interactivity. *IEEE Network*, 33(3), 58–64.
[38] Wan, J., Tang, S., Hua, Q., Li, D., Liu, C., & Lloret, J. (2017). Context-aware cloud robotics for material handling in cognitive industrial internet of things. *IEEE Internet of Things Journal*, 5(4), 2272–2281.
[39] Shamim Hossain, M., & Muhammad, G. (2019). An audio-visual emotion recognition system using deep learning fusion for a cognitive wireless framework. *IEEE Wireless Communications*, 26(3), 62–68.

Industrial Internet of Things 11

Applications and Challenges

Brijendra Singh
VIT Vellore, Tamilnadu, India

Contents

DOI: 10.1201/9781003244714-11

11.1 INTRODUCTION

Industrial Internet of Things (IIoT) is a network of sensors, intelligent objects, architecture, platforms, and applications that compromise various technologies and enable them to provide a communication environment. The main goal of Industrial IoT applications is to improve the availability of processors, adding intelligence to machines, affordability and enables sensors and other networking devices to communicate efficiently with real-time information. Industrial IoT applications have become very popular among industries and academia. Several sectors use the Industrial Internet of Things. Low-cost connected devices [1] play a vital role in IIoT applications. IIoT services are classified into two different categories as short-range and long-range connectives. Industrial IoT is used by various industries like automotive, agriculture, and healthcare. Predictive maintenance is one of the main advantages of using IIoT in automotive industries.

Key elements of Industrial IoT are assets management, artificial intelligence, big data analytics, safety improvement, cost-effectiveness, automated maintenance, enhanced connectivity, and cyber-physical systems. For better visibility of readers, we have demonstrated these critical elements in Figure 11.1. Assets management is a systematic approach to assess the value which comes from the group or entity from the entire development of the life cycle of the responsible project. IoT can help assess the physical devices and sensor's performance, its failure, and keep track of all these mishappenings to improve

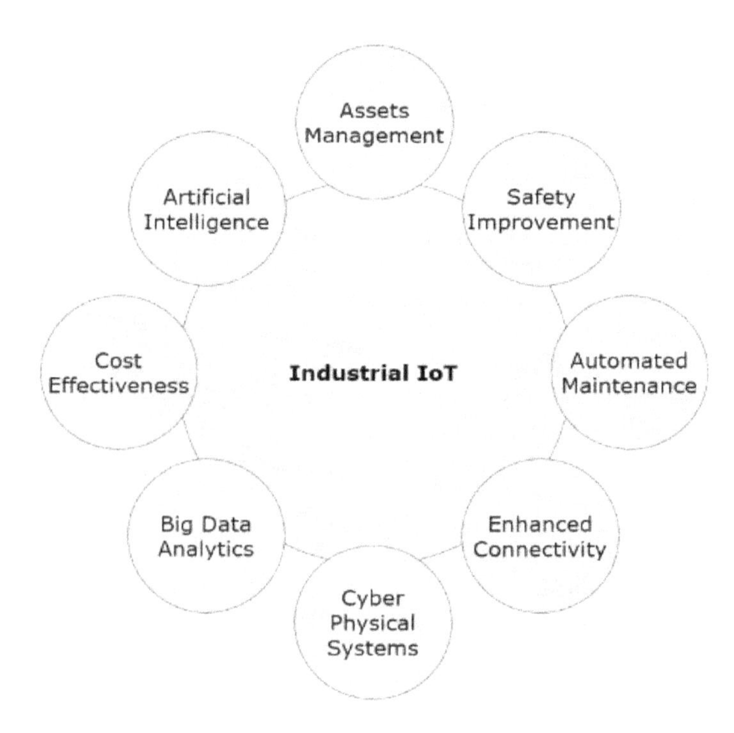

FIGURE 11.1 Industrial IoT key elements.

industrial processes and productivity. Artificial intelligence and big data analytics technologies add intelligence to industrial business processes. An AI-enabled framework [2] based on edge and cloud computing paradigms has proved its efficiency in maintaining the industrial processes. Safety improvements of human being assisted in Industrial IoT are essentially significant with automated and cost-effective solutions. Obtaining high network connectivity and optimizing network coverage are also essential parts of IIoT applications.

Moreover, cyber-physical systems are one of the core components in Industrial IoT systems as they integrate computational power and physical networking. Therefore, it is essential to understand how these components and different emerging technologies are being used in Industrial IoT applications with potential challenges. Therefore, the significant contribution of this book chapter is to provide a better understanding of related technologies, emerging applications in various sectors, and future directions to researchers solving research challenges in Industrial IoT.

The organization of the rest of this book chapter is as follows. The second section gives a detailed overview of significant technologies for Industrial IoT. In Section 11.3, we present the different architectural frameworks for Industrial IoT applications. Efficient and secure approaches in IIoT are presented in Section 11.4. The fifth section depicts the applications of Industrial IoT in various sectors. Implementation challenges incorporating Industrial IoT is presented in Section 11.6. Finally, the conclusion is presented in the last section.

11.2 MAJOR TECHNOLOGIES FOR INDUSTRIAL IOT

Technological interventions play a vital role in the successful implementation of Industrial IoT in different sectors. Integrating various technologies like big data analytics, cloud and fog computing, blockchain, machine/deep learning and 5G is vital in providing automation human-machine-centric applications in industries. Therefore, this section discusses all these major enabling technologies for the Industrial Internet of Things.

11.2.1 Big Data Analytics

IIoT systems use the integration of machine learning and parallel distributed systems such as cloud, grids, big data servers, and analytics. Edge devices that allow a local computer to connect and transmit information over the Internet in Industrial IoT applications usually generate massive amounts of data results in heavy traffic between networking, sensor devices, and cloud server. Due to this enormous amount of data generation, efficient data management is a challenge in IIoT applications. Therefore, there is a demand for big data processing, analytics, and management in Industrial IoT applications. Hadoop is an excellent platform to manage such a large amount of data in a distributed environment.

However, efficient storing, capturing, analyzing, sharing, and searching big data are still significant challenges.

11.2.2 Cloud and Fog Computing

Cloud computing integrated with IoT provides storage, processing power, and networking services in Industrial IoT applications. It offers various services at different levels as a platform, infrastructure, and software, enabling flexibility in business processes. It changes the way how people are doing business by interacting with various IT resources. Cloud computing is mainly used to control and monitor industrial processes, but it cannot monitor industrial machines because of unpredictable WAN delay [3]. Another limitation of the cloud is that it cannot guarantee dependability and trustworthiness. However, the limits of cloud computing, such as high delay, are overcome by introducing fog computing, which is a new standard to implement Industrial IoT solutions. To this end, various researchers [4–5] proposed different fog computing-based architectural platforms to meet Industrial IoT applications' needs.

11.2.3 Blockchain Technology

Industrial IoT applications play a vital role in Industry 4.0. However, existing solutions cannot deal with a malicious attack and a single point of failure. Because of the excellent security services of blockchain technology, it is a good idea to integrate blockchain with IIoT applications. Blockchain-enabled IIoT enhances industrial effectiveness by providing security and immutable storage for Industry 4.0 applications. Moreover, it offers a decentralized solution without the concerns of third-party involvement. Artificial intelligence has the capabilities to address significant issues in blockchain-enabled IIoT applications. Reinforcement learning approaches [6] solve such problems by minimizing block time and improving the throughput. However, high energy consumption demand and low throughput of blockchain technology are not suitable for power-restricted IoT devices. A credit-based proof-of-work method for IoT devices is developed [7]. A data authority management system is used to protect confidential data accessed from sensor devices. Directed acyclic graph architecture is used to satisfy the energy constraints for IoT devices. The proposed approach is implemented using PC and raspberry PI and tested using a case study of the smart factory. Future work can explore data quality control methods in blockchain and techniques to handle huge amounts of data.

11.2.4 Machine Learning

Automation is one of the crucial transformations in Industrial IoT applications. Predictive maintenance of industrial machines is one of the critical applications which can reduce cost and improve the effectiveness in IIoT systems. Machine learning algorithms play a vital role in achieving the same. Machine learning models are trained to detect the abnormal behavior of the machine and notify such behavioral changes. Machine learning techniques with big data analytics are used to perform cyber-attack analysis for safeguarding Industrial IoT systems. It is proved that it works well for anomaly detection [8] to identify such cyber-attacks well in advance. However, providing the industrial solutions to handle such a massive amount of data is still a challenge. It can solve it by providing automated solutions using deep learning approaches such as deep neural networks and deep reinforcement learning techniques.

11.2.5 5G Technology

Due the different requirements of Industrial IoT such as communication, security, and reliability, 5G wireless mobile technology plays a vital role as compared to 4G. 5G network has capabilities to support huge number of connections among different sensor objects. 5G-enabled technology helps industries to build smart factories which facilitates smart automation, augmented reality, and intelligent decision making for business processes. 5G can be used by manufacturing industries to enhance productivity and efficiency by introducing smart cameras in the assembly line. It allows efficient processing of massive amount of data for making fast decision on time. 5G has great potential to support cyber-physical manufacturing system and Industrial IoT. 5G technologies are widely used in industrial applications in network function virtualization, network slicing and multi access edge computing. There is no doubt that 5G support IoT industrial processes brings enhanced performance but still different challenges [9] remains same such as meeting various metrics requirements, interoperability, scalability, security and privacy requirements, effective device to device connectivity, effective network function virtualization solution, large-scale deployments, and utilization of advanced technologies.

11.3 ARCHITECTURAL FRAMEWORKS FOR IIOT APPLICATIONS

In general, the three-layer architecture is known well among researchers to enable Industrial IoT. Device, application, and connectivity layers play a vital role in different ways. The device layer in the Industrial IoT framework uses various sensor devices, control devices, actuators, processors, and controlling software. The application layer is responsible for processing the data collected from different devices to transform it into meaningful information for intelligent decision-making. Data analytics, cloud computing, blockchain, machine learning, and deep learning are a few powerful technologies that can deploy intelligent applications. Finally, the connection layer enables data to flow from various devices to the connection and application layers.

A 5C architecture [10] is described, which consists of five layers: configuration layer, cognition layer, cyber layer, conversion layer, and connection layer. Precise data acquisition is one of the essential functionalities in the connection layer. Further, selecting the best sensors for the data source and handling the heterogeneity among devices are critical tasks of this layer. Context-awareness from the collected data from different sensor devices is necessary functionality performed by the conversion layer. It will convert the raw data into valuable information and self-awareness to physical devices involved in Industrial IoT applications. The cyber layer gets information from all the machines in the network. It provides a central hub for the data for further processing to make intelligent decision making for all the devices. The cognition layer combines the information gathered from individual machines and integrated data sources to monitor the overall system. Big data analytics plays an essential role in this layer for intelligent decision-making. Finally, the configuration layer is used to provide feedback about the decisions made by the cognitive layer to the physical machines. This layer acts as an intelligent controlling system by offering the abilities of self-adaptiveness and self-control.

Integrating IoT for production and logistics has various challenges, like very long waiting times and a lot of energy wastage. To address these challenges, a framework is proposed [11] for intelligent modeling. The proposed framework for an intelligent production logistics system comprises three layers: self-organizing configuration, innovative production logistics system, and intelligent modeling. Production monitoring is performed smartly using various resources, sensors, manufacturing resources, and IoT technologies in the intelligent modeling layer. Real-time monitoring of smart devices is performed using dynamic behavior and cyber physical system monitoring. Hadoop clusters and data warehouse technologies are used for data management efficiently. Cloud computing is used in intelligent production logistics systems to facilitate necessary services. The self-organizing configuration layer is divided into three levels Analytical target cascading (ATH) hierarchy. The proposed ATC model is used to configure manufacturing elements and devices. Researchers can carry out future work to boost self-configuration models using advanced algorithms.

A healthcare monitoring framework based on the Industrial Internet of Things is proposed [12]. Patient data are collected through mobile devices and sent to a cloud server that healthcare professionals can access safely for efficient healthcare delivery. A complex industrial health IoT system enables various stakeholders to connect. Patient data safety is taken care by using this framework by accessing the confidential information by the authorized personnel. Further, cloud-based big data analytics techniques allow efficient storage, analysis, monitoring capabilities to medical data. The combination of cloud computing and big data analytics plays a vital role in Industrial IoT's success. Watermarking is a technique that is used to protect patient's critical information from forgery. Future work is suggested to use this framework for real-world environments and hospitals.

An advanced Industrial IoT solution is proposed [13] for intelligent factories to improve operational efficiency. The system is designed with IoT-based sensor devices to monitor the industrial machine in a pervasive environment. The performance of the proposed architecture is tested in a real environment. Performance shows promising results in terms of power consumption and latency. Further, this solution makes sure that each node is easily approachable with the help of pure IP-oriented methods with an acceptable delay. The proposed approach is feasible because operational time is extended to several years in an Industrial IoT environment.

Reference architecture to automate the industrial processes based on IoT is proposed [14]. This architecture uses a specific server for the Industrial IoT devices to become IIoT devices. This approach is helpful for researchers and practitioners, which helps speed up the commissioning process and reduce human mistakes due to the automation process. According to this reference architecture, the IoT device uses a particular function that exposes its input and output variables for the controller. However, this functional block is independent of the controller or any other applications. This information is used by plug and produces service for signal names matching from controllers.

An efficient resource service model [15] for IIoT applications is developed based on transparent computing, which helps manage and maintain various resources, data, and programs in a centralized manner. The multiple layers of the proposed model are sensing and aggregation, service and storage, interface, and management layers. This service-oriented framework for resource IIoT systems ensures reliability, efficiency, and safety of resources by reducing the development cost of the IIoT system is identified. For managing and controlling various resources, an integrative approach of IoT and cloud computing is widely used. However, this approach is not very satisfactory due to the heterogeneity of different Industrial IoT devices. Moreover, IIoT systems use several collection devices and sensor nodes, leading to high deployment and maintenance costs in a distributed environment.

A framework of industrial automation using the Internet of Things with the help of Raspberry PI is proposed [16]. Raspberry PI is used to collect the sensor data and act as a sensor and controller. This framework allows industries to remotely monitor and control various parameters using a mobile device in a real-world environment. Further, it facilitates industries to save workstation energy consumption using a blade aging system with the help of a web page. The advantages of the proposed approach are efficient and easy data collection from sensor devices. Furthermore, big data analytics techniques are used to analyze such enormous data for more intelligent decisions, strategically providing this data to the individual expert or personnel at the right time, enhanced performance by making a desirable decision for IIoT.

Industry 4.0 is revalorized with the idea of IIoTs, which enables all devices to connect with the Internet and share massive data. This enormous amount of data is sent to a cloud server for further analysis and fetches the meaningful information sold by the cloud service provider as a service. But short latency between data collection and output production [17] is not acceptable in industrial scenarios. Therefore, an approach to total latency for transferring the industrial data to a cloud server and sent it back to the industry is proposed. Results demonstrated that a simple solution could be effective if it is based on the MQTT protocol.

11.4 EFFICIENT AND SECURE APPROACHES IN IIOT

The Industrial IoT environment increases the demand for highly interconnected devices across different industries. However, it is always a big question mark that the access or the service of the device in the interconnected network is trustworthy or not. Implementing security techniques in the Industrial Internet of Things networks is always challenging as the number of devices increases with low computational capabilities. Moreover, it is noticed that there are more chances of security attacks at the same time. Hence, it requires different security approaches to prevent the devices from such attacks in Industrial IoT applications. Blockchain is proven to provide enhanced security because of its decentralized nature in a distributed environment. Towards this direction, various researchers proposed several secure methods to prevent devices from cyber-attacks in the IIoT environment. Table 11.1 presents a few such specific approaches to enhance the efficiency of Industrial IoT systems with their outcomes.

A secure framework approach [18] based on advanced technologies like blockchain and coalition formation theory is proposed. Additionally, for classifying benign and malicious attacks, a deep learning algorithm is promoted in an Industrial IoT environment. The efficiency of the proposed framework is tested through simulations. Similarly, another approach [19] based on reinforcement learning and artificial intelligence is proposed as an outcome of blockchain manager. The proposed blockchain manager is optimizable in real-time monitoring while considering security, latency, and cost. Another secure scheme [20] is presented as an outcome of lightweight identity-based generalize proxy signcryption capable of preventing security attacks and establishing secure communication among the people in the company if the concerned manager is out of the station. This scheme shows good performance in terms of communication cost and computational power.

A novel, trustworthy blockchain-enabled architecture [21] is proposed for IIoT. Distributed cloud radio and optical access network is used to ensure the privacy and low risk. The proposed architecture guaranteed confidentiality, efficiency, and credibility of the system. High credibility of the system is obtained by suggesting a tripartite authentication to access the device in IIoT networking. Further performance of the system is evaluated

TABLE 11.1 Secure Approaches with Their Outcomes

REF. NO.	*SECURITY APPROACHES*	*ADVANTAGES*	*OUTCOMES*	*YEAR*
40	Blockchain, deep learning classification	Malicious attack prevention	Security framework	2020
41	Deep reinforcement learning, Artificial Intelligence	Maximize security, minimum latency and cost	Intelligent blockchain manager	2021
42	Cloud computing, cryptography	Secure in terms of cyphertexts attack, Increased efficiency	Lightweight identity based generalized proxy signcryption scheme	2021
43	Distributed cloud radio, optical access network	Digital identity generation, anonymous access identification, trusted resource provisioning	Secure architecture	2021
44	Blockchain, real-time cryptographic algorithms	Secure and trustworthy operations	Blockchain based architecture	2021
45	Blockchain	Security and robustness of access control protocols	Framework	2021

in terms of identification cost, resource utilization, and mistrust rate. Another approach proposed [22] as a blockchain architecture that ensures secure and trustable operations in Industrial IoT applications and is implemented in a traditional food processing plant. The computational power of the proposed architecture is improved using real-time cryptographic algorithms. There is no doubt that blockchain solution is promising security, but the challenges are high latency rate and poor flexibility. Therefore, a novel access secure framework [23] is developed for 5 G-enabled blockchain applications. The three chain codes are designed, and the combination of these chain codes is utilized to support access control, access authorization, and behavior of IIoT devices. Experimental results of the proposed framework show low cost, high throughput, efficient resource utilization.

With the advancement in machine-to-machine communication, it is beneficial to incorporate it into Industrial IoT to exchange information automatically without human interaction. However, most M2M communication technology requires a high computational cost to include security aspects in it. Hence, a lightweight authentication method [24] is proposed for Industrial IoT. This lightweight security system provides Industrial IoT solutions with low computation cost, communication, and storage overhead. Further, an industrial-based IoT system is developed for lightweight and secure transmission using message queue telemetry transport protocol [25].

An M2M messaging system is proposed [26] for Industrial IoT architecture. A ZeroMQ based messaging system is found to be an efficient tool for data transmission and messaging purposes. It provides good connectivity, a rich sensing environment, and ubiquitous data access for IIoT applications. This ZMQ based data-oriented messaging system provides a detailed view of M2M-based IIoT applications. The proposed messaging technique is flexible with a different software platform and uses low-level UPD and TCP sockets. It can implement lightweight devices and powerful machines because of its unique features like compatibility with cross platforms, flexibility, and efficiency. Future research could optimize the ZMQ data-oriented messaging system to make it more efficient and suitable for various devices.

11.5 APPLICATIONS OF IIOT IN VARIOUS INDUSTRIES

Applications of Industrial IoT in different industries are comprehensive, and the spectrum of its advantages is enormous. Other industries benefited from Industrial IoT application are automotive, agriculture and healthcare to name a few. We have illustrated various applications of Industrial IoT in different sectors with its significant advantages in Table 11.2 for better readers' visibility.

11.5.1 Automotive Industry

Industrial IoT applications in the automotive industry lead to efficient monitoring and management of the entire automobile with the whole manufacturing process. The benefits are high production and better-quality products with low cost and less human intervention.

TABLE 11.2 Overview of Industrial IoT in Various Industries

REF. NO.	*INDUSTRIES*	*APPLICATIONS*	*MAJOR ADVANTAGES*	*YEAR*
20	Automotive	Autonomous electric vehicles	Efficient communication	2020
34	Automotive	Framework to predict device failure	Predictive based maintenance	2020
35	Automotive	Vehicle accident detection framework	Quick treatment	2020
26	Agriculture	IoT in precision agriculture	Real time monitoring, large production, water resource utilization	2015
22	Agriculture	Automated intelligence system using IoT	Less human effort, real-time field monitoring	2016
25	Agriculture	Automated irrigation framework	Efficient resource utilization	2015
36	Healthcare	Cloud assisted healthcare IoT system	Secure data management	2021
36	Healthcare	Cloud-assisted IoT based healthcare monitoring framework	Error reduction, efficient information access	2016

The major transformation in Industrial IoT from human-driven to self-driving cars will have a long impact on our life. With the growing number of vehicles, the term 'internet of vehicles' is evolving [27], providing a platform to communicate and transform information among automated cars.

Various researchers propose different applications towards the development of the Industrial Internet of Things. For example, a framework [28] to predict device failure in an industrial environment is proposed in condition and predictive-based maintenance. Another framework for detecting the accident in electric vehicles is proposed [29]. A unique device generates a quick response for the hospital and delivers location-based information about driver and car. Therefore, reducing the time spent in between accidents and the data passed to the hospital.

Robotics tools play a vital role in the intelligent industrial process. The convergence of smart robots in the production line and manufacturing process leads to less effort by human beings with their safety, high-quality production, and efficient delivery process to meet customers' requirements. However, one of the requirements and challenges in Industry 4.0 is transforming automated-based manufacturing into cyber-physical-based manufacturing processes. Cyber-physical system is essential as it provides the environment to connect humans and robots to communicate and exchange required information gathered from sensors. It enables industries to make intelligent decisions with the combined effort from both humans and robots. Cyber-physical systems allow comfortable to work with robots by providing intelligent sensing, communicating, wise decision making, and self-controlling capabilities. At the same time, it ensures the safety of human being in the working environment. However, various technical challenges associated with cyber-physical systems still need to be improved, like efficient communication between humans and machines, vagueness in the real-world data, and system performance measurement.

11.5.2 Agriculture Industry

Precision agriculture is one of the essential fields where IoT has successfully adopted, but only a proprietary solution is provided, leading to communication and compatibility issues [30]. However, the advantages of implementing wireless sensor networks (WSN) in precision agriculture are realized [31] to increase production, efficient use of water resources, and monitoring actual environmental conditions. Moreover, large applications in agriculture are developed and deployed to provide intelligent solutions to farmers. Plant and soil monitoring, animal monitoring, greenhouse environmental observation, precision agriculture, supply chain management, and food observation are essential applications in agriculture. An intelligent automated system using IoT [32] has been developed, which is used to make smart agriculture practice. An intelligent robot based on GPS technology is used in agriculture to perform different tasks by reducing human efforts. The system uses decision-making tasks based on real-time field monitoring information. Smart management warehouse is used to control temperature, humidity, and theft identification in the warehouse. Various technologies are used to develop the system: Wi-Fi, sensors, Internet, Zigbee, actuators, and Raspberry PI. Another system [33] for intelligent agriculture using IoT has been developed. Different features of the system include remote monitoring using GPS, sensors for temperature and moisture, leaf wetness, security, and irrigation control. Wireless Sensor Networks (WSN) is used to identify the goodness of soil and environmental factors affecting the crop. Various nodes are deployed at different locations in the field, which helps farmers increase crop productivity. Interaction between multiple nodes is accomplished using wireless networks. An automated field monitoring system is developed [34] for farmers based on different monitoring modules such as plant growth monitoring, reminders, irrigation planner, crop profit calculator, and knowledge base composed of varying crop details. This system is more efficient than traditional systems because of labor cost reduction and efficient use of water resources. An automated irrigation framework [35] is proposed for farmers as a resource utilizer. Temperature, humidity, and other environmental factors are easily recorded with the help of sensor technology in minimizing the farmer's efforts and errors in measuring those factors.

Another system is developed [36] to connect sensors to the cloud and the irrigation control system. Another IoT-based method [37] is proposed to facilitate farmers to monitor environmental parameters remotely for efficient crop production. IoT resources are utilized [38] in greenhouse horticulture to make the right decision at the right time. The system is designed based on sensors, electronic equipment, communication protocols, actuators, and different software operating in the cloud. Another system [39] is developed to monitor and discriminant using IoT system for insect pests and plant disease, which is the major problem in soil, plant cultivation, and environmental conditions. Apriori algorithm is used to generate association rules based on various environmental factors and monitors the outcomes. Finally, an automated and secure agriculture monitoring system [40] is developed using IoT to reduce human efforts using Arduino and Ethernet shield. Arduino is connected to agriculture appliances, and a connection is established between the Ethernet shield and site.

Wireless sensor networks are widely used in agriculture applications. It is used to monitor crops on a regular basis. Different protocols which are used to develop the system [41] are Wi-Fi, Zigbee, SigFox, LoRa, 3G/4G are used. Still, Zigbee and LoRa protocols were found to be best for agriculture because of less power consumption. A system is implemented [42] and used for monitoring purposes based on sensor networks and two different network protocols called tree-based collection and collection tree protocols. Wireless sensor networks (WSN), sensors, cloud computing, RFID, geographical systems, and Information Technology are essential to developing intelligent applications.

There is no doubt that a large number of applications in agriculture using IoT have been developed. But still, the challenge lies in monitoring soil quality, measurement of environmental conditions, and irrigation control mechanisms that affect crop quality and production [36]. Advantages of using IoT in agriculture are input improvements (e.g., soil, fertilizers, water, etc.), efficient crop production, food and environment monitoring and protection, rise in profit and sustainability. Furthermore, IoT facilitates farmers to supply crops directly to consumers in smaller and broader regions. Thus, it will reduce the complex process between consumers and producers. At the same time, cloud computing help farmers to provide necessary services at an affordable cost.

11.5.3 Healthcare Industry

Due to advanced technological development in the healthcare industry, the incorporation of IoT in healthcare provides safety and detection of severe diseases. The Internet of medical things combines robust sensors, wearable devices, communication networks, storage devices, interconnected apps, and people who work together as an intelligent system to improve the quality of healthcare delivery by reducing patient waiting time and intelligent decision making to provide an efficient treatment plan. Researchers have proposed different IoT-based healthcare-assisted methods for efficient patient care. A cloud-assisted healthcare IIoT system [43] is developed to ensure privacy and traceable fine delivery data delivery system. It is more efficient in encryption and decryption operations and tracing malicious attacks than the existing system. Performance evaluation shows the effectiveness and feasibility of the proposed solution. Another cloud-assisted IoT-based healthcare monitoring framework [44] is proposed where essential patient health information is collected from mobile devices and sensors. Further, this information is sent to a cloud server for processing and access by healthcare providers. Advanced analytical techniques like watermarking and signal enhancement are used to prevent identity theft or any accidental errors by physicians.

11.6 CHALLENGES IN INDUSTRIAL IOT

Different research challenges associated with Industrial IoT are identified as security and privacy, reliability, scalability, power efficiency, network management, interoperability, device diversity, and resource management. For better understanding for the readers, these challenges are demonstrated in Figure 11.2.

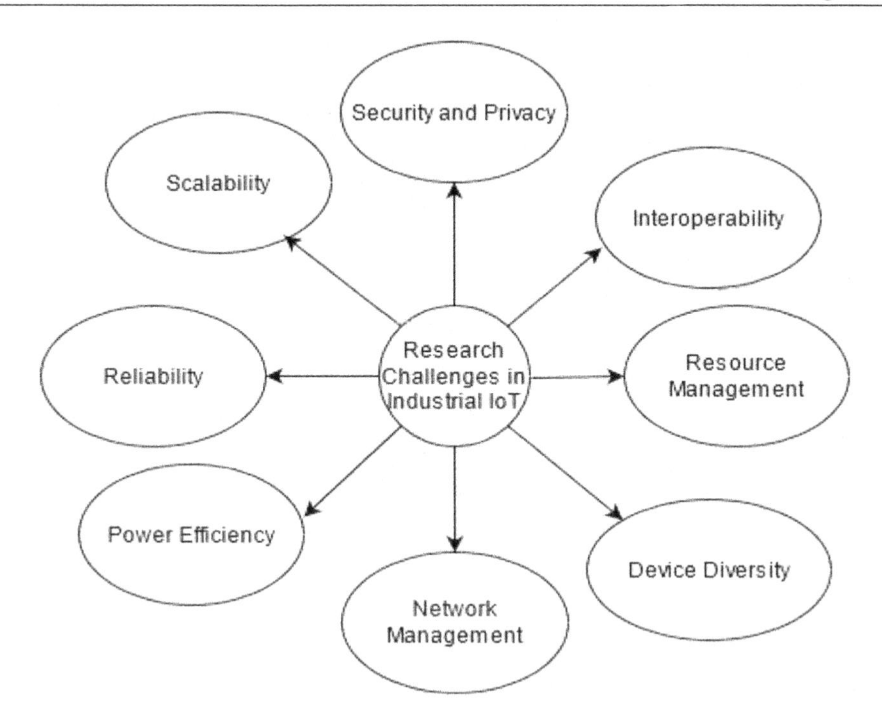

FIGURE 11.2 Challenges in Industrial IoT.

Various implementation issues [1] have to be addressed, like scalability, network management, cost, edge node dimensionality, power efficiency, and coverage. Further, other challenges [16] in IoT automation are identified as data security, data integrity, privacy, interoperability, trust, and scalability. M2M communication [26] plays a crucial role in the implementation of Industrial IoT applications. M2M-enabled technology connects sensors, actuators, other computational devices, and exchanges necessary for information and messages. The challenges involved in M2M technologies are flexibility, compatibility, and efficiency.

Different challenges [7] in IIoT applications are identified as privacy, access control, dependability, and scalability. Further, the challenges [45] in developing Industrial IoT systems are identified in a distributed environment. Integration of diverse hardware and software platforms in IIoT is one of the significant challenges. Further, interoperability with existing infrastructure is another challenge. IIoT enabled solutions must have availability, reliability, maintainability, safety, and security, making them dependable. Industrial and academia collaboration is needed to figure out more centric problems and restrictions faced by industries to incorporate IIoT systems in their business processes.

The heterogeneity of various devices and sensors is the major challenge [15]. Other challenges are limited storage capacity and computing power, problems in data sharing among multiple devices. These sub-challenges turn to significant challenges like resource management, integration of various IoT devices, and interconnectivity. Providing security and safety to IIoT applications in practice is one of the fundamental challenges.

5G mobile technology has a great potential to address different communication requirements in Industrial IoT applications. But there are some challenges associated with 5G implementation in IIoT applications are identified by the researchers [46] are providing QoS metrics, interoperability, meeting privacy and security requirements, scalability, effective device to device solutions, exploring the capabilities of cloud edge computing, standardization, cost effective solutions, utilization of emerging technologies and deployment of private campus networks.

Now we will discuss various challenges in Industrial IoT applications in more detail in different subsections as follows.

11.6.1 Security and Privacy

Security and privacy are the primary concerns and should be guaranteed while implementing IoT in the industrial sector. Software used in IIoT systems needs to protect the used devices and machines in the network from malicious attacks. Security challenges [47] in IIoT systems are data confidentiality, cyber-physical system integrity, pivotal pairing devices, and secure device management. Advanced data encryption and network security techniques ensure data confidentiality in the industrial and intelligent factory environment. Most of the devices in the Industrial IoT environment are considered cyber-physical devices, and the system should maintain the integrity of such devices. A device called prover helps verify device integrity to avoid unwanted and unauthorized modification of the software. Pairing new devices securely with the existing IIoT devices is a crucial challenge and must be addressed through cryptographic mechanisms. A large deployment of industrial and sensor devices securely is also a significant challenge in IIoT environments. Few approaches like firmware updates, live result monitoring, sending commands, and solving problems on demand may be helpful for the successful and secure deployment of IIoT devices.

11.6.2 Scalability

Scalability is a key challenge due to the increased number of sensor devices, machines, and smart factories in Industrial IoT applications. Scalability issues come into the picture for various reasons, like many sensor devices generating a massive amount of continuous and high-frequency data. Scalability issues can be addressed using modern management techniques in the system design at ground level. However, various approaches were proposed by the researchers to provide scalable solutions for Industrial IoT applications. For example, a scalable and secure intelligent transportation system is proposed [48]. Data mining, deep learning, and blockchain technologies were used to support scalable and secure frameworks. Data mining is used to extract urban traffic patterns. Deep learning is used to merge these patterns extracted from each site using the keras package, and blockchain addresses security issues. Another scalable solution for fleet monitoring and visualization for smart machine maintenance using the middleware cloud platform is proposed [49] in a mission-critical industrial system.

11.6.3 Interoperability and Device Diversity

The primary issue in Industrial IoT applications is to achieve interoperability between different devices. Interoperability between the devices is essential to transmit and receive the data between the cloud server and the system. Therefore, standard protocols and interface between devices to communicate each other is necessary. But in a real industrial environment, most devices use diverse protocols and interfaces that seem challenging to integrate and configure. However, various solutions convert devices with standard interfaces and protocols. For example, an ethernet gateway is used to convert all protocols to standard protocols. Different researchers provided solutions for interoperability in an Industrial IoT environment. SEMIoTICS architectural framework is proposed [50], which addresses some of the key challenges such as connectivity and interoperability of the system. Another interoperable approach for Industrial IoT is proposed [51] based on the integration server. Finally, more interoperable approaches are recommended for the successful integration of IoT in industrial applications.

11.6.4 Reliability

Reliable data is the key success factor in Industrial IoT applications for successful digital transformation. The challenge lies in the heart of a huge amount of data collected through industrial manufacturing processes to make intelligent and informed decisions. Data reliability depends on the selection of proper sensors and the use of appropriate data analytic techniques. The backend of the system must be reliable for the success of any front-end application in an Industrial IoT environment. Defective devices due to different environmental factors in Industrial IoT applications have a negative impact on the adoption of IoT by industries. Other approaches are required to ensure fast network recovery due to unwanted failure. At the same time, the success of reliable Industrial IoT applications depends on solid backend infrastructure. However, various researchers have proposed reliable Industrial IoT solutions for successful IoT adoption. For example, a reliable aware contention-free protocol is proposed [52] for automated RFID monitoring in an Industrial IoT environment.

11.6.5 Resource Management

Efficient resource management is the critical challenge to satisfy the end-user experience. Due to IoT devices' diverse and distributed nature, it is recommended to pay more attention to resource management. Resource management becomes critical when an Industrial IoT environment deals with cloud computing for storage and data processing. Efficient resource allocation is essential to maintain the quality of service requirements in an IoT networking environment. Resource management can be viewed from different perspectives, such as monitoring, scheduling, controlling, and discovery. Efficient resource utilization is important in the cloud computing environment, as it leads to reduced operational costs and efficient energy utilization. Diversities among devices and the remoteness between them make resource management difficult and should be addressed with high importance. However, different resource management techniques are proposed by researchers in the past for improving the overall performance of the system and productivity. For example, a multi-virtual resource utilization approach for integrated satellite-terrestrial Industrial IoT network is proposed [53] using a matching design that considers the resource cube algorithm. Another approach based on deep reinforcement learning [54] is proposed for an efficient real-time resource decision-making system in an Industrial IoT environment.

11.6.6 Network Management and Power Efficiency

Device failure in the industrial network is the major challenge as it leads to high energy consumption, wastage of computing resources such as CPU, memory, and network bandwidth. Appropriate strategies like design protocols and frameworks in network management lead to good network performances and energy-efficient solutions in Industrial IoT systems. However, designing low-power network management is challenging due to heterogeneous devices, different network topologies, and a lack of resources. In addition, scalability, fault tolerance, quality of service, energy efficiency, and security are key challenges [55] in networking management. Various energy-efficient frameworks based on network management designed by researchers are classified as a cloud-based framework, network management protocols, SDN-based framework, machine learning-based, and semantic-based framework.

Low power wide area network (LPWANs) has shown its popularity after the growth of IoT in various fields. LPWANs are used in multiple domains effectively. However, it has the potential to revalorize Industrial IoT also. An application is developed [56] to demonstrate the performance of LPWANs in indoor industrial monitoring. Some of the parameters of LPWANs for industrial monitoring identified are the number of nodes, coverage radius, message length, and transmission period. Further, the behavior of LPWANs for industrial monitoring is compared with the IEEE802.15.4 network. Results show the best performance of the LPWANs network in terms of reliability, timeliness, and energy consumption. Various platforms [1] to support LPWAN are LoRaWAN, Sigfox, and weightless.

11.7 CONCLUSION

In the era of automation and technological advancement of applications, it is essential to meet customer requirements to improve the reputation of the organizations. Therefore, there is a shift to move from traditionally based systems to Industrial IoT-based approaches to improve business processes as the demand for such systems is high nowadays. For understanding the importance of IIoT applications in the future, this book chapter gives an overview of various technologies and Industrial IoT applications in the automotive, agriculture, and healthcare sectors. It also identifies critical challenges associated with IIoT applications.

Industry 4.0 plays a vital role in satisfying business requirements. However, there are some potential challenges in the industrial automation process like security and privacy, interoperability, network performance, resource management, scalability, and trust must be addressed at high importance for successfully implementing IIoT in different sectors. Therefore, industries need to adopt future technologies like blockchain, deep learning, big data analytics, and fog computing to implement Industrial IoT in future applications successfully.

REFERENCES

1. Sanchez-Iborra, R., & Cano, M. D. (2016). State of the art in LP-WAN solutions for industrial IoT services. *Sensors*, *16*(5), 708.
2. Sun, W., Liu, J., & Yue, Y. (2019). AI-enhanced offloading in edge computing: When machine learning meets industrial IoT. *IEEE Network*, *33*(5), 68–74.
3. Breivold, H. P., & Sandström, K. (2015, December). Internet of things for industrial automation – Challenges and technical solutions. In *Proceedings 8th IEEE International Conference on Internet Things (iThings)* (pp. 532–539). IEEE.
4. Pop, P., Zarrin, B., Barzegaran, M., Schulte, S., Punnekkat, S., Ruh, J., & Steiner, W. (2021). The FORA fog computing platform for industrial IoT. *Information Systems*, *98*, 101727.
5. Cerina, L., Notargiacomo, S., Paccanit, M. G., & Santambrogio, M. D. (2017, September). A fog-computing architecture for preventive healthcare and assisted living in smart ambients. In *2017 IEEE 3rd International Forum on Research and Technologies for Society and Industry (RTSI)* (pp. 1–6). IEEE.
6. Jameel, F., Javaid, U., Khan, W. U., Aman, M. N., Pervaiz, H., & Jäntti, R. (2020). Reinforcement learning in blockchain-enabled IIoT networks: A survey of recent advances and open challenges. *Sustainability*, *12*(12), 5161.
7. Huang, J., Kong, L., Chen, G., Wu, M. Y., Liu, X., & Zeng, P. (2019). Towards secure industrial IoT: Blockchain system with credit-based consensus mechanism. *IEEE Transactions on Industrial Informatics*, *15*(6), 3680–3689.
8. Zolanvari, M., Teixeira, M. A., Gupta, L., Khan, K. M., & Jain, R. (2019). Machine learning-based network vulnerability analysis of industrial internet of things. *IEEE Internet of Things Journal*, *6*(4), 6822–6834.
9. Varga, P., Peto, J., Franko, A., Balla, D., Haja, D., Janky, F., . . . Toka, L. (2020). 5G support for industrial IoT applications – Challenges, solutions, and research gaps. *Sensors*, *20*(3), 828.
10. Li, J. Q., Yu, F. R., Deng, G., Luo, C., Ming, Z., & Yan, Q. (2017). Industrial internet: A survey on the enabling technologies, applications, and challenges. *IEEE Communications Surveys & Tutorials*, *19*(3), 1504–1526.
11. Zhang, Y., Guo, Z., Lv, J., & Liu, Y. (2018). A framework for smart production-logistics systems based on CPS and industrial IoT. *IEEE Transactions on Industrial Informatics*, *14*(9), 4019–4032.
12. Hossain, M. S., & Muhammad, G. (2016). Cloud-assisted industrial internet of things (iiot) – enabled framework for health monitoring. *Computer Networks*, *101*, 192–202.
13. Civerchia, F., Bocchino, S., Salvadori, C., Rossi, E., Maggiani, L., & Petracca, M. (2017). Industrial internet of things monitoring solution for advanced predictive maintenance applications. *Journal of Industrial Information Integration*, *7*, 4–12.
14. Koziolek, H., Burger, A., & Doppelhamer, J. (2018, April). Self-commissioning industrial IoT-systems in process automation: A reference architecture. In *2018 IEEE International Conference on Software Architecture (ICSA)* (pp. 196–19609). IEEE.

15. Li, W., Wang, B., Sheng, J., Dong, K., Li, Z., & Hu, Y. (2018). A resource service model in the industrial IoT system based on transparent computing. *Sensors*, *18*(4), 981.
16. Merchant, H. K., & Ahire, D. D. (2017). Industrial automation using IoT with raspberry pi. *International Journal of Computer Applications*, *168*(1), 44–46.
17. Ferrari, P., Sisinni, E., Brandão, D., & Rocha, M. (2017, September). Evaluation of communication latency in industrial IoT applications. In *2017 IEEE International Workshop on Measurement and Networking (M&N)* (pp. 1–6). IEEE.
18. Sharma, M., Pant, S., Kumar Sharma, D., Datta Gupta, K., Vashishth, V., & Chhabra, A. (2021). Enabling security for the industrial internet of things using deep learning, blockchain, and coalitions. *Transactions on Emerging Telecommunications Technologies*, *32*(7), e4137.
19. Al-Marridi, A. Z., Mohamed, A., & Erbad, A. (2021). Reinforcement learning approaches for efficient and secure blockchain-powered smart health systems. *Computer Networks*, *197*, 108279.
20. Hussain, S., Ullah, I., Khattak, H., Khan, M. A., Chen, C. M., & Kumari, S. (2021). A lightweight and provable secure identity-based generalized proxy signcryption (IBGPS) scheme for industrial internet of things (IIoT). *Journal of Information Security and Applications*, *58*, 102625.
21. Yang, H., Bao, B., Li, C., Yao, Q., Yu, A., Zhang, J., & Ji, Y. (2021). Blockchain-enabled tripartite anonymous identification trusted service provisioning in industrial IoT. *IEEE Internet of Things Journal*, *9*(3), 2419–2431.
22. Latif, S., Idrees, Z., Ahmad, J., Zheng, L., & Zou, Z. (2021). A blockchain-based architecture for secure and trustworthy operations in the industrial internet of things. *Journal of Industrial Information Integration*, *21*, 100190.
23. Feng, Y., Zhang, W., Luo, X., & Zhang, B. (2021). A consortium blockchain-based access control framework with dynamic orderer node selection for 5G-enabled industrial IoT. *IEEE Transactions on Industrial Informatics*, *18*(4), 2840–2848.
24. Esfahani, A., Mantas, G., Matischek, R., Saghezchi, F. B., Rodriguez, J., Bicaku, A., . . . Bastos, J. (2017). A lightweight authentication mechanism for M2M communications in industrial IoT environment. *IEEE Internet of Things Journal*, *6*(1), 288–296.
25. Katsikeas, S., Fysarakis, K., Miaoudakis, A., Van Bemten, A., Askoxylakis, I., Papaefstathiou, I., & Plemenos, A. (2017, July). Lightweight & secure industrial IoT communications via the MQ telemetry transport protocol. In *2017 IEEE Symposium on Computers and Communications (ISCC)* (pp. 1193–1200). IEEE.
26. Meng, Z., Wu, Z., Muvianto, C., & Gray, J. (2016). A data-oriented M2M messaging mechanism for industrial IoT applications. *IEEE Internet of Things Journal*, *4*(1), 236–246.
27. Vaidya, B., & Mouftah, H. T. (2020). IoT applications and services for connected and autonomous electric vehicles. *Arabian Journal for Science and Engineering*, *45*(4), 2559–2569.
28. Ciancio, V., Homri, L., Dantan, J. Y., & Siadat, A. (2020). Towards prediction of machine failures: Overview and first attempt on specific automotive industry application. *IFAC-PapersOnLine*, *53*(3), 289–294.
29. Dashora, C., Sudhagar, P. E., & Marietta, J. (2020). IoT based framework for the detection of vehicle accident. *Cluster Computing*, *23*(2), 1235–1250.
30. Ojha, T., Misra, S., & Raghuwanshi, N. S. (2015). Wireless sensor networks for agriculture: The state-of-the-art in practice and future challenges. *Computers and Electronics in Agriculture*, *118*, 66–84.
31. Savale, O., Managave, A., Ambekar, D., & Sathe, S. (2015). Internet of things in precision agriculture using wireless sensor networks. *International Journal of Advanced Engineering & Innovative Technology*, *2*(3), 1–5.
32. Gondchawar, N., & Kawitkar, R. S. (2016). IoT based smart agriculture. *International Journal of Advanced Research in Computer and Communication Engineering (IJARCCE)*, *5*(6), 177–181.

33. Suma, D. N., Samson, S. R., Saranya, S., Shanmugapriya, G., & Subhashri, R. (2017). IOT based smart agriculture monitoring system. *International Journal on Recent and Innovation Trends in Computing and Communication*, *5*(2), 177–181.
34. Mohanraj, I., Ashokumar, K., & Naren, J. (2016). Field monitoring and automation using IOT in agriculture domain. *Procedia Computer Science*, *93*, 931–939.
35. Kansara, K., Zaveri, V., Shah, S., Delwadkar, S., & Jani, K. (2015). Sensor based automated irrigation system with IOT: A technical review. *International Journal of Computer Science and Information Technologies*, *6*(6).
36. Rukhmode, S., Vyavhare, G., Banot, S., Narad, A., & Tugnayat, R. M. (2017) IOT based agriculture monitoring system using wemos. In *International Conference on Emanations in Modern Engineering Science and Management (ICEMESM-2017)* (Vol. 5, No. 3, pp. 14–19). Yashika publications.
37. Sarkar, P. J., & Chanagala, S. (2016). A survey on iot based digital agriculture monitoring system and their impact on optimal utilization of resources. *Journal of Electronics and Communication Engineering (IOSR-JECE)*, 1–4.
38. Carrasquilla-Batista, A., Chacón-Rodríguez, A., & Solórzano-Quintana, M. (2016, November). Using IoT resources to enhance the accuracy of overdrain measurements in greenhouse horticulture. In *Central American and Panama Convention (CONCAPAN XXXVI), 2016 IEEE 36th* (pp. 1–5). IEEE.
39. Wang, X. F., Wang, Z., Zhang, S. W., & Shi, Y. (2015, October). Monitoring and discrimination of plant disease and insect pests based on agricultural IOT. In *International Conference on Information Technology and Management Innovation (ICITMI 2015)* (p. 112–115). Atlantis-Press
40. Mahajan, T. (2016). IOT based agriculture automation with intrusion detection. *International Journal of Scientific and Technical Advancements*, *2*(4), 269–274.
41. Jawad, H. M., Nordin, R., Gharghan, S. K., Jawad, A. M., & Ismail, M. (2017). Energy-Efficient wireless sensor networks for precision agriculture: A review. *Sensors*, *17*(8), 1781.
42. Liqiang, Z., Shouyi, Y., Leibo, L., Zhen, Z., & Shaojun, W. (2011). A crop monitoring system based on wireless sensor network. *Procedia Environmental Sciences*, *11*, 558–565.
43. Sun, J., Chen, D., Zhang, N., Xu, G., Tang, M., Nie, X., & Cao, M. (2021). A privacy-aware and traceable fine-grained data delivery system in cloud-assisted healthcare IIoT. *IEEE Internet of Things Journal*, *8*(12), 10034–10046.
44. Hossain, M. S., & Muhammad, G. (2016). Cloud-assisted industrial internet of things (iiot) – enabled framework for health monitoring. *Computer Networks*, *101*, 192–202.
45. Iwanicki, K. (2018, July). A distributed systems perspective on industrial IoT. In *2018 IEEE 38th International Conference on Distributed Computing Systems (ICDCS)* (pp. 1164–1170). IEEE.
46. Varga, P., Peto, J., Franko, A., Balla, D., Haja, D., Janky, F., . . . Toka, L. (2020). 5G support for industrial IoT applications – Challenges, solutions, and research gaps. *Sensors*, *20*(3), 828.
47. Sethi, R., Bhushan, B., Sharma, N., Kumar, R., & Kaushik, I. (2021). Applicability of industrial IoT in diversified sectors: Evolution, applications and challenges. In *Multimedia Technologies in the Internet of Things Environment* (pp. 45–67). Springer.
48. Belhadi, A., Djenouri, Y., Srivastava, G., & Lin, J. C. W. (2021). SS-ITS: Secure scalable intelligent transportation systems. *The Journal of Supercomputing*, 1–17.
49. Moens, P., Bracke, V., Soete, C., Vanden Hautte, S., Nieves Avendano, D., Ooijevaar, T., . . . Van Hoecke, S. (2020). Scalable fleet monitoring and visualization for smart machine maintenance and industrial IoT applications. *Sensors*, *20*(15), 4308.
50. Petroulakis, N. E., Lakka, E., Sakic, E., Kulkarni, V., Fysarakis, K., Somarakis, I., . . . Waledzik, K. (2019, June). Semiotics architectural framework: End-to-end security, connectivity and interoperability for industrial iot. In *2019 Global IoT Summit (GIoTS)* (pp. 1–6). IEEE.

51. Karaagac, A., Verbeeck, N., & Hoebeke, J. (2019, April). The integration of LwM2M and OPC UA: An interoperability approach for industrial IoT. In *2019 IEEE 5th World Forum on Internet of Things (WF-IoT)* (pp. 313–318). IEEE.
52. Zhai, C., Zou, Z., Chen, Q., Xu, L., Zheng, L. R., & Tenhunen, H. (2016). Delay-aware and reliability-aware contention-free MF – TDMA protocol for automated RFID monitoring in industrial IoT. *Journal of Industrial Information Integration*, *3*, 8–19.
53. Chen, D., Yang, C., Gong, P., Chang, L., Shao, J., Ni, Q., . . . Guizani, M. (2020). Resource cube: Multi-virtual resource management for integrated satellite-terrestrial industrial IoT networks. *IEEE Transactions on Vehicular Technology*, *69*(10), 11963–11974.
54. Zhang, W., Yang, D., Haixia, P., Wu, W., Quan, W., Zhang, H., & Shen, X. (2021). Deep reinforcement learning based resource management or DNN inference in industrial IoT. *IEEE Transactions on Vehicular Technology*, *70*(8), 7605–7618.
55. Aboubakar, M., Kellil, M., & Roux, P. (2021). A review of IoT network management: Current status and perspectives. *Journal of King Saud University-Computer and Information Sciences* (In press).
56. Luvisotto, M., Tramarin, F., Vangelista, L., & Vitturi, S. (2018). *On the Use of LoRaWAN for Indoor Industrial IoT Applications*. Wireless Communications and Mobile Computing, 2018.

Machine Learning in IoT-Based Ambient Backscatter Communication System

12

Shivani Chouksey[1], Tushar S. Muratkar[2], Ankit Bhurane[3], Prabhat Sharma[4], and Ashwin Kothari[5]

[1,3,4,5]Visvesvaraya National Institute of Technology, Nagpur, India; [2]Indian Institute of Information Technology, Nagpur, India

Contents

DOI: 10.1201/9781003244714-12

12.1 INTRODUCTION

The most basic definition of the IoT is that it connects a wide range and number of devices so that they can communicate with one another over the Internet [1]. Globally, there are around 21.5 billion networked devices, with the number predicted to rise to 30.9 billion by the year 2025. Sharing information data between multiple connected devices in the IoT is typically accomplished using a variety of wireless standards including Wi-Fi (IEEE-802.11), LRP-UWB-PHY (low rate pulse-ultra wide band-physical IEEE-802.16), Bluetooth (IEEE-802.15.1), wi-max (IEEE-802.16), Bluetooth Low Energy, Zigbee (IEEE-802.15.4), etc., but their power consumption is quite high [2].

Connecting such large number of devices is one of the primary communication challenge for fifth generation (5G) IoT communication system. One method to overcome this power consumption issue is to use a Backscatter Communication (Back-Com) system, which consumes much less power. Back-Com is a communication method based upon the concept of modulated signals that reflects back from an RF receiver.

There will be an increasing demand for efficient spectrum usage and network infrastructure as emerging technologies such as the self-driving vehicles, IoT, and intelligent sensors gain popularity. In smart agriculture, an IoT network can measure water absorbed by crops, plant disease and pest management, and fertilization use to enhance productivity.

While designing such networks, three factors must be considered: energy efficiency, spectrum efficiency, and cost efficiency [1]. There are two major issues with wireless sensors or tags.

- **Source of energy:** In sensors or tag devices, most popular source of energy are batteries but they have a limited lifespan and must be recharged or replaced on a regular basis [3].
- **Radio frequency (RF) component costs:** In compared to baseband circuits, few addition RF components like amplifiers and oscillators are required, which results in increasing the system's cost.

ABCS is emerging as one of the more innovative and low-power alternatives in this environment. We can also use ML to detect information that has been backscattered from the tag to the reader.

The use of ML to retrieve the backscattered signal from the tag to the receiver has been investigated. In supervised learning, the operator supplies a predetermined data set with intended inputs and outputs, and the system must figure out how to get to those inputs and outputs, whereas in unsupervised learning, the ML algorithm searches for patterns in the data. And the reinforcement learning is used to enhance the efficiency of backscatter system.

12.2 BACKSCATTER COMMUNICATION AND ITS TYPES

Radio Back-Com was initiated during World War II, when radio waves were sent from some radar devices and backscatter from an oncoming aircraft was used to determine whether it was a 'friend or enemy'. Harry Stockman was the first to establish the concept of Back-Com in 1948 [4]. Since then, there has been ongoing research on the subject, as well as an examination of RFID (Radio-frequency identification) products for supply chain and identifying applications. From 1990 to 2000 a well-known application of Back-Com was Electronic Toll Collection (ETC), which allows vehicles on busy highways to pay the money for toll price very quickly with the help of sensors without stopping the vehicle [5]. Over the past ten years, extensive research on coding methods, link budgets, channel fading and modeling and multi-antenna strategies, among other things has been done on radio Back-Com systems. Other examples are sensor networks and IoT, because of their capacity to save energy and cost effectiveness. Furthermore, the privacy and security of the backscatter physical-layer systems have attracted significant attention [6].

Back-Com is a type of wireless communication that allows a passive device to transmit data bits employing the signal it receives. It connects two wireless nodes and establishes communication between them. Back-Com is a radar detection technique that detects the presence of an object in the radar range. A Back-Com system must have three basic components: (1) transmitter, (2) receiver, and (3) carrier source/emitter [2]. An RF signal source called as a carrier emitter. Backscatter transmitters are essentially tags that modulate and reflect the carrier emitter's signal to the backscatter receiver (reader). The reader decodes information from the backscattered signal that is sent back from the tag.

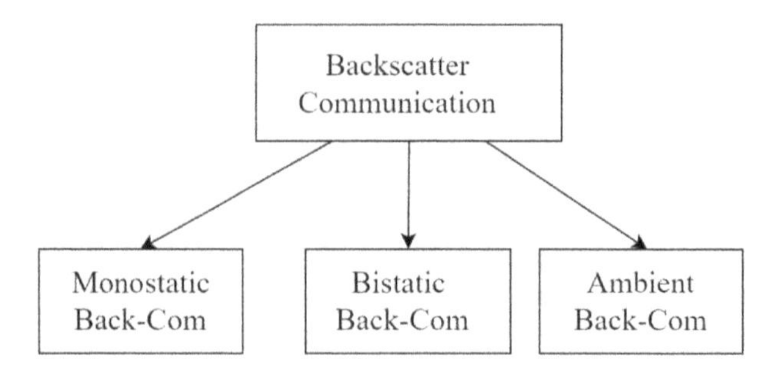

FIGURE 12.1 Types of backscatter communication.

Back-Com can be classified into three major types:

1. **Monostatic Back-Com**
2. **Bistatic Back-Com**
3. **Ambient Back-Com**

Low-power IoT devices are often used in industry and agriculture to monitor humidity and pressure in the air using environmental sensing application of the Back-Com. Some of the most recent applications of Back-Com include healthcare, environmental monitoring, and ad-hoc communication. It can assist designers create wireless sensors that don't require batteries and have end-to-end power requirements in microwatts or less. A backscatter temperature sensor samples at 1 Hz and operating in a loop to sense and transmit with no other overheads (e.g., no receive circuit, protocol overhead, etc.) would consume less than 10 W of power, allowing such sensors to operate on harvested energy constantly. Micro-implants and on-body temperature sensors have already been used with backscatter tags in the healthcare industry. During the uplink data transmission in a BackCom system, the reader may absorb reflected signals from many transmitters. As a result of this collision or interference in the uplink data may occur. To avoid collisions in a Back-Com network, uplink transmissions are commonly multiplexed in the time-domain or frequency-domain [7].

Back-Com is an effective approach for Internet-of-Things (IoT) applications that rely on passive sensing. Backscatter uses less energy by transferring carrier production to the reader and only using power to clock the RF transistor of the reader. To obtain the desired performance, the two parameters, namely the transmit time and the number of receiving antennas, can be mutually adjusted. Continuous data transfer from a range of data rich sensors, including cameras and microphones, becomes immensely effective with high-speed and extremely low power communications, allowing complex distributed sensing applications to be accomplished.

12.2.1 Monostatic Back-Com

It is the simplest backscatter system as shown in Figure 12.2. It has tag and reader only. In Monostatic Back-Com system the reader sends the carrier signal to tag which then

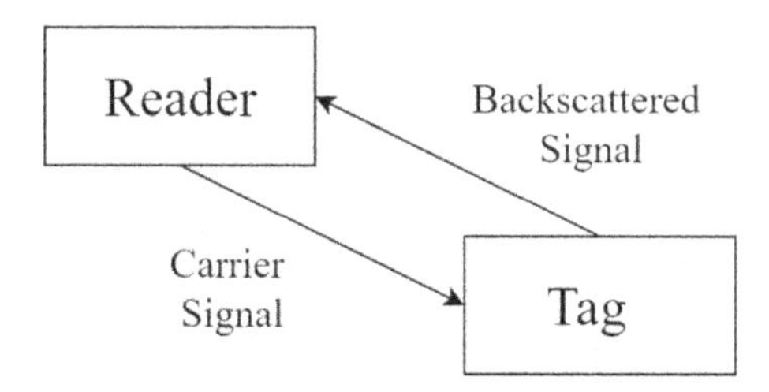

FIGURE 12.2 Monostatic Back-Com.

backscattered by tag after modulating it. Because the carrier wave is supplied by a reader and a sensor only needs to modulate the signal to pass on information, backscatter requires relatively minimal circuitry, eliminating traditional radioactive components that require a significant amount of power [8].

The tag can be of two types:

1. Active device
2. Passive device

An active tag has a dedicated battery for its power requirements, whereas a passive tag is powered by the harvested energy [9]. The tag is consists of microchip antennas and Coil Antennas. In monostatic Back-Com the reader also works the carrier emitter send the carrier signals towards the tag and then the microchip antenna at the tag processes these signals inside before sending them back to the reader. Engineered impedance mismatching at the antenna input side is used to reflect the RF signal [7]. If x(n) = 0, the tag's impedance changes, allowing just a small amount of s(n) energy to be reflected, whereas if x(n) = 1, the tag's impedance changes, allowing the signal to be scattered to the reader. Because the backscatter receiver and RF carrier signal emitter are on the same device(reader), this form of Back-Com system suffers from round trip path loss. Another disadvantage of this monostatic Back-Com system is that it is influenced by the doubly near-far problem, which states that the RF source can extract lesser energy and achieve low throughput to users who are farther away as compared to those users who are near. Hence this technique is ideal for communicating across small distances.

12.2.2 Bistatic Back-Com

Unlike a monostatic communication network, the carrier signal in a bistatic communication network is sent by a third device rather than the reader. To establish connection between the tag and the reader, an additional device (RFID) is required. The radio frequency identification transmits the carries signal to the tag and reader, which is then modulated by the tag with the information bit and backscattered to the reader. Tag is a passive device and it uses the harvested energy from RFID for this operation. The only difference between monostatic Back-Com and bistatic Back-Com is, in a monostatic communication

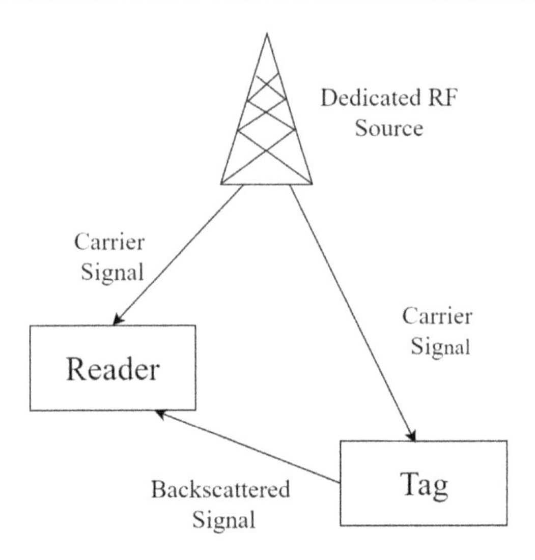

FIGURE 12.3 Bistatic Back-Com.

system, the information receiver and carrier emitter are the same node, but in a bistatic Back-Com system, they are two separate nodes.

A tag's core consists of microchip and coil antennas. These antenna get energy as well as RF signals from the dedicated RFID source. The microchip internally processes the signals before returning them to the reader. The RFID system, in addition to being power intensive, has three primary drawbacks: (1) an external power supply powers the backscatter transmitter, which is heavy and costly (e.g., a reader); (2) when a backscatter transmitter receives a request from the reader then only it begins the communication; (3) a nearby active reader could interfere with the signal backscattered by the tag. We need to build the network with sensors and actuators that can work for long periods of time while consuming very little power to overcome these limits and fully realize the potential of IoT. This could be achieved by ABCS.

12.2.3 Ambient Back-Com

ABCS enables wireless nodes to harvest energy and communicate with other devices, the harvested energy is used to keep the circuit running. This makes the system require a very small amount of energy [10]. The ambient radio frequency (RF) signal present in the surrounding received by the tag as well as the reader and the tag transmit its information bit modulating these RF signals, the tag (transmitter) broadcasts the information bit either 1 or 0 by switching the antenna impedance to reflecting and nonreflecting modes respectively without requiring any additional dedicated infrastructure [11] [6].

The research is based on the existence of RF waves in the immediate vicinity. The tag communicates with the reader by generating energy from received ambient RF signals and then modifies it with the information bit and reflects the surrounding RF signals towards the reader.

The main role of the reader is to identify the tag's backscattered signal. As the reader receives two signals from the tag and the ambient source, the direct signal from the ambient source is stronger than the tag's backscattered signal. The detection of the backscatter bit has took a deal of time and resources. This system is advantageous as it reduces the power consumption, being battery-free can be of smaller size improves system mobility and effectively reduces the cost of wireless communication. It can be used for various purpose as a sensor devices for example: large sized IOT networks, defense system, medical devices and patient monitoring, environment monitoring automobiles and smart homes. In today's era of developing technology, there is the high demand of devices having smaller size while consuming very less amount of power. In ABCS, we can achieve this by utilizing the existing radio frequency signals which are specifically used for TV, mobile phones, etc. This system consumes very little power and can be called a passive network.

Ambient Back-Com network has two main advantages: (1) it does not require additional dedicated infrastructure for the carrier transmission while using the energy present in the air; (2) radio frequency components are minimized. Ambient Back-Com can be considered as a typical spectrum sharing model that consists of two communication system. The Legacy systems and Back-Com system, in both of these systems the tag backscatter the carrier signal at the identical frequency band as of radio frequency carrier ambient signals.

In practice, tags and sensors are powered by batteries that are not be suitable for a wide range of IoT applications because of their limited power supply. Ambient Back-Com system is an unique green communication mechanism that has recently been offered as a feasible solution to the battery problem in IoT devices. The differential encoding technique is used to detect the signal for an ABCS, which eliminates the need of channel estimation [12]. One of advantage of ABCS is that it assumes the channel to be perfect and eliminates the channel state information, as estimating the channel would be a cumbersome task for the following reasons:

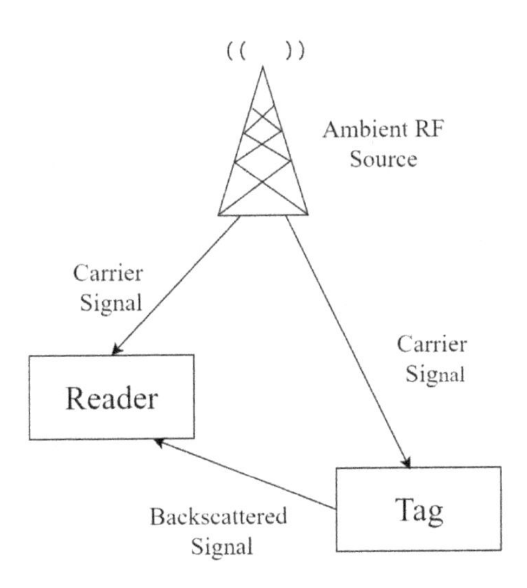

FIGURE 12.4 Ambient Back-Com.

1. The channel characteristics for broadcasting '0' bits differ from those for transmitting '1' bits.
2. Both the tags and the readers are unaware of the carrier waves.
3. Tags are typically incapable of transmitting training symbols due to their simplistic design and limited collected power [3].

Minimum Mean Square Error detectors are used for the detection of backscatter signal.

12.2.3.1 Minimum Mean Square Error (MMSE) Detector

In a Bayesian setting, MMSE refers to estimating with a quadratic cost function. The Bayesian approach to estimating is based on the assumption that we have some information of the parameter that has to be estimated. For example, we may already be aware of the parameter's range of values or estimated the parameter already that we want to update upon the occurrence of a new observation; or we may already be aware of the signal's statistics. Let Y be the random quantity to be estimated by a constant value c. We have to find the value of c such that the second moment of difference y-c (i.e., error) come out to be minimum.

$$e = E(y-c)^2 = \int_{\infty}^{-\infty} (y-c)^2 f(y)\, dy \tag{12.1}$$

Error 'e' is dependent on the value of c, to obtain minimum value second moment we will find the first derivation of error with respect to constant c and equate it to 0.

$$\frac{de}{dc} = \int_{\infty}^{-\infty} 2(y-c) f(y) dy = 0 \tag{12.2}$$

$$c = E(y) = \int_{\infty}^{-\infty} y f(y) dy \tag{12.3}$$

As we note here the optimal criteria for MMSE is the average value of random vector y. A MMSE estimation is a method to estimate the reduced mean square error (MSE) of the dependent variable, which is a typical a measure of quality of the estimator. The MMSE estimation is determined by averaging the latest known parameters. Designing of such estimator is confined in a small space because calculating the posterior average is difficult task.

12.3 MATHEMATICAL ASPECTS OF ABCS

Assuming the ambient RF signal from the source is s(n), and channel coefficients h_{st}, h_{tr}, h_{sr} for ambient source to backscatter transmitter, backscatter transmitter to backscatter receiver and ambient source to backscatter receiver respectively. The information signal

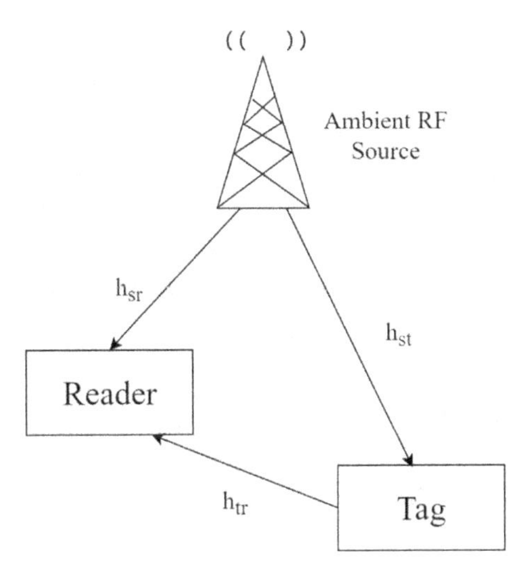

FIGURE 12.5 Ambient Back-Com system model.

from a backscatter transmitter(tag) is x(n), y(n) be the received signal at the backscatter reader, and w(n) is additive white gaussian noise AWGN.

Let c(n) be the signal from ambient source to the backscatter transmitter:

$$c(n) = h_{st}s(n) \tag{12.4}$$

Signal backscattered from backscatter transmitter towards backscatter receiver:

$$g(n) = h_{st}s(n)x(n) \tag{12.5}$$

Where x(n) is 0 or 1 for n=1,2,3, . . . N.
Received signal at backscatter receiver:

$$y(n) = h_{sr}s(n) + \alpha h_{st}h_{tr}s(n)x(n) + W(n) \tag{12.6}$$

Alpha is the reflection coefficient ranges between 0 and 1:

$$y(n) = h_{sr}s(n) + w(n)\, for\, x(n) = 0 \tag{12.7}$$

$$y(n) = h_{sr}s(n) + \alpha h_{st}h_{tr}s(n) + w(n)\, for\, x(n) = 1 \tag{12.8}$$

12.4 MACHINE LEARNING MODELS IN ABCS

ML is the study of learning systems based on research. ML is a highly interdisciplinary subject that draws on statistics, engineering, cognitive science, probability theory, theory of optimization, computer science, and a variety of other fields of science and mathematics. To get started with ML, you'll need a basic understanding of linear algebra, calculus, probability, and statistics.

Although the words 'machine learning' and 'artificial intelligence' are sometimes interchanged, ML is actually a sub-field of artificial intelligence. ML is sometimes considered as analytical method or modeling method based on prediction process. At a fundamental level, ML works predefined algorithms to analyze input value in order to obtain the output data which belongs in a reasonable range of desired output data. When a new data is given as input to these algorithms, they reads the pattern and improve their processes in order to improve efficiency, culminating in the gradual development of 'intelligence'. So the machine's purpose is to observe the pattern in the given input to output data set and learn how to create the output for the next input. This output could be a classification or regression (items belongs different classes or a numeric output). In ambient Back-Com, ML-based approaches can be utilized to detect the backscattered signal at the receiver.

We'll assume reader gets two energy levels that correspond to 0 and 1 bits. The tag delivers known data to the reader in order for the system to learn input bits, which is called featured data for ML methods. ML can take three different approaches depending on the properties of the input:

1. **Supervised ML**
2. **Unsupervised ML**
3. **Reinforced ML**

In supervised ML, user provides input and output data set to the machine, and the machine must figure out an algorithm to find pattern in data sets to obtain the output and for given input. The algorithm analyzes patterns in data upon multiple iterations, learns and makes predictions while knowing the correct output for the input. In supervised learning, the ML algorithm finds for patterns in the input to output of data with some given data sets, whereas in unsupervised learning, the machine looks for patterns in the input

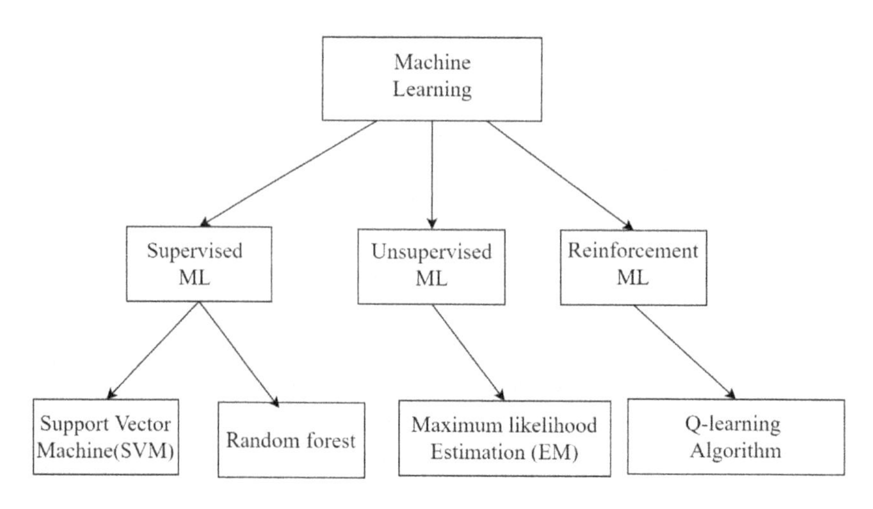

FIGURE 12.6 Machine learning methods.

to output data without any data sets. There is no help available in the form of an answer key or an actual operator. Instead, the machine explores available data for correlations and connections. The ML algorithm is indeed capable of evaluating massive data sets and responding to them in a timely manner in this process. These algorithm try to put everything in order. The algorithm attempts to put everything in its proper place. The algorithm tries to organize the data so that its structure can be defined in some way. This algorithm uses average grouping the data into clusters or organizing it into more logically. Reinforcement learning refers to structured learning approaches in which a ML algorithm is given a set of actions, parameters, and end results. While the creation of the algorithm, the ML attempts to test a wide range of options and possibilities, analyzing and monitoring each output to establish the best one. It tells the computer to learn by trial and error. This process applies what it has learnt latest and applies it to the current environment in order to get the best possible outcome.

12.4.1 Supervised Machine Learning

This technique highlights the property of a machine to read the tag signal by first removing the direct ambient signal and finding the correlation in the remaining signal with the estimated ambient signal. To recognize ambient Back-Com tag symbols, supervised ML methods such as SVM and random forest are used to divide input features into two groups based on the learned model generated from labeled training data. The reader is trained to build an ideal hyperplane from the training data for the SVM algorithm. The following received signal can be separated into two groups according to the ideal hyperplane, and the matching tag symbols can then be decoded. The random forest approach requires the receiver to construct several decision trees from training data using different samples and beginning variables. By averaging the output of these decision trees, the tag symbols may be decoded.

12.4.1.1 Support Vector Machine (SVM)

SVM is an algorithm for optimizing a mathematical function into graphical relation to the input values. SVM learning, or ML with maximization (support) of separation margin (vector), is a potent classification approach that has been used for cancer genomics categorization or sub-typing [13]. SVM is a basically two-class classifier that linearly separates the input parameters x_i with the maximum feasible margin using an optimum hyperplane. The maximum margin hyperplane parallel goes through the intersection of two parallel hyperplanes that distinguish different classes of training data, with the margin equal to the distance between the separating hyperplane and the maximum margin hyperplane. The purpose of such algorithm can be determined without reading the equation. Let the vector w defines the direction of the hyperplane, b is a scalar quantity that defines the distance between the hyperplane and the point in the data set. The separations of two hyperplanes is defined as:

1. For the data set $y_i = -1$

$$w^T x_i + b \leq -1 \tag{12.9}$$

2. For the data set $y_i = +1$

$$w^T x_i + b \geq +1 \tag{12.10}$$

The hyperplane is generally denoted as (w,b), the hyperplane equidistant from both the clusters gives the maximum margin for the clusters. We may write it as:

$$x \in S : w^T x_i + b = 0 \tag{12.11}$$

where x belongs to the given data set.

The distance from given data set point to the hyperplane of (w,b) is:

$$d = \frac{| w^T x_i + b |}{\| w \|} \tag{12.12}$$

Assuming the nearest data point to the hyperplane is $\frac{1}{\| w \|}$ The margin would be:

$$\rho = min(x_i, y_i \in S) \frac{| w^T x_i + b |}{\| w \|} = \frac{1}{\| w \|} \tag{13}$$

Clearly the distance between two such data points which are separated by are 2*ρ (i.e., $\frac{2}{\| w \|}$) and maximizing this margin can be done by minimizing of $\frac{1}{2} \| w \|$ This maximum distance in the data points separated by the hyperplane can be done only if the given data set is classified linearly into two groups. However in most of the cases the classification is not linear. SVM offer balanced predicted performance, even in studies with small sample sets, because of their relative simplicity and flexibility in handling a variety of classification problems [14].

12.4.1.1.1 Separating Hyperplane

In separating hyperplane the given data is represented geometrically, which makes it easy to create cluster and predict the expression for the function. A hyperplane is created separating the two. When the new input comes in, the algorithm predicts on which side of the hyperplane it should lie, above the hyperplane or below it.

The tag bit is decoded using a Support Vector Machine (SVM), and the received signal is divided into distinct groups using a hyperplane. The hyperplane is constructed using known information bits while training the ML system.

12.4.1.1.2 Maximum Margin

It is not unique in high-dimensional space to classify items as points and create a hyperplane to separate them. However, many of these kinds exists in SVM to separate the objects. In maximum margin hyperplane, the line in which gives the maximum distance while separating them would be selected. The theorems from statistical learning also supports the theory. This approach improves the prediction efficiency of the SVM.

12.4.1.1.3 Soft Margin

Soft margin describes the tradeoff between the violating point in the hyperplane and margin size. In some cases, the object resides on the opposite side on the hyperplane; there needs a soft margin up to which the objects are allowed in the SVM. Soft margin defines how far from the hyperplane the point can reside in the opposite classes. The SVM should not allow too much of these points.

12.4.1.1.4 Kernel Function

A mathematical strategy for separating inseparable things by increasing the dimension. Separation of such classes is challenging in SVM in this scenario, because the objects are distributed randomly and jumbled between the two groups. The kernel function gives the SVM a dimension and finds the hyperplane that classifies the items and divides them into two groups. Increasing the dimension, however, may also increase complexity. The method focuses on the most common practical issues. Cross validation can be used to find an optimal kernel, but it is time demanding and does not guarantee that the kernel function will perform with fresh input items.

SVM is the simplest approach. It is also quick when dealing with big amounts of data. SVM is used in a variety of molecular biological applications, including DNA, protein sequence, human gene expression, bone marrow, and cancer classification. However, SVM has several flaws. SVM has only one disadvantage. SVM is used to solve binary classification issues. There must be a parallel number of SVM necessary for more than two classes, as determined by the number of classes. For example, to classify X Y and Z classes, 3 SVM ('Is it X?', 'Is it Y?', and 'Is it Z?') has to be defined.

12.4.1.2 Random Forest

Random forest is a computational technique for both classification as well as regression problems. For huge data sets, it operates swiftly. The decision tree is the foundation of random forest. In this method the data separated by creating a tree-like structure called as decision tree keeping the training data set in its branches, each of which is further branched, until a leaf node is reached. The decision tree is a decision assistance technique that is constructed in the shape of a tree. Root node, leap node, and decision node are the three pieces that make up a decision tree. These nodes represent the characteristics that will be utilized to forecast the information bit from the tag. The decision tree is a classification technique which maps the input to the output of the given data sets. In this tree the root node is corresponding to a question related to the feature of the input data which then further move to it left sub-tree if the answer is true and right sub-tree when the answer is negative, these nodes now act as a root node and repeats the same process until the terminal node is reached. In this way the decision tree is formed using the given data sets. Such algorithms are called greedy algorithms.

12.4.1.2.1 Algorithm 1 – Decision Tree Algorithm

1. **Initializing:**
   $\hat{a}_0$ as a root node
2. For loop up to i=1 to N
3. **Assign:** $\underset{(\hat{a},\hat{b})}{\arg\min}\, \overline{F}(\hat{a},\hat{b})$ to $(\hat{a}_i,\hat{b}_i)$
4. Split the tree to $(\hat{a}_i,\hat{b}_i)$

5. **End for Loop**
6. Return root node as a_{i+1}.

The decision tree starts with the node a_0 as a root node with question b_0, which then splits into left node as $(a_1, b_1)_L$ and right node as $(a_1, b_1)_R$ keep splitting the tree further until the leaf node, and F(a,b) is the measure of impurity after optimal splitting. $\bar{F}(\hat{a},\hat{b})$ is the measure of decrease in impurity after node splitting. When the value of $\bar{F}(\hat{a},\hat{b})$ is maximum then it is said that optimum splitting has been achieved.

$$\bar{F}(\hat{a},\hat{b}) = F(a) - \left[\eta_{\hat{a},\hat{b}} F(\hat{a},\hat{b})_L + (1-\eta_{\hat{a},\hat{b}}) F(\hat{a},\hat{b})_R\right] \tag{12.14}$$

where $\eta_{\hat{a},\hat{b}}$ is the probability that sample value move to the left node or the right node.

Assuming the probability of a sample from node a_1 belonging to k class is equal to $p_k(\hat{a})$. Then the impurity of the node F(n) is defined as:

$$F(n) = -\sum_{k=1}^{2} p_k \log_2 p_k(\hat{a}) \tag{12.15}$$

There can be decision trees in the Random forest, which are used to predict the output of the given input. Each decision in the forest uses the subset of the given input data set. Let T is the number of decision tree used, S_t be the subset used in decision and S be the set of given input data.

12.4.1.2.2 Algorithm 2 – Random Forest Algorithm

1. For loop up to i=1 to T
2. Create a training data set S_t
3. Build decision tree T_i out of S_t
4. Produce ensembled tree as output
5. End for loop
6. Majority voting of decision trees

Final output of the random Forest is decided by the majority voting of the decision trees. Predictor for the Random Forest method is defined as

$$H(x) = \underset{(y \in Y)}{\arg\max} \sum_{t=1}^{T} I(h_t(x) = y) \tag{12.16}$$

where I(.) denoted the indicator function, $h(x, \theta_k)$ is the ensemble for 't' tree and θ_k is the independent and identically distributed random vector [15]. Ensemble learning is used in this method, a technique that combines a large number of classifiers to solve the problem. This pattern continues until a leaf node is reached. It makes use of ensemble learning, which is a technique for solving complicated problems by combining several classifiers. Majority voting of the decisions trees predicts the outcome in most cases. While increasing the number of trees increases the precision, it also increases the computation cost [13]. It decodes the tag bit by using a multiply choice strategy for the received data and then averaging it.

12.4.2 Unsupervised Machine Learning

Dealing with strong direct-link interference from the RF source is one of the most difficult aspects of ambient backscatter. Treating the interference as noise and demodulating the backscattered information using energy detection was a simple approach. Using unsupervised learning approaches, the energy set elements are divided into two clusters that correspond to two different types of transmission bits either 0 or 1 from the tag. For these elements of the energy set in the cluster, a point is calculated, which a machine learns without obtaining supervised target outputs or rewards from its environment; it is said to be unsupervised learning. Given that the machine receives no feedback from its environment, it may be impossible to deduce what it could be capable of learning. However, a framework for unsupervised learning is based on the assumption that the purpose of the machine function is to generate characteristics of the input that can help in decision making, prediction of the future inputs, transferring the inputs to another machine efficiently, and so on. Unsupervised learning is be defined as the recognition of patterns in the data that will be above the level of pure unstructured noise. Clustering and dimensionality reduction are two simple instances of unsupervised learning. Almost every unsupervised learning research may be considered as a model that gives probability of the data being learned. Even if the machine is not supervised and compensated, developing a model that shows the probability distribution for a new input based on previous inputs is a beneficial approach.

The k-means technique, which is an unsupervised clustering algorithm, is strongly linked to the mixture of Gaussian model: assuming all Gaussian models have the same proportional covariance matrix to the identity matrix.

This model's maximum likelihood parameters could be estimated using the iterative technique known as EM, which we'll go through next.

When we apply the limit $\sigma \to 0$, the resulting algorithm is identical to the k-means method.

12.4.2.1 Maximum Likelihood Estimation (EM)

EM is a technique for parameter estimation of a probability distribution based on observed data. This is accomplished by optimizing a likelihood function such that the observed data is the most anticipated under the given statistical model.

Consider a model with output variables 'y', latent variables 'x', and parameters θ.

For any data point, we can lower bound the log likelihood as follows:

$$l(\theta) = log\, p(y \mid \theta) = log \int \left[p(x, y \mid \theta)dx\right] \tag{12.17}$$

When number of input is very large, we can say that the output follows Gaussian distribution with mean μ_k and variance σ_k^2 where k can be either 1 or 0 corresponding to the information bit from the tag. For backscatter communication we consider the probability of transmitted bit either 0 or 1 is equal which is 1/2. Now the probability y_i is given as:

$$P(y_i) = \sum_{k=0}^{1} \pi_k N\left(y_i \middle| \mu_k, \sigma_k^2\right) \tag{12.18}$$

where k could be either 0 or 1 and $N\left(y_i|\mu_k,\sigma_k^2\right)$ is the Gaussian distribution with mean μ_k and variance σ_k^2 we would estimate the parameters using EM algorithm based on the energy set of the signal. The log-likelihood function is given by:

$$\text{In } p\left(\mathbf{Y}|\mu,\sigma_k^2\right)=\sum_{i=0}^{l}\text{In}\left(\sum_{k=0}^{1}\frac{1}{2}N\left(y_i|\mu_k,\sigma_k^2\right)\right) \tag{12.19}$$

where $\mathbf{Y}=[y_1, y_2, \ldots, y_l]$ and μ_k σ_k^2 are the unknown parameters. As the log likelihood function is a non-convex function, it is difficult to estimate the parameters. A better approach is to find the probability that in which of the class does y_i belongs. It is given as:

$$p\left(z_k=1|y_i\right)=\frac{N\left(y_i|\mu,\sigma_k^2\right)}{\sum_{j=0}^{1}N\left(y_i|\mu_j,\sigma_j^2\right)} \tag{12.20}$$

where z_k is the latent variable such that $z_k \in (0, 1)$. The parameter mean μ_k and variance σ_k^2 can be estimated through simple iterative methods.

The mean is expressed as:

$$\mu_k=\frac{1}{N_k}\sum_{i=1}^{l}p\left(z_k|y_i\right)y_i \tag{12.21}$$

and the variance is expressed as:

$$\sigma_k=\frac{1}{N_k}\sum_{i=1}^{l}p\left(z_k|y_i\right)\left(y_i-\mu_k\right)^2. \tag{12.22}$$

where $N_k=\sum_{i=1}^{l}p\left(z_k|y_i\right)$

12.4.2.2 Algorithm 3 – Maximum Likelihood Estimation Algorithm for Gaussian Mixture Model

1. **Initialize:**
   μ_k, σ_k, $m = 1$, $\epsilon = 10^{-6}$
2. Calculate the probability of y_i belonging k^{th} cluster
3. Update the parameter μ_k,σ_k
4. Update m to $m + 1$
5. **Repeat Until**
   $\left|ln(y\,|\propto_m,\sigma_m)-\ln p(y\,|\propto_{m-1},\sigma_{m-1})\right|<\epsilon$

In the preceding algorithm, parameter are initialized with some initial value; we can estimate the belonging of y_i in either k=0 or k=1 by selecting the maximum value of the probability $p(z_k|y_i)$, that is if $p(z_k = 0|y_i) > p(z_k = 1|y_i)$ then the y_i belongs to the cluster ‘0’ or else $p(z_k = 1|y_i) > p(z_k = 0|y_i)$ then the y_i belongs to the cluster ‘1’. This process is repeated for multiple iterations. There is a tradeoff in transmission rate of tag and the value spreading gain. When the value of spreading is large we use central limit theorem and assume that the output energy set has Gaussian mixture distribution. But when the transmission rate is higher the spreading gain must be less and does not obey the Gaussian mixture distribution. Considering case which unity spreading gain. In this case the distribution is chi-square mixture model with mean λ and σ^2 as its parameters. For x(n)=0 the energy set follows non-central chi-square with mean $\lambda_0 = (h_{sr}s)$ and for x(n)=1 the energy set follows non-central chi-square with mean $\lambda_0 = (h_{sr}s + \alpha h_{st} h_{tr} s)$.

Now the probability y_i is given as:

$$P(y_i) = \sum_{k=0}^{1} \pi_k \phi\left(y_i \mid \lambda_k, \sigma_k^2\right) \tag{12.23}$$

where k could be either 0 or 1 and $\phi\left(y_i \mid \lambda_k, \sigma_k^2\right)$ is the distribution for non-central

Chi-square with mean λ_k and variance σ_k^2

Chi-square distribution:

$$\phi\left(y_i \mid \lambda_k, \sigma_k^2\right) = \frac{1}{2\sigma_k^2} \exp\left(\frac{-y_i + \lambda_k}{2\sigma_k^2} J_0 \frac{\sqrt{y_i \lambda_k}}{\sigma_k^2}\right) \tag{12.24}$$

where J_0 is the Bessel function of zero-th order. According to the model of non-central chi-square distribution the variance due to noise of both the clusters is same $\left(\sigma^2 = \sigma_0^2 = \sigma_1^2\right)$ and the log-likelihood for the chi-square distribution is given as:

$$\ln p\left(\mathbf{Y} \mid \lambda, \sigma_k^2\right) = \sum_{i=0}^{l} \ln\left(\sum_{k=0}^{1} \frac{1}{2} \phi\left(y_i \mid \lambda_k, \sigma^2\right)\right) \tag{12.25}$$

where $\mathbf{Y} = \left[y_1, y_2, \ldots, y_l\right]$ and $\lambda_k\, \sigma^2$ are the unknown parameters. The probability that in which of the class does y_i belongs is given as:

$$p\left(z_{ik}\right) = \frac{\phi\left(y_i \mid \lambda_k, \sigma^2\right)}{\sum_{j=0}^{1} N\left(y_i \mid \lambda_j, \sigma^2\right)} \tag{12.26}$$

where z_k is the latent variable such that $z_k \in (0,1)$. The parameter mean λ_k and variance σ_k^2 can be estimated through simple iterative methods. Mean is expressed as:

$$\lambda_k = \left(\frac{\sum_{i=1}^{l} p\left(z_{ik}\right)\sqrt{y_i}}{\sum_{i=1}^{l} p\left(z_{ik}\right)}\right)^2 \tag{12.27}$$

And variance is expressed as:

$$\sigma^2 = \frac{\sum_{i=1}^{l}\sum_{k=0}^{l} p\left(z_{ik}\right)\left(\sqrt{y_i} - \sqrt{\lambda_k}\right)^2}{\sum_{i=1}^{l}\sum_{k=0}^{l} 2p\left(z_{ik}\right)} \tag{12.28}$$

12.4.2.3 Algorithm 4 – Maximum Likelihood Estimation Algorithm for Chi-Square Mixture Model

1. **Initialize:**
   $\lambda_k, \sigma_k, m = 1, \epsilon = 10^{-6}$
2. Calculate the probability of y_i belonging to k^{th} cluster
3. Update the parameter λ_k, σ_k
4. Update m to m+1
5. **Repeat Until**
   $\left| ln(y \mid \lambda_m, \sigma_m) - \ln p(y \mid \lambda_{m-1}, \sigma_{m-1}) \right| < \epsilon$

In the preceding algorithm parameter are initialized with some initial value, we can estimate the belonging of y_i in either k=0 or k=1 by selecting the maximum value of the probability $p(z_{ik})$, that is if $p(z_{ik} = 0) > p(z_{ik} = 1)$ then the y_i belongs to the cluster '0' or else $p(z_k = 1|y_i) > p(z_{ik} = 0)$ then the y_i belongs to the cluster '1'. The algorithm converges more quickly when the probability of overlapping of the clusters is lesser or zero. EM has wide range of application in latent variable models, despite the fact that it is not always the quickest optimization method.

Additionally, keep in mind that the likelihood can have multiple local optimum values, out of which EM will converge on one of them, which may or may not be the global optimum value [16].

12.4.3 Reinforced Machine Learning

Backscatter devices must possess learning capabilities in order for upcoming 5G IoT networks to become a powerful platform with smart sensing devices that operate in a quick learning manner. The goal of this research is to employ reinforcement learning to improve backscatter network performance.

One of the three main ML models is reinforcement learning, which includes both supervised and unsupervised learning. ML branch deals with how intelligently the agents will react to a given problem in order to achieve maximum reward. In reinforcement learning, the machine communicates to the surroundings through producing actions. These tasks have an environmental impact, which results in scalar incentives for the machine. The aim of such machine is to enhance its perform in such a way that it earns maximum rewards over time. Reinforcement learning is somewhat similar to decision theory in statistics and management science, as well as control theory in engineering. The underlying problems investigated in above domains are often compared theoretically, and getting similar answers, distinct elements of the problems are frequently addressed. Apart from Back-Com system, reinforcement learning have been recently developed in wireless networks in many publications.

An agent investigates and exploits its surroundings in exchange for a positive or negative reward in reinforcement learning. The agent receives a positive reward for a successful action and a negative reward for a failed activity. The agent takes a greedy approach. In the ideal strategy, the agent receives positive reward when the SINR (Signal to Interference Noise Ratio) achieved through the algorithm is equal to the desired effective SINR. The agent receives a negative reward when the achieved SINR is less than the desired SINR. Generally, these reward is defined by program implementation. A reinforcement learning-based system was used to generate power allocation instructions, which enhances capacity, reduces interference and improves energy usage of backscatter networks. Low-power IoT networks can benefit from combining reinforcement learning with Back-Coms to provide much-needed knowledge. In moving data from one location to another, reinforcement learning uses substantially less mean energy per bit. This variation could be explained by a decrease in the mean distance between backscatter devices, allowing reinforcement learning to perform better even in low density. The reinforcement learning technique dramatically reduces interference for high density of backscatter devices, making it a feasible alternative for dense and large scale networks. The energy spent per bit falls as the Mean distance between the backscatter transmitter and the receiver decreases because of the reduced device path loss.

12.4.3.1 Q Learning Algorithm

The Q-learning methodology is used to reduce network interference and increase the received SINR of the backscatter network Q-learning discovers an optimal solution in terms of maximizing the expected value of the total incentives across all iterative steps, initializing with the current state. In the backscatter power allocation context, the algorithm begins by determining exploration rate, state, action, decay rate, and exploration thresholds. In the ambient Back-Com simulation, the backscattered data is first supplied to the reader. Each time slot is examined to see if the goal effective SINR threshold has been met or if the transmit time interval has been exceeded. The exploration rate is compared to a uniform distribution random variable. The Q-value maximizing action is chosen if the value of the random variable is greater than the exploration rate. While modifying the records, the step is done and the reward is decided. The status is updated at the terminal point, and the procedure continues with next step doing the same operations. As a result of this procedure, the best sequence is generated [17].

12.4.3.2 Algorithm 5 – Q Learning Algorithm

1. **Input:**
   Target SINR value as γ_0, Initially computed effective SINR $\bar{\gamma}_{init}$
2. **Define:**
   S=power allocation states.
   A=power allocation actions.
   w=exploration rate.
   d=decay rate.
   w_{min}=minimum exploration threshold.
3. **Initialize:**
   $\bar{\gamma} = \bar{\gamma}_{init}$, $s = 0$, $a = 0$, *and* $t = 0$.
4. **For loop**
   Start the simulation and transmit the backscatter signal
5. **while**
   $\bar{\gamma} \leq \bar{\gamma}_0$ or $t \leq \tau$
   $w = max(w.d, w_{min})$
   Select r randomly from uniform distribution

7. **if** $r \leq w$ Select a random action
8. **else**
   Select action to maximize the reward
9. **end if**
10. Calculate the reward after the action
11. Check the impact on next step.
12. Update the state s.
13. **end while**
14. **end for**
15. **Output:**
    Optimized power allocation actions for different target SINR.
16. **End procedure**

12.5 APPLICATION OF BACK-COM

In ABCS the tag uses the harvested energy from the ambient RF signal and reflects the modulated signal to the nearest reader, which could be on the same device or any terminal node like mobile, laptop, or other cellular device. The source is generally a base station in cellular network and tags are the IoT devices [18].

In ABCS, backscatter transmitters can operate automatically with little human interaction. The advancement of ambient backscatter technology has vast implications for data transmission. ABCS can be used in a variety of fields, including smart life, logistics, and medical biology.

12.5.1 Biomedical Applications

Small and long-lasting communication devices are required for biomedical applications such as wearable and implanted health monitoring. These requirements are fulfilled with ABCS. Some of the wearable biomedical application includes (1) smart fabric, which is used to record the heart activity and breathing rate; (2) smart watch, which is very popular nowadays, it can also be used to monitor the heart and breathing rate of a person wearing it; (3) smart shoes, that is sensors and ambient backscatter modules are used to build a pair of shoes. The sensor in each shoe performs independent duties, such as counting steps and heart rate, while the ambient backscatter modules are used to coordinate two shoes. The results of the experiments show that the proposed platform works well in real-world situations. But they are limited for the speed of moving object or person increases.

12.5.2 Smart World

There are variety of areas where ABSC has improved the standard of living, for example in smart houses, a large number of sensor transmitters can be put in a smart house at various locations, such as inside walls, furniture, and floors. These sensors are battery efficient and can operate for long time without maintenance or an additional energy source. Other applications include harmful gas detection, such as gas, smoke, and CO, as well as movement monitoring and surveillance. Smart card application is also a very well-known application for ABCS, in which sensors are embedded on a card to transfer the data.

12.5.3 Logistics

ABCS can have many applications depending on its low cost implementation. In grocery stores it can be used to maintain the items, to check the available quantity, to refill the stock, or to locate the items without much human effort within a few seconds. Each object has a unique identifying number and is coupled with a backscatter transmitter [19].

12.6 CONCLUSION

IoT is expected to connect billions of devices, bringing digital change to a wide range of industries, including agriculture, homes, and healthcare. Back-Com appeals to a wide range of wireless sensors because it combines energy collection with low-power communication. Backscatter has emerged as a significant competitor for other wireless systems because of its capacity to transmit electricity while also delivering ultra-low power wireless back-haul. Ambient backscatter is a new method of wireless communication with economic promise as well as a number of issues [20]. Ambient backscatter systems enhance the overall IoT architecture in two ways: first, they enable local access the sensors which consume very low power, and second, they work well with large-scale networks and are more reliable as compared to the other wireless IoT systems, at nearly negligible energy and complexity costs [21]. ML methodologies to extract the backscattered signal from the tag to receiver have been studied. The user provides the machine with a known data set with desired inputs and outputs, and the system must figure out how to get to those inputs and outputs in supervised learning, but in unsupervised learning, the machine hunts for patterns in the data. Reinforcement learning can help improve the performance of backscatter networks in ambient Back-Com.

REFERENCES

[1] Q. Zhang, H. Guo, Y.-C. Liang, and X. Yuan, "Constellation learning-based signal detection for ambient backscatter communication systems," *IEEE Journal on Selected Areas in Communications*, vol. 37, no. 2, pp. 452–463, 2018.

[2] T. S. Muratkar, A. Bhurane, and A. Kothari, "Battery-less internet of things – a survey," *Computer Networks*, vol. 180, p. 107385, 2020.

[3] S. Ma, G. Wang, R. Fan, and C. Tellambura, "Blind channel estimation for ambient backscatter communication systems," *IEEE Communications letters*, vol. 22, no. 6, pp. 1296–1299, 2018.

[4] T. Hong, C. Liu, and M. Kadoch, "Machine learning based antenna design for physical layer security in ambient backscatter communications," *Wireless Communications and Mobile Computing*, vol. 2019, 2019.

[5] B. Ji, B. Xing, K. Song, C. Li, H. Wen, and L. Yang, "The efficient backfi transmission design in ambient backscatter communication systems for iot," *IEEE Access*, vol. 7, pp. 31 397–31 408, 2019.

[6] H. Guo, Q. Zhang, S. Xiao, and Y.-C. Liang, "Exploiting multiple antennas for cognitive ambient backscatter communication," *IEEE Internet of Things Journal*, vol. 6, no. 1, pp. 765–775, 2018.

[7] S. Zeb, Q. Abbas, S. A. Hassan, A. Mahmood, R. Mumtaz, S. H. Zaidi, S. A. R. Zaidi, and M. Gidlund, "Noma enhanced backscatter communication for green iot networks," in *2019 16th International Symposium on Wireless Communication Systems (ISWCS)*. IEEE, 2019, pp. 640–644.

[8] P. Zhang, J. Gummeson, and D. Ganesan, "Blink: A high throughput link layer for backscatter communication," in *Proceedings of the 10th International Conference on Mobile Systems, Applications, and Services*. Association for Computing Machinery, 2012, pp. 99–112.

[9] G. Wang, F. Gao, R. Fan, and C. Tellambura, "Ambient backscatter communication systems: Detection and performance analysis," *IEEE Transactions on Communications*, vol. 64, no. 11, pp. 4836–4846, 2016.

[10] D. Li and Y.-C. Liang, "Adaptive ambient backscatter communication systems with mrc," *IEEE Transactions on Vehicular Technology*, vol. 67, no. 12, pp. 12 352–12 357, 2018.

[11] X. Wen, S. Bi, X. Lin, L. Yuan, and J. Wang, "Throughput maximization for ambient backscatter communication: A reinforcement learning approach," in *2019 IEEE 3rd Information Technology, Networking, Electronic and Automation Control Conference (IT- NEC)*. IEEE, 2019, pp. 997–1003.

[12] Q. Liu, S. Sun, X. Yuan *et al.*, "Ambient backscatter communication-based smart 5g iot network," *EURASIP Journal on Wireless Communications and Networking*, vol. 2021, no. 1, pp. 1–19, 2021.

[13] S. Huang, N. Cai, P. P. Pacheco, S. Narrandes, Y. Wang, and W. Xu, "Applications of support vector machine (svm) learning in cancer genomics," *Cancer Genomics & Proteomics*, vol. 15, no. 1, pp. 41–51, 2018.

[14] D. A. Pisner and D. M. Schnyer, "Support vector machine," in *Machine Learning*. El- sevier, 2020, pp. 101–121.

[15] Y. Hu, P. Wang, Z. Lin, M. Ding, and Y.-C. Liang, "Machine learning based signal detection for ambient backscatter communications," in *ICC 2019–2019 IEEE International Conference on Communications (ICC)*. IEEE, 2019, pp. 1–6.

[16] Q. Zhang and Y.-C. Liang, "Signal detection for ambient backscatter communications using unsupervised learning," in *2017 IEEE Globecom Workshops (GC Wkshps)*. IEEE, 2017, pp. 1–6.

[17] F. Jameel, W. U. Khan, S. T. Shah, and T. Ristaniemi, "Towards intelligent iot networks: Reinforcement learning for reliable backscatter communications," in *2019 IEEE Globecom Workshops (GC Wkshps)*. IEEE, 2019, pp. 1–6.

[18] S. Zhou, W. Xu, K. Wang, C. Pan, M.-S. Alouini, and A. Nallanathan, "Ergodic rate analysis of cooperative ambient backscatter communication," *IEEE Wireless Communications Letters*, vol. 8, no. 6, pp. 1679–1682, 2019.

[19] N. Van Huynh, D. T. Hoang, X. Lu, D. Niyato, P. Wang, and D. I. Kim, "Ambient backscatter communications: A contemporary survey," *IEEE Communications Surveys & Tutorials*, vol. 20, no. 4, pp. 2889–2922, 2018.

[20] W. Zhao, G. Wang, S. Atapattu, C. Tellambura, and H. Guan, "Outage analysis of ambient backscatter communication systems," *IEEE Communications Letters*, vol. 22, no. 8, pp. 1736–1739, 2018.

[21] R. Duan, X. Wang, H. Yigitler, M. U. Sheikh, R. Jantti, and Z. Han, "Ambient backscatter communications for future ultra-low-power machine type communications: Challenges, solutions, opportunities, and future research trends," *IEEE Communications Magazine*, vol. 58, no. 2, pp. 42–47, 2020.

Index

X

Z

CPSIA information can be obtained
at www.ICGtesting.com
Printed in the USA
BVHW050511060323
659649BV00002B/95